Human Evolution and Development

Human Evolution and Development

Textbook for Life Sciences

Nico M. van Straalen and Dick Roelofs

Amsterdam University Press

Originally published as: Nico M. van Straalen and Dick Roelofs, *Evolueren wij nog? Alles wat je wilt weten over ontwikkeling en evolutie van ons lichaam* © Amsterdam University Press, 2017 [ISBN 978 94 6298 130 0]

Translated by Translation Kings

Expanded and updated by the authors

The publication of this book was supported by the Education Directorate, Faculty of Science, Vrije Universiteit Amsterdam

Cover design: Janine Mariën
Lay-out: Crius Group, Hulshout

ISBN	978 94 6372 920 8
e-ISBN	978 90 4854 397 7
NUR	922

© N.M. van Straalen, D. Roelofs / Amsterdam University Press B.V., Amsterdam 2019

All rights reserved. Without limiting the rights under copyright reserved above, no part of this book may be reproduced, stored in or introduced into a retrieval system, or transmitted, in any form or by any means (electronic, mechanical, photocopying, recording or otherwise) without the written permission of both the copyright owner and the author of the book.

Every effort has been made to obtain permission to use all copyrighted illustrations reproduced in this book. Nonetheless, whosoever believes to have rights to this material is advised to contact the publisher.

Table of Contents

Preface   7

1. The story of our ancestors   11
   The revolutionary innovation: walking upright   11
   How old is that fossil?   14
   The hominin tree   17
   The earliest hominins   20
   The heyday of the ape-men   22
   The first *Homo*   25
   Towards modern times — *Homo neanderthalensis*   32

2. From ovum to human   37
   Heterochrony and Haeckel's law   37
   Cleavages and germ layers   41
   Axes to provide direction   48
   Model animals in developmental biology   52
   The molecular toolkit for development   59
   New axes for limbs   64

3. Our tinkered body   69
   Tinkers, watchmakers and a Boeing 747   69
   The naked human   73
   Adaptations to bipedalism in the locomotor apparatus   77
   Gill slits, larynx and middle ear   84
   The intestines and the lung   87
   Heart and urogenital system   91
   Evolution of the brain   99

4. There must be differences   109
   Giant leaps, neutral fluctuations or gradual adaptation?   109
   The emergence of variation   114
   Equilibrium between allele and genotype frequencies   124
   Neutral evolution   135
   Geographical distance causes genetic differences   140
   On top of genetics   142

5. The past in the present   151
   Phylogenetic reconstruction   151
   The molecular clock   161
   Out of Africa or multiregional evolution?   164
   Migrations in all directions   170
   Hybridisations between ancient humans   176

6. The cultural human   183
   Prehistoric tools and cave drawings   183
   The Neolithic transition   192
   Language: early or late?   198
   Living in groups: altruistic behaviour   201
   Cultural evolution   208

7. Do humans still evolve?   215
   Quantitative characters and heritability   215
   Ecogeographic variation in human body form   221
   Evolution of biomedical traits   224
   Evolution of the life cycle   229
   Partner choice and sexual selection   230
   Evolutionary medicine   235

Epilogue   241

Further reading   245

Primary literature   249

Credits   275

Index   281

Preface

No scientist will doubt nowadays that humans are a biological species, subject to the same evolutionary mechanisms as all other biological species, but how that evolutionary process took place exactly and which selective forces made us into what we are now, is still largely uncertain. Over the last ten years, however, research on human evolution has rapidly gained momentum. This was caused in the first place by the discoveries of a large number of new fossils. In 1964, a 'golden period' started for human palaeontology, and a clear understanding of human ancestry began to take shape. Since 2002, a large number of spectacular discoveries added to our knowledge, thoroughly adjusting our idea about the diversity of our ancestors. Many of those discoveries were and continue to be published in newspapers or otherwise reach the general public. Over the last years, however, the number of species has increased such that for laymen it has become almost impossible to maintain a proper overview. One of the objectives of this book is to present the abundance of extinct ape-men and humans in a coherent story.

Secondly, modern biotechnology has allowed us to map the DNA of humans in such a detailed way that important conclusions can be reached about our evolutionary history. The genetic variation among humans, in conjunction with the places where they live, the languages they speak and the cultures they share, are an invaluable source of knowledge for evolutionary reconstruction: a large part of our evolutionary history is stored in the current DNA. Due to the enormous development of DNA research and the possibility to sequence and compare entire genomes of hundreds of people simultaneously, we are now able to tell much more about human evolution than a number of years ago.

Thirdly, it has also proven to be possible to isolate DNA from fossil material, provided that it is not too old (currently to approximately 400,000 years). In 2010, sequencing of ancient DNA led to the astonishing conclusion that humans must have crossbred with Neanderthals. For us, teachers of evolutionary biology, this was one of the most shocking events, because for many years we had taught our students that humans and Neanderthals were two biologically separated species.

Lastly, we emphasise that over the last years evolutionary biology has become integrated with developmental biology. A new area of expertise is developing, referred to as 'evo-devo'. Every animal develops from a fertilised ovum, via the embryonic and foetal stages into a mature and reproducing individual. The genetic machinery managing this development is beginning

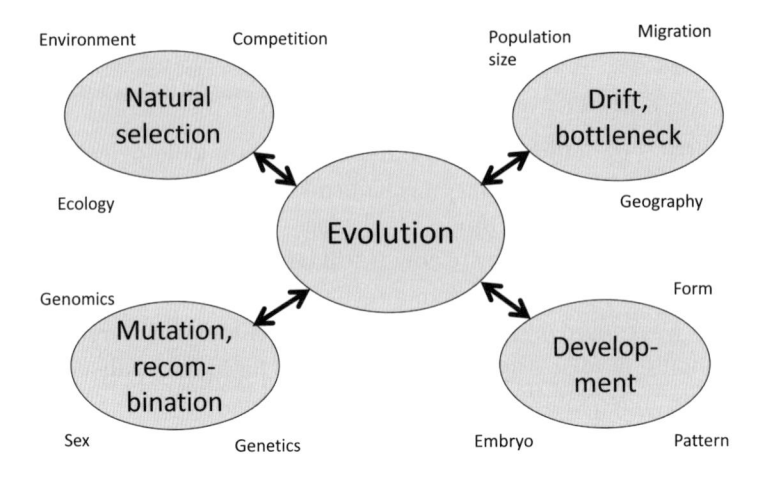

```
           Environment    Competition       Population    Migration
                                              size
              ╱─────────╲                   ╱─────────╲
             │  Natural  │                 │   Drift,  │
             │ selection │                 │ bottleneck│
              ╲─────────╱                   ╲─────────╱
        Ecology      ↘         ↙       ↘        ↙   Geography
                      ╱─────────╲
                     │ Evolution │
                      ╲─────────╱
        Genomics     ↗         ↖       ↗        ↖     Form
              ╱─────────╲                   ╱─────────╲
             │ Mutation, │                 │  Develop- │
             │  recom-   │                 │   ment    │
             │  bination │                  ╲─────────╱
              ╲─────────╱
            Sex        Genetics        Embryo        Pattern
```

to be understood for an increasing number of model species. Evolution is a process of change in the way genes regulate that development and thus alter the external appearance of a species.

Evolution is, in fact, the result of four different drivers (see figure). The basis for evolution lies in genetics. Changes in the hereditary material (mutations) take place continuously, recombine between individuals as a consequence of sexual reproduction, and are passed on to offspring. Mutation and recombination generate the variation that becomes available for selection. Natural selection subsequently ensures that mutations that provide beneficial properties to the individual are maintained and spread within the population at the expense of less beneficial variants. In a small population, mutations can also settle due to genetic drift and bottleneck effects, which is the subject matter of neutral evolution theory. Finally, embryonic development has a strongly integrating and canalising effect on the possibilities for evolution, as any mutation should fit into the genetic machinery directing an organism's body plan before it can become manifest as an external feature available for selection.

Our book aims for an integrative approach of evolutionary biology, considering mutation, recombination, natural selection, neutral evolution and developmental processes in mutual interdependence. Our focus on developmental biology distinguishes this book from other books on human evolution. The appearance of a new and unique developmental plan, a human body, which differentiates in many respects from the great apes, yet is so similar, demands an explanation in terms of evo-devo. Our body turns out to have been tinkered in a wonderful way, with all kinds of inconveniences and relicts, which are still evident and can only be explained from our evolutionary history.

PREFACE

This book is the result of years of teaching evolution to students of biology, biomedical science and health sciences at the Vrije Universiteit Amsterdam. As part of the course, the students were to apply their obtained knowledge by writing an essay, under the theme 'Do humans still evolve?' This assignment time and again resulted in numerous unexpected and original ideas. This book is our own answer to that question.

1. The story of our ancestors

Around 7 million years ago a radical new evolutionary lineage emerged among the African great apes: upright walking, ape-like creatures began to explore the savannah instead of spending their lives in trees. This can be regarded as one of the most important events in evolutionary history, because eventually this lineage would produce a species that was to fundamentally change the earth's face in all corners of the world: humankind. In this chapter we discuss the different ape-men and humans, no less than 26 species that lived within a period of 7 million years, from the end of the Miocene up to modern day. All those species have gone extinct, except one. We, *Homo sapiens*, are the only survivors; from the others we know nothing but fossils and artefacts. How do all these species cohere, which ones can be considered our ancestors and how did they live? These fascinating questions will be discussed in this first chapter.

The revolutionary innovation: walking upright

In biological terms, humans are primates, mammals and vertebrates. Biologists classify species in accordance with a hierarchical system of genera, families, orders, classes and phyla. By using prefixes such as 'super', 'sub', and 'infra' additional categories can be inserted and the classification can be made more precise (table 1.1).

Many of the characteristics of our body remind us that we are apes, mammals and vertebrate animals. These characteristics will be discussed in chapter 3. In this chapter we focus on the group of Homininae, technically a subfamily of the Hominidae family, which, apart from humans, also includes chimpanzee, bonobo, and gorilla. *Homo sapiens* is the only surviving species of the subfamily Homininae (table 1.1).

The defining feature of the Homininae is *bipedalism*, upright walking. That must have been tremendously beneficial, given the fact that no less than 26 species emerged, while much less divergence took place in the chimpanzee lineage. Yet the exact benefit of a life of upright walking cannot be indicated that easily. Various theories were formulated about this. Often it was argued that upright walking freed the hands, allowing them to become available for taking food to a home base. Scientists have also claimed that walking on two legs is beneficial as it enhances the overview of the surroundings, due to which predators are more easily spotted. In

Table 1.1: Systematic classification of *Homo sapiens*

Phylum	Chordata	Chordates
Subphylum	Vertebrata	Vertebrates
Class	Mammalia	Mammals
Subclass	Placentalia	Placental mammals
Order	Primates	Primates (prosimians and monkeys)
Suborder	Haplorhini	Tarsiers and monkeys
Infra-order	Catarrhini	Old World monkeys
Superfamily	Hominoidea	Apes
Family	Hominidae	Human, chimpanzee and gorilla
Subfamily	Homininae	Ape-men and humans*
Genus	*Homo*	Humans*
Species	*Homo sapiens*	Human

*including extinct representatives

addition, walking on two legs is a very efficient way of moving forward if you do not want to proceed too fast. Finally, biologists have indicated that a straight-up posture reduces the quantity of absorbed solar radiation, while the upper side of the body is able to cool off by making use of the air flow (fig. 1.1). Of all these theories, the latter, stressing a better heat regulation in a hot climate, seems to be the most logical.

The transition from complete *arboreality* (confinement to trees) to a fully terrestrial way of life was already somewhat initiated with the apes. Chimpanzee and gorilla, the species most related to humans, habitually

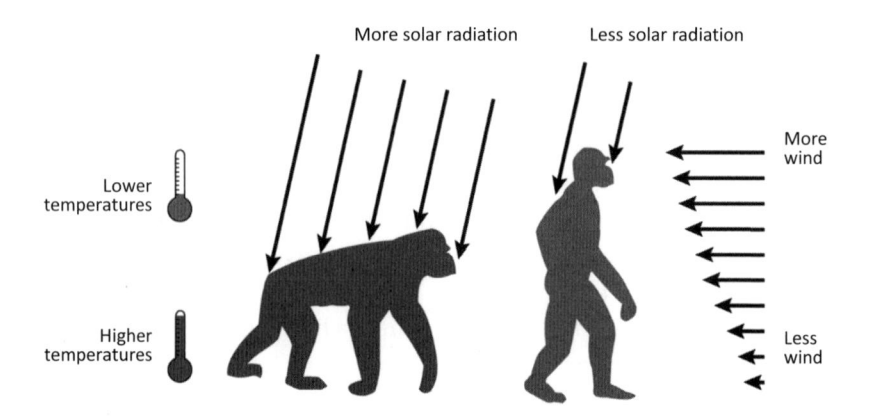

Fig. 1.1: Upright walking of the hominins is seen as an adaptation to the warm African savannah climate. The heat regulation of bipedal hominins was better than that of knuckle-walkers, because they absorbed less sun radiation and were better able to make use of the cooling airflow.

lean on their middle knuckles (the knuckle walk). It is highly likely that this is also the way of locomotion from which bipedalism derived. Strong evidence for this is the presence of a cavity in the distal part of the radius, which is typical for knuckle-walkers. Due to this cavity, the connection between the forearm and the metacarpals is additionally solid and rigid. Humans lack such a cavity – on the contrary, our wrist is quite flexible. A cavity in the radius does, however, occur in the well-known hominin fossil of *Australopithecus afarensis*. The fact that this hominin still had a remainder of a knuckle-walk adaptation indicates that upright walking gradually evolved.

All fossil hominins older than 2 million years are found in Africa, in the Great Rift Valley, an area that runs from the Red Sea via Ethiopia and the African lakes to South Africa (fig. 1.2). The valley is an elongated depression in the landscape, caused by opposite drift of the eastern and western part of the African continent. On no other continent old hominin fossils are found, so it is likely that we should locate the emergence of the upright walking hominin ancestors in Africa, most likely eastern Africa. There are, however, two exceptions, namely the fossils of *Sahelanthropus tchadensis* (6.5 million years old) and of *Australopithecus bahrelghazali* (3.5 million years old).

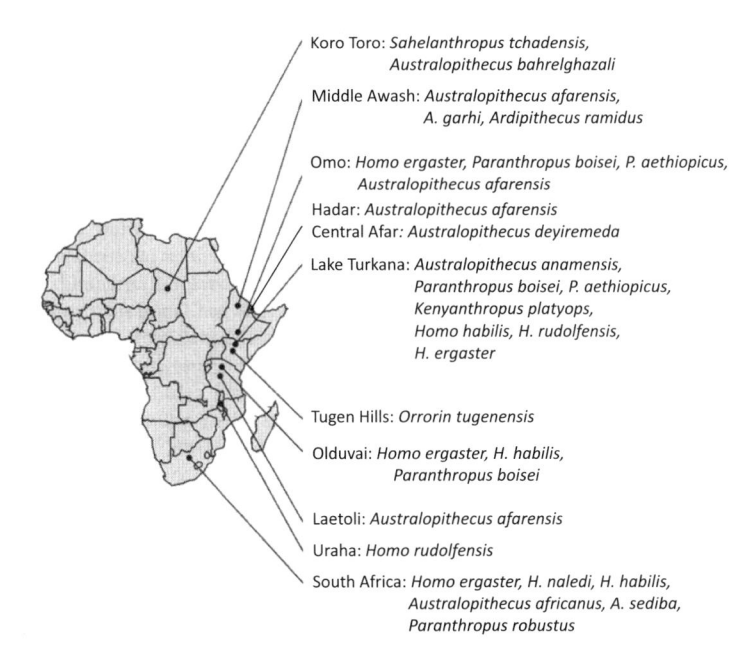

Koro Toro: *Sahelanthropus tchadensis, Australopithecus bahrelghazali*
Middle Awash: *Australopithecus afarensis, A. garhi, Ardipithecus ramidus*
Omo: *Homo ergaster, Paranthropus boisei, P. aethiopicus, Australopithecus afarensis*
Hadar: *Australopithecus afarensis*
Central Afar: *Australopithecus deyiremeda*
Lake Turkana: *Australopithecus anamensis, Paranthropus boisei, P. aethiopicus, Kenyanthropus platyops, Homo habilis, H. rudolfensis, H. ergaster*
Tugen Hills: *Orrorin tugenensis*
Olduvai: *Homo ergaster, H. habilis, Paranthropus boisei*
Laetoli: *Australopithecus afarensis*
Uraha: *Homo rudolfensis*
South Africa: *Homo ergaster, H. naledi, H. habilis, Australopithecus africanus, A. sediba, Paranthropus robustus*

Fig. 1.2: The main hominin sites (except of *Homo sapiens*) in Africa. All hominin fossils older than 2 million years are from Africa, and virtually are all found in the Great Rift Valley.

These were found in Central Africa, in the Republic of Chad. How this adds up to the idea that the Great Rift Valley is the cradle of mankind remains uncertain to this day. Possibly the first hominins' distribution range was more extended than we assume nowadays.

From approximately 2 million years ago we also find hominin artefacts and fossils outside Africa, the oldest sites being Dmanisi (Georgia, 1.8 million years) and Shangchen (China, 2.1 million years). The ancient hominins in these areas are classified as *Homo erectus*. Also *H. sapiens* itself originated in Africa, 250,000 to 350,000 years ago, as genetic research indicates (see chapter 5). Between 180,000 and 90,000 years ago *H. sapiens* first dispersed to the Arabian peninsula, followed by a second dispersal, about 60,000 years ago, to the Middle East, from which Asia and Europe were colonised about 45,000 years ago and the Americas about 20,000 years ago. We will look at these dispersals in more detail in chapter 5. For the moment it suffices to state that, for the largest part of their existence, hominins lived only in Africa.

How old is that fossil?

Palaeontology is the science that deals with the study of fossils. For a palaeontologist, knowledge of rocks, sediments and geological processes is essential, so palaeontologists can be most often found in an earth science environment. Palaeontology involves three main aspects. Firstly, the fossil itself: what it looks like, which species it belongs to, *etc*. Here, also knowledge of the fossilisation process is required. The soft parts of an organism usually degrade quite quickly, but the harder parts could become mineralised and, as such, be preserved. In this process, dissolved minerals in groundwater or pore water penetrate the biological material, which results in precipitates. The fossil obtains the colour and the structure of its environment. In fact, a good fossil is a piece of rock that has retained the shape of the original biological material but consists of the minerals in which it is present.

Fossils are usually cleared in pieces from a deposition and subsequently a part of the skeleton (skull, pelvis *et al*.) is reconstructed by putting the pieces together. Previously, the missing parts would be filled with plaster. Nowadays models are made by use of computer programs, allowing a 3D print of the fossil to be made. Scanning methods and three-dimensional image formation help to give palaeontologists a good impression of an object, and enable them to make all kinds of measurements, allowing a statistical comparison with other fossils.

Secondly, palaeontology is about the context in which the fossil was found, particularly the layer, the stratigraphy, other fossils and artefacts, residues of plants, and the spatial arrangement of all these materials. During the discoveries in the nineteenth century, scientists paid less attention to the context of the fossil. Nowadays valuable information about the lifestyle or the habitat of a species can be obtained by describing the fossil in connection with its environment.

Thirdly, palaeontologists are obviously interested in the age of the fossil. Usually not the fossil itself is dated, but the layers between which it was found. It is hard to establish a date for fossils that were not discovered between two datable layers. This often applies to material that was found in caves, for instance, the fossils of the hobbit (*Homo floresiensis*) and those of *Homo naledi* were initially reported without reliable dates.

The chance of things to fossilise is very slim. Ideally, straight after death, a layer of sediment or volcanic ash covers the body, without having been torn to pieces by wild animals. The best fossils we have are of big animals and animals with a calcified skeleton, like vertebrates, snails, crustaceans, echinoderms, and similar, while the fossil record of plants, insects and unicellular organisms is meagre. Also the lifestyle of the species plays a role. We have, for instance, not been able to gather many fossils of primates because almost all of them live in trees and they are not covered by sediment directly after they die. This is contrary to the hominins, who lived on the ground.

Teeth play a special role in the fossil record. Since teeth consist of a hard material, *dentin*, covered by a still harder layer of *enamel*, they are often found in fossil collections and sometimes they are the only finds. Compared to apes, hominin teeth tend to have a rather thick layer of enamel, and enamel thickness has increased further in some lineages, such as the robust Australopithecines (see below). In the lineage leading to *H. sapiens* enamel thickness also increased.

A large number of dating methods are available for estimating the age of a fossil. They all have a clock mechanism, such as radioactive decay or accumulation of damage, and a designated time at which the clock started running, a set point or zero mark. For example, after a volcanic eruption, in the descending ash no gaseous elements are present, due to the high temperature, thus all gas that is now being found in a sedimentary rock has accumulated as a result of radioactive decay after the volcanic eruption. Each dating method has its own time frame, which is determined by the speed of the applied 'clock' (for instance, the half-life of an unstable isotope).

For the study of human evolution, the *potassium-argon* method is highly important, due to the precision of the measurement and the suitable time

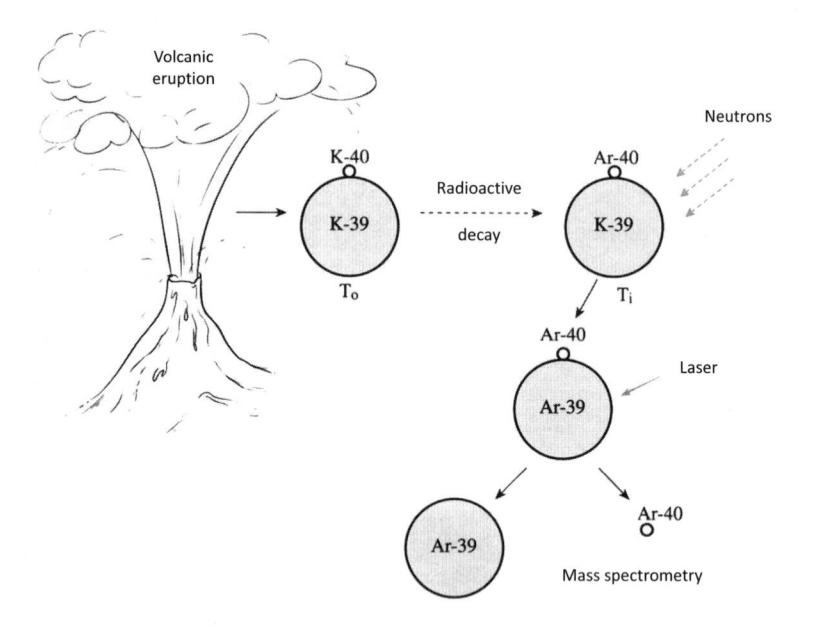

Fig. 1.3: Principle of ^{39}Ar/^{40}Ar-dating. Volcanic ashes rich in potassium (e.g. feldspar) contain an amount of unstable ^{40}K, which, over the years, decays to ^{40}Ar with a half-life of 1,252 Ma. The ^{40}Ar to ^{39}Ar isotope ratio provides an estimate of the time that has passed since the formation of the ash, when it was devoid of ^{40}Ar.

frame (fig. 1.3). Radiocarbon dating is only suitable for recent fossils (*H. sapiens* and Neanderthal), as the relatively short decay time of ^{14}C (half-life 5,730 years) does not allow dating biomaterials older than 100,000 years. In potassium-argon dating, the accumulation of ^{40}Ar in potassium-rich sedimentary rock is studied. ^{40}Ar is the decay product of the unstable potassium isotope ^{40}K, which has a half-life of approximately 1,250 million years. The concentration of ^{40}Ar gas measured in a piece of rock is compared with the original quantity of ^{40}K, which is estimated from the concentration of ^{39}K. This in turn is defined after all the ^{39}K present in a crystal is converted to ^{39}Ar, by use of neutron activation (fig. 1.3). The major advantage of this method is that the isotope ratio ^{40}Ar/^{39}Ar can be determined with a high level of accuracy in just one mass spectrometry run.

Following geologists' conventions, the age of a layer or a fossil is expressed in thousands or millions of years before present, abbreviated to ka (*kilo-annum*) BP or Ma (*mega-annum*) BP. These abbreviations will also frequently be used in this book. To be more precise on the recent past, archaeologists have agreed that the year 1950 should be regarded as the 'present' in 'before present', the year in which Willard Libby discovered radiocarbon dating.

The hominin tree

A coherent picture of the different ape-men and humans, according to the most recent scientific insights, is shown in fig. 1.4. Almost all relationships are still debated, but some principles are now becoming clear. Firstly, the tree is characterised by many ramifications and dead-end lines, a process of evolution by *cladogenesis* (table 1.2). This means that new species emerge from side branches, while the main branch continues or dies out. The opposite process is *anagenesis*. In this case the main branch evolves in a straight line into a new species, without splitting; only one species is alive at a time. For a long time, from the days of Darwin up to the 1960s, it was assumed that human beings had developed in a continuous line, through successive species. Neanderthals were considered an ancestor or primitive form of *H. sapiens* and the possibility that it might have been an extinct sister species was neglected. Nowadays we are aware that cladogenesis, with its numerous dead-end lines, is the most common mechanism in the evolution

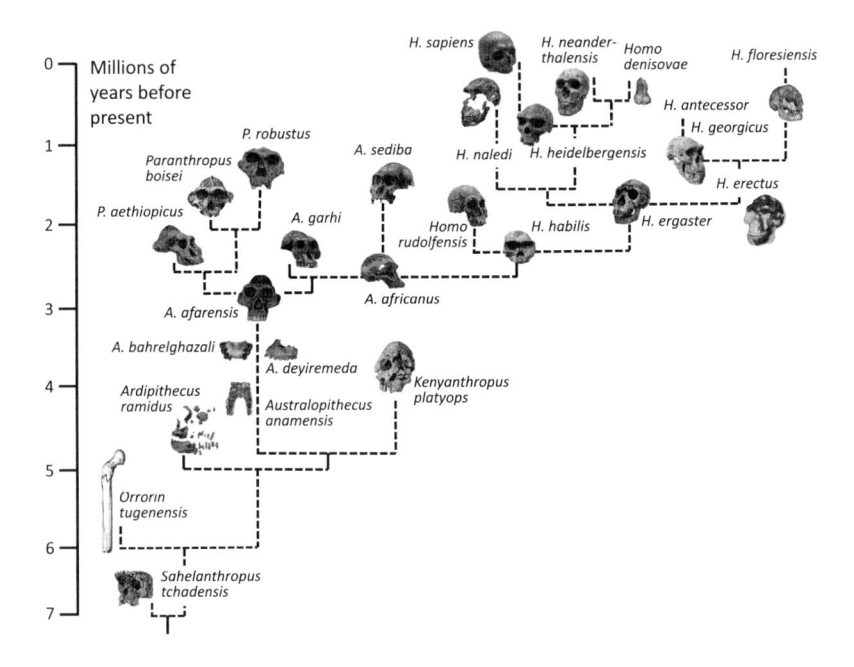

Fig. 1.4: Schematic evolutionary tree of the hominins, based on reported fossils. In the fossil record some 26 hominins (usually designated as species) lived from around 7 million years ago that, apart from *Homo sapiens*, have all gone extinct. The tree is characterised by a large number of branches and extinct sidelines. For the sake of clarity dashed lines in this figure suggest connections between species, but essentially nearly all relationships are still disputable and uncertain. Note that also the distinction between species is debated in many cases.

Table 1.2: Cladogenesis and anagenesis

Process	Definition	Examples
Cladogenesis	Origin of new species due to splitting of lineages; multiple species live simultaneously, of which some become extinct	The split between robust Australopithecines (genus *Paranthropus*) and the gracile *Australopithecus* species
Anagenesis, also called phyletic evolution	Origin of new species due to a lineage gradually converting into a new form; only one species is present at a time	The evolution of *Ardipithecus* (5 to 4 million years old) towards *Australopithecus* (4 to 3 million years old)

of the hominins. Still, anagenesis cannot be excluded completely, especially not in early hominin evolution. According to the American palaeontologist Tim White, anagenesis was the most important process between 5 and 3 million years ago, in the evolution of *Australopithecus* from *Ardipithecus*.

Another remarkable phenomenon is the split-off of *robust* species, with a broadly flaring zygomatic arch, large molars, enormous jaws and a striking crest on top of the skull (fig. 1.5). Of this type of hominin three species are known, which can be counted among the genus *Paranthropus* (literally 'side-men'). The large, bony ridge on top of the skull, called a *sagittal crest*, is an attachment site for the jaw muscle, musculus temporalis, which must

Fig. 1.5: The fossil skull of *Paranthropus aethiopicus*, one of the 2.5-million-year-old robust Australopithecines. Notable are the widely flaring cheekbones (zygomatic arch), the small brain and the huge sagittal crest on top of the skull. The *Paranthropus* species were dedicated herbivores with heavy jaws and represent an extinct sideline in hominin evolution.

have been huge. The *Paranthropus* species most likely were herbivores that, with their flat molars with thick enamel and enormous jaws, were very well able to grind leaves and nuts that were not easy to digest, hence the popular name: Nutcracker Man. The other evolutionary lineage is characterised by lighter skulls with a small sagittal crest or only a longitudinal vault or keel at the centre of the head. The species belonging to this lineage are called *gracile*.

That robust species could exist next to gracile ones for a prolonged period of time raises the question on differences in their lifestyle. Those were probably in the food they ate: the robust species specialised in plant material whereas the gracile species became omnivores.

The side-by-side evolution of different species in the same habitat (*sympatric speciation*) is an evolutionary mechanism that does not occur very often. The reason is that reproductive isolation (absence of cross-breeding between two groups) isn't likely to occur if individuals live in the same place and keep meeting each other. In order to form two separate species, reproductive isolation is required. Usually a possibility for new species to evolve is presented when populations divide (split up), start living in different environments, and adapt to different circumstances, while isolated from each other. This is called *allopatric speciation*. Examples of both evolutionary mechanisms, sympatric and allopatric speciation, can be found in the hominin tree (table 1.3).

The robust species didn't make it; they all became extinct. Possibly they were adversely affected by their own food specialisation. Elsewhere in evolutionary history we observe that specialists also find difficulty in coping with changing conditions. Humans descend from the gracile *Australopithecus* species, an

Table 1.3: **Different modes of speciation**

Process	Definition	Examples
Sympatric speciation	Origin of new species while ancestral and descendant species keep occupying the same habitat	Robust species (genus *Paranthropus*), gracile *Australopithecus* and early *Homo* species living closely together in East Africa
Allopatric speciation	Origin of new species after a group separates and occupies new territory	Evolution of *Homo erectus* in Asia from *Homo ergaster* in Africa
Parapatric speciation	Origin of new species on the borders of the distribution range, caused by local evolutionary pressures different from the centre of the range	Evolution of *Homo neanderthalensis* in Europe and Asia from *Homo heidelbergensis* during the ice ages

evolutionary lineage of opportunistic, omnivorous ape-men who were better able to deal with the changes in the African climate, 2.5 million years ago.

The earliest hominins

We know little about the first period of the hominins, 7 to 4 million years ago, because little fossil material remained. The best-known species is *Ardipithecus ramidus* ('Ardi'), 4.4 Ma old. The first discovery, in Ethiopia, was reported in 1994, but Tim White described the species in detail much later, in 2009. Another, older incomplete collection of fossil bones, 5.5 Ma old,

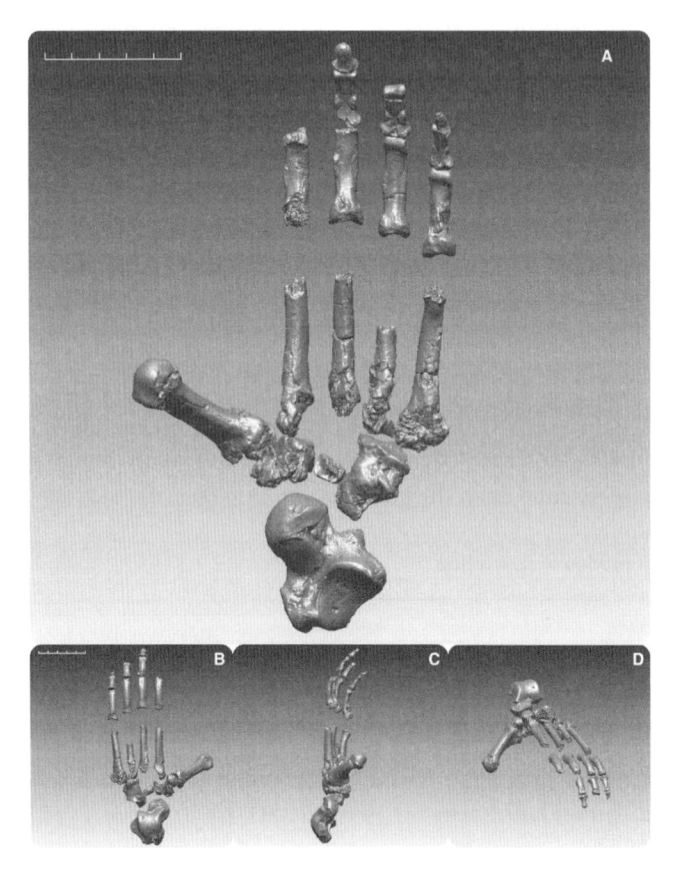

Fig. 1.6: The fossil foot bones of *Ardipithecus ramidus*, an early representative of the Homininae, 4.4 million years old, discovered near Aramis, Ethiopia. Notable is the wide-spread, apparently opposable big toe, which indicates a lifestyle that partly took place within the trees. The fossil lacks a heel bone and some of the phalanges. A bottom view, B top view, C side view, D oblique top view.

was initially described as a separate species, *Ardipithecus kadabba*, but this is now usually considered an early form of *A. ramidus*. The skeleton of Ardi shows obvious signs of upright walking, but the foot is still very primitive. It is a prehensile foot with a widely spread big toe, which is typical for an arboreal lifestyle (fig. 1.6). The upright gait on two legs was probably not as far advanced. The brain volume was still hardly larger than that of the chimpanzee. In all respects it is a primitive hominin, one that probably started to explore the savannah, yet was still also active in the trees. Some authors therefore dispute that Ardi is one of the hominin species.

Other early hominins are *Orrorin tugenensis* and *Sahelanthropus tchadensis*. Of *Orrorin* merely an incomplete femur and some molars were found, so it is impossible to determine what this species looked like. The upper part of the thighbone though is very human-like and clearly indicates that they were walking upright. Brigitte Senut and Martin Pickford presented *Orrorin* in the year 2000, which is why sometimes it is also called 'Millennium Man'. The presentation initially generated a lot of discussion as Pickford and Senut argued that *Orrorin*, 6 Ma old, was the direct ancestor of *Homo*, which would move the transitional genus *Australopithecus* to a sideline. Nowadays, however, *Orrorin* is generally accepted as an early hominin.

Sahelanthropus is an entirely different story. It is the oldest hominin to have been found so far, 6 to 7 million years old, described in 2002 by Michel Brunet and his staff. It was discovered in Chad, so not in East Africa, but in Central Africa. In addition, the skull has a number of typical characteristics, such as a narrow, yet flat face and an extremely heavy, bony arch above the eye sockets (*torus supraorbitalis*), which continues from left to right. The position of the foramen magnum at the bottom of the skull, normally a good indicator for walking on two legs, is unclear as the fossil is severely deformed at the underside. As a consequence, some palaeontologists believe that *Sahelanthropus* is not at all a hominin, but an unknown gorilla species. Most textbooks, however, place *S. tchadensis*, along with *Ardipithecus* and *Orrorin*, at the base of the hominin tree. We also did this in fig. 1.4, although we as well are in doubt about the hominin status of *Sahelanthropus*.

All three early hominins fall within the time frame that is predicted from the genetic divergence between human and chimpanzee. As will be shown in chapter 5, it is possible to estimate, from the DNA of living species, the time that has passed since they had a common ancestor. These estimates vary from 6.5 to 7.5 million years (although some estimates go down to 12 Ma). Fossils older than 7 million years and belonging to the hominins have never been found, so in this regard the genetic and palaeontological data are properly corresponding.

The heyday of the ape-men

The second period of the hominins, from 4 to 1.5 million years ago, can be considered as the heyday of the ape-men. Many species emerged, living side by side, yet with a different ecology. The best-known one is *Australopithecus afarensis*, which was discovered in 1974 in Ethiopia and described by Donald Johanson in 1978. The world-famous fossil was given the nickname 'Lucy', due to the song 'Lucy in the Sky with Diamonds', on the album *Sergeant Pepper's Lonely Hearts Club Band* by the Beatles, which was played all the time in the camp. Lucy is the ancestral fossil most appealing to the imagination, as it is so complete. About 20 different *afarensis* specimens were found later, giving us a good picture of what she may have looked like.

It quickly became apparent after the first discovery that different *A. afarensis* fossils appeared to have different body sizes, which was attributed to *sexual dimorphism*: the males were about 30% larger than the females. Initially, this conclusion seemed questionable, yet the *communis opinio* among palaeontologists is now that the female Lucy was much shorter than her male peers. The large dimorphism is remarkable as modern humans are hardly dimorphic anymore. There is also no evidence for sexual dimorphism in the older species *Ardipithecus ramidus*, so dimorphism may be a peculiarity of the Australopithecines.

What could be the meaning of the disappearance of sexual dimorphism in the later hominins? Outside hominins, sexual dimorphism especially occurs with species that live in harem groups where one male dominates multiple females. Among mammals (for instance, gorillas, baboons, deer, lions, walruses, and the like) there is a significant positive correlation between the size of the harem and the degree of sexual dimorphism. Given that the sex ratio even in the case of harems is usually 1, in these populations a majority of the males has no access to females. This causes fierce competition between males and creates a strong evolutionary pressure in favour of a large and imposing body, often reinforced, for instance, by enlarged canines. This is called *intra-sexual selection*. The disappearance of sexual dimorphism could mean that in the course of their evolution hominins would more rarely live in harems, and their behaviour became more monogamous, like that of the current gibbons.

Another remarkable aspect of *A. afarensis* is that it shows clear signs of a transitional species that has a number of characteristics left in its body corresponding with the ancestors, while modern characteristics are also present. This is called *mosaic evolution*. During the evolution of a body plan not everything changes at once, as a result of which each species shows a

Table 1.4: Primitive and derived characteristics of Australopithecus afarensis

Plesiomorphic (primitive) characteristics	Small brain volume, prognathic face, canines protrude from the dental arch, diastema present
	Chest funnel-shaped, shoulder joint oriented towards skull
	Short legs, curved phalanges, hallux abduction 20-30 degrees
Apomorphic (derived) characteristics	Pelvis shortened and widened, bowl-shaped, tilted, iliac arch bending backwards, inclination angle of femoral neck smaller than 90 degrees
	Valgus angle of the knee joint about 10 degrees

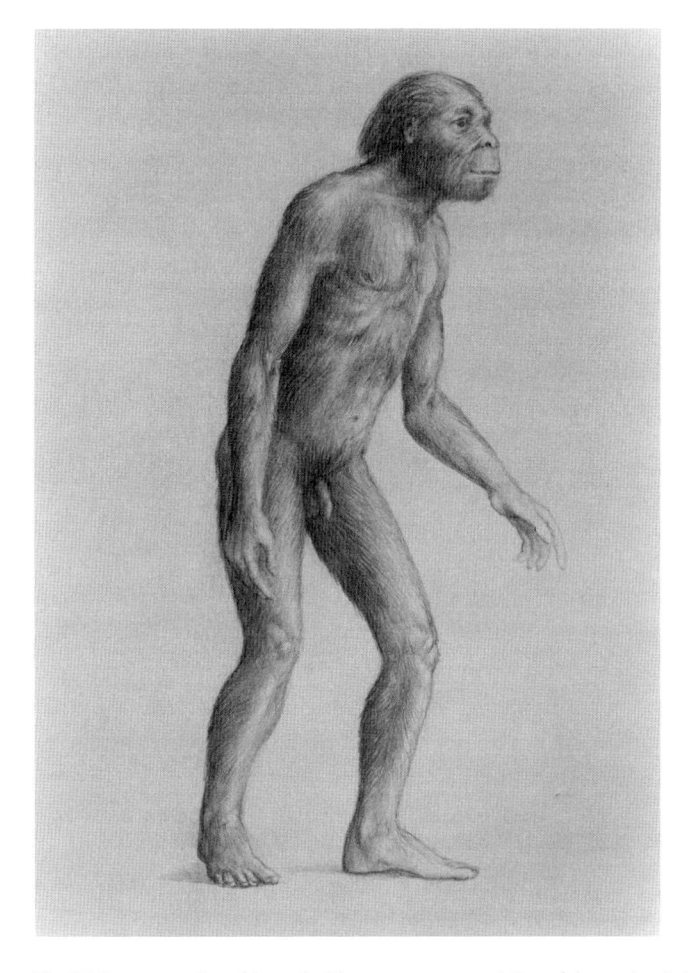

Fig. 1.7: Reconstruction of *Australopithecus*, a group consisting of six species, living in Africa between 4 and 2 million years ago. Although *Australopithecus* did walk upright, not all parts of the body were fully adapted to that, and the brains were still small.

mixture of primitive (*plesiomorphic*) and derived (*apomorphic*) characters. In table 1.4 an overview is provided of the plesiomorphic and apomorphic traits of *A. afarensis*.

Apart from *afarensis* six other more or less familiar hominins are considered part of the genus *Australopithecus*: *A. anamensis, A. africanus, A. bahrelgazali, A. garhi, A. sediba* and *A. deyiremeda*. Of the latter species, reported in 2015, as yet only parts of the jaws and some molars have been found. Also, *A. bahrelgazali* and *A. garhi* are rather incomplete.

A. africanus is the best known of all these species. Raymond Dart described the first fossil of *A. africanus* as early as 1924. It was discovered in a quarry near the town of Taung in South Africa. The genus name *Australopithecus* in fact means 'ape from the south' and *A. africanus* was the first to be called this. Together with the skull, part of the brain was also fossilised, which is why the Taung fossil plays a role in the discussion about the evolution of language, as we will see in chapter 6. Another particularity is that the fossil is of a child, which is why the discovery is known as the *Taung Child*. Back in 1924, there was a lot of confusion about Dart's discovery and for a long time *A. africanus* was not accepted as part of hominin ancestry. Now, when added to similar fossils, we know that it is a typical *Australopithecus*, slightly less robust than *A. afarensis* and, given the dating, to be regarded as an evolutionary descendant of *afarensis* (*cf.* fig. 1.4).

In 2010 a new *Australopithecus* was added to the already known species: *A. sediba*, discovered in the Malapa cave near Johannesburg in South Africa, and described by Lee Berger. The striking thing about this species is its late dating, namely 1.8 Ma BP. In those days, *Homo habilis* was already present in East Africa. As *A. sediba* turned out to have a number of modern characteristics, the discovery led to the speculation that not *H. habilis*, but *A. sediba* is the ancestor of the later hominins like *H. erectus*. Other than that, *A. sediba* strongly resembles *A. africanus* and some authors believe that *sediba* is a late *africanus*. We have implemented this interpretation in fig. 1.4.

Finally, we mention another mysterious species, which is certainly part of the heyday of the ape-men, but whose position in the ancestry is absolutely obscure: *Kenyanthropus platyops*, presented by Maeve Leakey in 2001. The name means 'flat-faced man of Kenya', referring to the distinctively flat face, a feature that reminds us of the later species *Homo rudolfensis*. Some authors therefore consider *K. platyops* an ancestor of *H. rudolfensis,* who was renamed *Kenyanthropus rudolfensis* for that purpose. *K. platyops* was the contemporary of *A. afarensis*, which suggests that it derives from an older hominin, yet which remains uncertain. In fig. 1.4 we present it as a descendant of the early hominins, but the precise position is by no means obvious. *K. platyops* does, however, fit into the view that between 4 and 1.5 million years ago different Australopithecines lived side by side in Africa.

With three *Paranthropus* species, seven *Australopithecus* species, *K. platyops* and probably another two *Homo* species (*H. habilis* and *H. rudolfensis*, which will be discussed below), it is justified to call the period of 4 to 1.5 million years ago the heyday of the ape-men.

The first *Homo*

The transition of *Australopithecus* to the genus *Homo* has always been a point of discussion in human palaeontology. The oldest species called *Homo* is *H. habilis* (Handy Man), presented as such in 1964, by Louis Leakey, together with two other famous palaeontologists, Philip Tobias and John Napier. The material, found in the Olduvai Gorge in Tanzania, consisted of remains of hand and foot bones, an upper jaw and part of a skull, 1.8 million years old. Later, a complete skull was found. The fossils were associated with stone tools, found at the same location, hence the name *H. habilis*. But the brain capacity of the skull was cause for concern, as it was remarkably small, no more than 640 cm³. Many scientists felt that a hominin could only be regarded *Homo* if the brain volume exceeded 750 cm³ (the so-called *Cerebral Rubicon*). The acknowledgement of Leakey's discovery required the genus *Homo* to be redefined: the ability to produce tools rather than the brain volume should be decisive.

After multiple fossils were discovered at the same location, a second problem arose: the variation in the shape of the skulls turned out to be significant. That is why the fossils were divided into two species: *H. rudolfensis* with a flat face, larger brain, and a somewhat robust appearance, and *H. habilis* in the strict sense, with a smaller brain volume, yet less robust (fig. 1.8).

The dispute about *H. habilis* is still ongoing today. Several scientists prefer to remove *habilis* from the genus *Homo* and rename it *Australopithecus habilis*, while *H. rudolfensis* should be called *Kenyanthropus rudolfensis* (*cf. K. platyops*). This suggestion is substantiated by the discovery in 2015 of a fossil lower jaw of 2.8 Ma old, which is reckoned with *H. habilis*, though also showing features of *Australopithecus*. In addition, old stone tools were discovered in Kenya, no less than 3.3 million years old, thus from the age of *Australopithecus*, although *H. habilis* was defined as the first user of stone tools.

It seems that at the time of early *Homo* there was no fixed relationship between the appearance of a species and its ecology. The different ecological features, for instance, expansion of the diet, the use of stone tools and extension of the life cycle, seem to appear like packages in different species. Perhaps

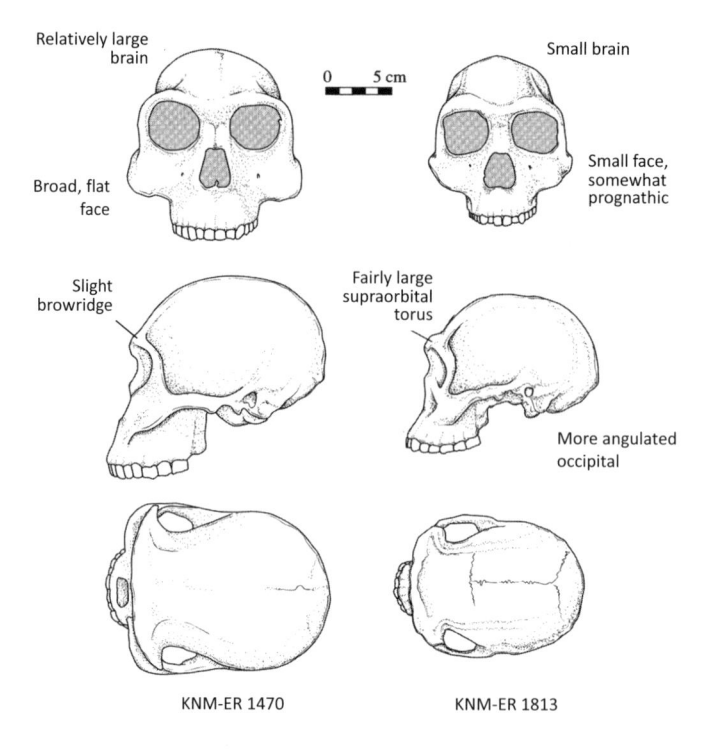

Fig. 1.8: Two groups of fossils, indicated as KNM-ER 1470 and KNM-ER 1813, found in Kenya, used to be both regarded as *Homo habilis*. Nowadays KNM-ER 1470 is renamed *Homo rudolfensis* and KNM-ER 1813 *Homo habilis* in the strict sense, the ancestor of *H. ergaster*. If that is a fact, *H. rudolfensis*, sometimes called *Kenyanthropus rudolfensis*, would be a dead-end sideline. But other interpretations exist, such as that KNM-ER 1813, due to its small brain size, should be called *Australopithecus habilis*, and *A. sediba*, rather than *A. habilis* would be the ancestor of *Homo*.

this ecological approach, suggested by Bernard Wood in 2014, fifty years after Leakey's publication, can put some order in the *habilis* complex. If even *Australopithecus* was able to produce primitive hand axes, there is also no more reason to call *H. habilis* '*Homo*', which would clear the way for thoroughly redefining the early *Homo* complex, which perhaps should not be called *Homo* after all! As yet, such a radical action to stir things up has not been suggested.

While the *Homo habilis* complex is obviously in disarray, there is little doubt about the status of what comes after. *Homo erectus* is a genuine *Homo*. Thus we are, 1.5 Ma BP, at a tipping point in human evolution: no more ape-men or anthropoids of doubtful status, but real humans. *H. erectus* is associated with a large number of innovations: a leap forward in brain volume, fully upright walking with all associated adaptations, the use of a new type of hand axe, life in a home base, the practising of hunting, and, quite remarkable, large-scale migrations out of Africa.

Fossil remains of *H. erectus* are not only connected with Africa, but also with Indonesia (Java Man), China (Peking Man) and Georgia. The Dutch palaeontologist and physician Eugène Dubois found the first hominin fossil outside Europe in 1891, near the town of Trinil, along the Solo River in central Java (fig. 1.9, fig. 1.10). The remains consisted of a calvarium, a femur and several molars. Dubois named it *Pithecanthropus erectus*, upright walking ape-man. From the femur it was obvious that the species was bipedal, while the calvarium suggested a rather primitive skull.

The original calvarium and the femur of Java Man can be admired in Naturalis National History Museum in Leiden, The Netherlands. It is certainly worth a visit, as it is one of the very few hominin fossils of which the original is being exhibited. *H. erectus* is characterised by a rather heavy eyebrow ridge (torus supraorbitalis), a longitudinal vault on top of the skull (*sagittal keel*), a somewhat protruding (*prognathic*) face, the absence of a chin and a brain volume that is clearly larger than that of *Australopithecus*.

When Dubois returned to Europe with his *Pithecanthropus* fossil, it emerged that only few scientists were able to assess the value of his discovery. In those days the only human fossil that it could be compared with was that of the Feldhofer Neanderthal. While most palaeontologists at the time considered Neanderthal an ancestral, primitive form of *H. sapiens*, Java Man was hard to place. In addition, its age was also uncertain. Only in 2014 a reliable estimation was made. Shells from the Dubois collection were dated,

Fig. 1.9: Portrait of Eugène Dubois, the Dutch physician who discovered the remains of Java Man, initially called *Pithecanthropus erectus*, later renamed *Homo erectus*.

Fig. 1.10: View on the Solo River from the museum near Trinil, Central Java. On the right bank is the excavation site where, in 1891, Dubois found Java Man.

by use of the K/Ar method, at 540-430 ka; it is highly likely that also Java Man fits into that time frame.

In 1923 an important discovery was added to the fossil collection of Asian hominins: Peking Man, excavated at Zhoukoudian, near Beijing. Over a number of years several skulls and many teeth and molars were found there, which altogether provide a complete picture. Also near the town of Sangiran on Java, in 1934, the German palaeontologist Ralph von Koenigswald discovered new fossils from approximately the same time. Although the Chinese fossils seemed a bit more modern than the Javanese, it slowly became obvious that they all belonged to a single species, with a wide natural range in Asia: *Homo erectus*.

After the discoveries in Asia, fossils were also found in Africa that could be classified as *H. erectus*. The best known is Nariokotome Boy, an almost complete skeleton of a young man, found in 1984 near Turkana Lake in Kenya. Later, however, it was decided that the differences between the African *erectus* and the Asian Java Man and Peking Man were too large. The African *erectus* was subsequently renamed *H. ergaster*, whereas the name *H. erectus* was still reserved for the Asian forms. Lately, the situation has become even more complicated. In 2015, the South African palaeontologist Lee Berger (who also reported *A. sediba*) published the discovery of an extensive fossil collection in a hard-to-reach chamber of the Rising Star cave system near Swartkrans, South Africa. This collection consists of no less than 1,550 specimens, parts of skulls as well as postcranial bones, which belong to at least fifteen different individuals. The species were given a name, *Homo naledi*. In 2017 a similar collection of fossils was reported from another chamber in the same cave system, and this material was dated using various methods that converged to a range of 236-335 ka BP. This surprisingly young age, overlapping with the rise of *Homo sapiens* in East Africa, suggests that *H. naledi* is a primitive form of *H. ergaster* that survived for a long time in South Africa, isolated from any other hominin lineage. The location where the fossils were found is equally enigmatic. Did *H. naledi* deliberately carry the bodies of the deceased, crawling through narrow holes in the pitch dark to bury them in a remote chamber of the cave? We don't know.

Outside Africa, between 1990 and 2013, *H. erectus* discoveries were made in Georgia, near the town of Dmanisi. These fossils, not surprisingly indicated as 'Georgia Man', or *H. georgicus*, comprise five well-preserved skulls with significant mutual differences in brain volume. The remarkable thing is that they are older than the Asian *erectus* fossils (up to 1.8 million years). Some of the skulls are also substantially more primitive; they are even closer to *H. habilis* than to the Asian *H. erectus* or the African Nariokotome Boy. Finally, fossils were also discovered in Spain, in an excavation near Atapuerca, which can be classified within the *erectus* complex, although they have been described as a separate species, *H. antecessor*.

In fig. 1.11 results are shown of a multivariate statistical analysis by the team of David Lordkipanidze, the project leader of the Dmanisi excavation. This graph shows that the evolution of skull morphology can be depicted along two axes: increase of brain volume, making the skull more globular, and diminishment of the protruding 'snout' (reduction of *prognathism*), making the face more flat (*orthognathic*). These two trends are only loosely correlated: the level of prognathism significantly differs between fossils as well as among skulls with similar brain volume. The *erectus* fossils do not seem to break down into two or three clear clusters. For that reason Lordkipanidze considered all of them to be *H. erectus*, a species that, according to him, must have been subject to a large degree of intraspecific variation. However, other palaeontologists, *e.g.* Ian Tattersall, strongly deny this proposal.

Whether *H. erectus* is one species, or should be subdivided into five separate species (*erectus*, *ergaster*, *georgicus*, *antecessor* and *naledi*), is a matter of taste. The biology does not provide a solution, because cross-breeding

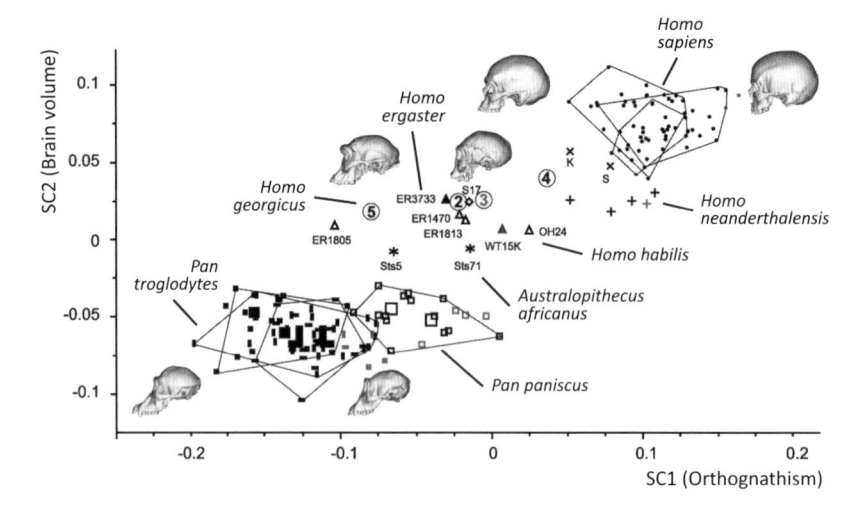

Fig. 1.11: Representation of a multivariate statistical analysis of morphological skull traits of a number of *Homo* species, compared with chimpanzee (*Pan troglodytes*) and bonobo (*Pan paniscus*). The analysis shows that changes occur along two axes: increase of brain volume and decrease of prognathism.

cannot be established and the fossils are too old to isolate DNA from. It is a matter of how much morphological variation one allows within a species and how consistent the differences are between the delineated species. The various positions of palaeontologists can be categorised as splitters and lumpers (table 1.5).

As the last species arising from the *erectus* complex, we mention the discovery of the enigmatic *H. floresiensis*. In 2004, Australian scientists found a very peculiar hominin with a tiny brain (in the range of *Australopithecus*), on the Indonesian island Flores, in the Liang Bua cave. The fossil was very short, which made him known worldwide as the 'hobbit'. But the

Table 1.5: Distinction of hominin species from the perspective of splitters and lumpers

Species names according splitters	Intermediate position	Species names according to lumpers
H. erectus, H. ergaster, H. georgicus, H. antecessor, H. naledi	H. erectus, H. ergaster	H. erectus
H. habilis, H. rudolfensis		H. habilis
A. africanus, A. garhi, A. sediba	A. africanus, A. sediba	A. africanus
A. afarensis, A. anamensis, A. bahrelgazali, A. deyiremeda	A. afarensis, A. anamensis	A. afarensis

most striking feature was the dating of 60,000 years; it could have been a contemporary of *H. sapiens*, when *H. erectus* was long extinct.

A fierce debate developed around the hobbit. Some scientists claimed that the small brains were a pathological defect, a case of microcephaly. Others, the Indonesian palaeontologist Teuku Jacob in particular, stressed that the small body fits in the variation of indigenous peoples on the Indonesian islands, like, for instance, the pygmies of Rampasasa, a village on Flores close to Liang Bua. The hobbit could therefore simply be *H. sapiens*. Nevertheless, this does not explain its small brain volume. The most likely explanation is still that Flores Man is a descendant of *H. erectus* who lived in isolation for a long time, in a small population, and was subject to dwarfism. This frequently occurs on islands and can also be seen in mammals (for instance, the extinct pygmy elephants of Sicily, Malta and Crete). Present-day pygmies on Flores indeed show signatures of positive selection for small body size in their genomes.

The Dutch palaeontologist John de Vos pointed out in 2007 that the hobbit is not only small, but also shows juvenile characteristics. The large eye sockets, as well as other parts of the skull indicate this. This would make the evolution of the hobbit an example of *neoteny*: retention of juvenile characteristics in the adult stage. As the evolution of *H. sapiens* itself also shows neotenic characteristics, it would explain why the skull of the hobbit looks so human. In chapter 2 we will further discuss the importance of neoteny and other forms of heterochrony in human evolution.

In 2016, the interpretation that *H. floresiensis* is a dwarfed descendant of *H. erectus* was confirmed by the description of a partial mandibula and a series of molars, found in an excavation near Mata Menge, 60 kilometres east of Liang Bua on Flores, headed by the Dutch-Australian scientist Gerrit van den Bergh. These fossils are much older than the hobbit of Liang Bua, an estimated 700,000 years, which makes it impossible that they belong to *H. sapiens*. The conclusion seems inevitable that in the Malay Archipelago, *H. erectus* was subject to a form of island dwarfism, which we have seen in mammals, and that some of those dwarf populations lived on in isolation for many years, until the time that modern *H. sapiens* arrived there.

Towards modern times

The hominin fossil that was first found is the last to be discussed here, as it is so recent: Neanderthal or *Homo neanderthalensis*. In 1856, in a quarry called Feldhofer, in the valley of the Düssel River (locally called Neander, after Joachim Neander, a popular composer of church hymns and psalms from the seventeenth century), labourers found a collection of petrified bones and skull parts that seemed, in all respects, to belong to a primitive human. The quarry workers took the material to a local teacher, Johan Fuhlrott, who in turn contacted an anthropologist, Hermann Schaaffhausen, stationed in Bonn. Schaaffhausen wrote a publication about it in 1858. He called Neanderthal a separate species, *Homo neanderthalensis*, distinct from *H. sapiens*. It is remarkable that this publication appeared virtually at the same time as the famous book by Charles Darwin, *On the Origin of Species* (1859). In his book, however, Darwin makes no reference whatsoever to Neanderthals, and in one of his later books, the *Descent of Man*, published 1871, he mentioned Neanderthals just once, observing that 'some skulls of very high antiquity, such as the famous one of Neanderthal, are well developed and capacious'. Darwin seemed to consider Neanderthal as a primitive form of *H. sapiens*. Even Thomas Huxley, in those days the most fervent supporter of Darwin's theory of evolution, stressed the continuity between Neanderthal and humans and was not willing to regard the Feldhofer fossil as a separate species. History proved Huxley and Darwin wrong; the interpretation of Schaaffhausen was correct.

One of the reasons that Neanderthal was misunderstood was the fact that the scientific world in those days was dominated by the German physician Rudolf Virchov. Around 1860, he claimed that the Neanderthal fossil belonged to a Cossack who had sunk in a swamp with horse and all, while battling Napoleon. The Cossack allegedly suffered from rachitis and as a result of all the frowning in pain, the heavy eyebrow ridge would have appeared. In his very entertaining book from 2015, the American palaeontologist Ian Tattersall described how the 'rickety Cossack' should serve as a warning example of the errors that palaeontologists occasionally make when their discoveries do not fit into the leading framework at the time.

As more Neanderthal fossils were being discovered, it became obvious that Neanderthal was a modern hominin who, however, in many respects deviated consistently from modern humans. The skulls of Neanderthal fossils have a brain volume approximately similar to ours (even somewhat larger on average), yet the cranium is more curved backwards, instead of upwards, as a result of which the forehead is low. Furthermore, the brow ridge is heavier, while the face beneath the nose points somewhat forward. During the bone practical in the authors' course on human evolution, students have no trouble whatsoever telling a Neanderthal skull from an *H. sapiens* skull (fig. 1.12). Looking at the rest of the body we see that the Neanderthal skeleton is clearly more robust than that of humans: relatively heavy bones, a short shinbone relative to the thighbone, and a sturdy posture (fig. 1.13). The form of the body shows similarities with that of the current Sami, Koryak and Greenlanders, suggesting that Neanderthal was a cold-adapted form of human. Neanderthals lived from 600,000 years ago until 30,000 years ago, during the Pleistocene, when ice ages dominated the appearance of the earth in Europe and northern Asia.

Our view of Neanderthals has been changing constantly over the years. Around 1920, the French palaeontologist Marcellin Boule described a Neanderthal fossil, found in the French town of La Chapelle-aux-Saints, as a savage that was barely able to walk upright. For many years this image dominated the opinion of the general public about Neanderthals, but it was completely unjustified. On the other hand, sometimes a romantic picture exists of Neanderthals being people like us; making music and having a culture that included worshipping of cave bears. Neither this is likely to be correct. The idea that Neanderthals worshipped cave bears is based on discoveries in the Drachenloch cave in Switzerland, described between 1917 and 1923 by Emiel Bächler. It is most likely that the large numbers of cave bear fossils, which were found there, as well as in other mid-European caves, mixed with Neanderthal fossils, are a consequence of material flushed in from the surrounding landscape and do not result from deliberate human action. The modern image of Neanderthals is that of a human species, well adapted to its environment, with his own culture of stone tools, and a characteristic way of using the space around him, but who did not create cave drawings, probably didn't know any language and was the lesser compared to *Homo sapiens* in terms of cognition.

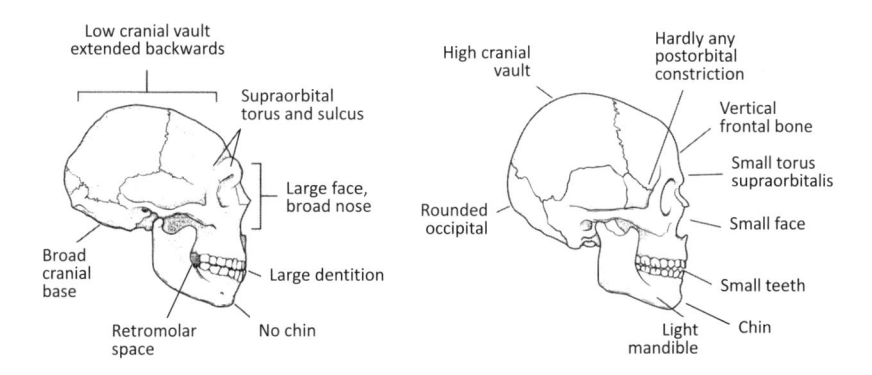

Fig. 1.12: Lateral view of the skulls of Neanderthal and *H. sapiens*, with a number of diagnostic features, which demonstrate that morphologically the two were quite distinct species.

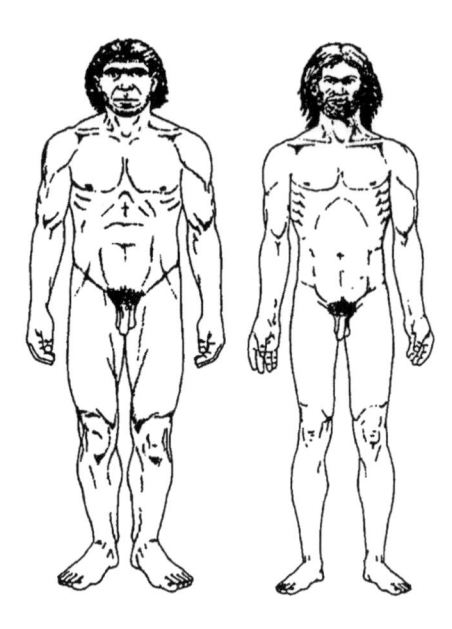

Fig. 1.13: Reconstruction of Neanderthal (*Homo neanderthalensis*, left) and Cro-Magnon man (*H. sapiens*, right), who both lived in Europe, 40,000 years ago. The Neanderthal's build was heavier and stockier, reflecting adaptation to the cold European climate of the Pleistocene.

In one significant respect we should be thankful to Neanderthal: he was probably the one who invented the use of fire. Under the influence of the British primatologist Richard Wrangham, for years it was assumed that the use of fire to cook food dates back to *Homo ergaster*, 2 million years ago. However, this reasoning was mainly based on the emergence of hunting and not so much on archaeological evidence. Looking at the remains of fire pits, such as heated pebbles, charred wood and scorched bones, anthropologist Wil Roebroeks from Leiden University concluded, in 2011, that the habitual use of fire in Europe couldn't date further back than 300 to 400 ka. This conclusion is supported by work conducted at Tabun Cave in Israel. If true, it would imply that Neanderthals, originating in southern Europe 600 ka ago, during the cold Pleistocene climate, didn't have any fire yet and only later learned to control the use of fire. Also Roebroeks' analysis implies that *H. sapiens*, unless he reinvented the art of fire making, learned or copied it from Neanderthals.

The modern vision is that Neanderthals and *H. sapiens* are sibling species. They did not succeed one another, but descended from the same ancestor and lived side by side in the Middle East, Europe and the nearby Asia, from approximately 90,000 to 30,000 years ago. The common ancestor that we share with Neanderthal is *Homo heidelbergensis*, a hominin that was initially described on the basis of a discovery at Heidelberg (the 'Mauer jaw'), but

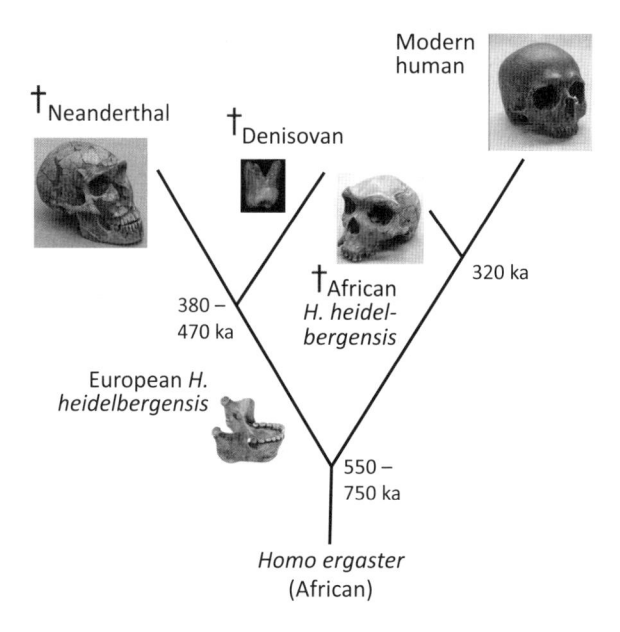

Fig. 1.14: Evolutionary relations between the three recent human species, Neanderthal, Denisova Man and *Homo sapiens*. It is assumed that all three of them are descendants of *Homo heidelbergensis* (Heidelberg Man). DNA data show that Neanderthals and Denisovans were more related to each other than each of them to modern man.

whose origin is in Africa. The 'Florisbad skull' found by T.F. Dreyer in 1935 near Bloemfontein, South Africa, and dated 259 ka old, may also be included in *H. heidelbergensis*, although it was originally described as a separate species, *Homo helmei*. *H. heidelbergensis* resembles Neanderthal, with an at least as heavy, robust skull, but has a few characteristics (*e.g.* a lightly sagittal keel) that remind us of his ancestor *H. ergaster*. Our evolutionary reconstruction thus runs as follows: in Africa, from *H. ergaster* evolved a new species, *H. heidelbergensis*, that migrated to Europe and gave rise to Neanderthal. *H. heidelbergensis* also lived on in Africa, and later gave rise to a new lineage, from which *H. sapiens* emerged, followed by *H. heidelbergensis* becoming extinct in both Europe and Africa, and later the extinction of *H. neanderthalensis* (fig. 1.14).

One more species must be added to this story, the Denisovans. This hominin does not yet have a scientific name because it is only known for a phalanx (finger bone) and a molar. But from these fossils DNA has been isolated that clearly indicates that it concerns a species related to Neanderthals as well as to us. In 2010, the group of Svante Pääbo, a world-renowned investigator of ancient DNA at the Max Planck Institute in Leipzig, published this shocking discovery. The phalanx, which was initially assumed to be of a Neanderthal, was found in Denisova Cave in the Altai mountain range in Siberia. Due to the frozen condition, the DNA was well conserved. On the basis of their nuclear DNA, Denisovans are regarded as a sibling species of Neanderthals. The split between the two is dated at approximately 400,000 years ago, probably in Europe or nearby Asia, in any case before modern humans appeared in Africa (fig. 1.14).

The Denisovan molar is extremely large, which suggests that a very robust human is involved here. A glimpse of the skull morphology may be obtained from two incomplete crania excavated in 2017 by a Chinese team in Xuchang, in the province of Henan. Both in terms of dates (105-125 ka) and in terms of morphology, these skulls could very well have been ascribed to Denisova Man, however, the Chinese team prefers to consider them primitive *H. sapiens*, a view in line with the theory of multiregional evolution, discussed in chapter 5. We have tentatively labelled Denisova Man as '*Homo denisovae*' in fig. 1.4, but this name is not yet accepted.

Homo sapiens itself is approximately 320 ka years old and originated in Africa. There is an overwhelming amount of genetic evidence to support an African origin. It is also well in line with the fossils. Old hominin fossils that can still be attributed to *H. sapiens* are remains that were found near Omo, Ethiopia, presented in 1967 by Louis Leakey. These rather incomplete skull parts are estimated at 196,000 years old. Later, Tim White found a quite complete skull of 160,000 years old near the town of Herto, also in Ethiopia. These first African people are indisputably *H. sapiens*, although they still have a few primitive characteristics that remind us of *H. heidelbergensis*. Recently, fossils discovered by the French palaeontologist Jean-Jacques Hublin, in Jebel Irhoud, a site in Morocco, dated at 315 ka BP, were also presented as *Homo sapiens*, although the cranial shape of these hominins is still more primitive, rather Neanderthal-like.

At this point we stop the story of our ancestors. *Homo sapiens* itself, its migrations, tools and culture, will be further discussed in detail in later chapters. We mentioned all 26 species of hominins known to date, and discussed some of them at length. Maybe we leave the reader confused by the large number of scientific names. We recommend that they take another look at fig. 1.4 and envisage the journey once again. That will provide a good basis for the following chapters. We have hardly discussed the changes in physical appearance of the hominins and how some of those changes can be explained as adaptations to bipedalism. This will also be discussed in detail in the next chapters.

2. From ovum to human

If we want to understand how human beings came about, from a great ape walking on its knuckles, via an upright walking ape-man, up to a modern large-brained human, we should first become acquainted with the evolutionary mechanisms underlying those changes. The body plan of a species, which involves such things as the number of legs, the position of the eyes, the kind of respiratory system, the arrangement of internal organs, *etc.*, is established in its development. Evolutionary changes in the body plan are primed by mutations in the genes that regulate development. The collective set of developmental genes is sometimes called the toolbox of developmental biology. It is remarkable that the structure of the tools has seen little change in the course of evolution and the genes involved are fairly similar across species. Thus, crucial developmental processes of humans can be studied equally well in invertebrates. In this chapter we will discuss the principles of embryology and developmental genetics on the basis of a number of widely used model animals.

Heterochrony and Haeckel's law

In the nineteenth century the Estonian biologist Karl Ernst von Baer developed a zoological theory that included four laws of development, which still can be considered the foundation of comparative embryology today. He argued that during its development, each embryo first shows the general features of the group to which it belongs, and in its later stages shows more specialised features that discriminate it from other species in the group. Consequently, the similarity among embryos of a certain group is greatest in the early stages while later stages show more and more divergence. The laws of Van Baer were the basis for the concept of the *phylotypic stage* or phylotypic period, which was formulated by a number of biologists between 1960 and 1980. The phylotypic stage is defined as an early embryonic form that all representatives of a phylum go through. You can tell the phylum to which a species belongs from its phylotypic stage. Vice versa, showing such a stage is proof that the animal belongs to the pertinent phylum.

According to the Swiss zoologist Denis Duboule, the resemblance among the embryos of a phylum can be shown in the form of an hourglass (fig. 2.1). In the developmental phases before the phylotypic stage, the embryos differ due to differences in reproduction biology, particularly the amount

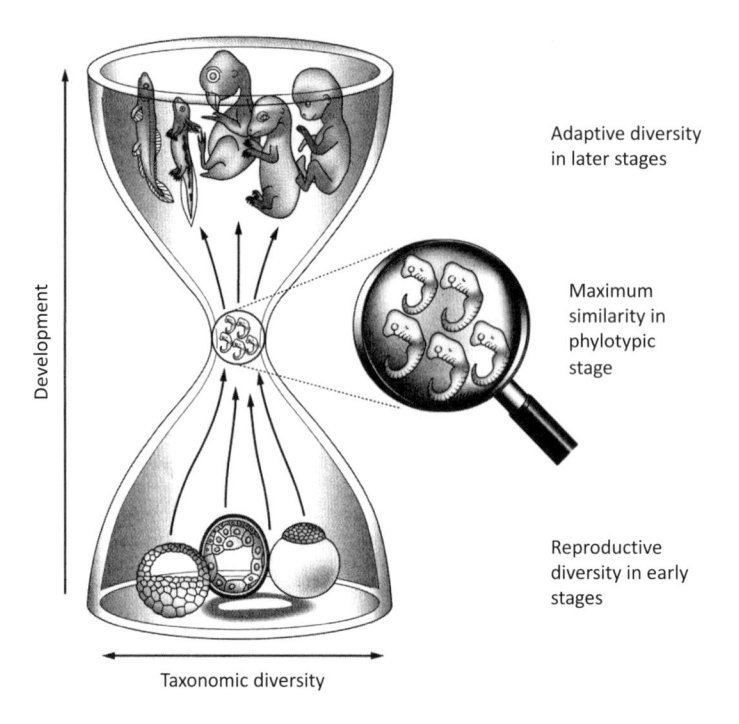

Fig. 2.1: Hourglass model for the morphological diversity of animal groups during their development, as suggested by Denis Duboule. The model indicates a developmental window of minimal morphological diversity due to developmental constraints in which the various members of a phylum all look alike. This 'phylotypic stage' characterizes the phylum.

of yolk included in the egg and the process of the first cleavages. In the phases after the phylotypic stage, morphological diversity increases due to all kinds of ecological specialisations. In between lies a developmental phase in which the coherence in the developmental programme is so strong that few evolutionary changes are being tolerated, as a result of which the resemblance with the common ancestor is maximal. In the case of Chordata (including vertebrates) this phylotypic stage consists of a 'tadpole' or '*pharyngula*'. Humans go through such a stage, with gill arches and a tail, as we will learn later in this chapter.

Ernst Haeckel (1834-1919), an influential and versatile German biologist, expressed the coherence between development and evolution even more strongly. Haeckel was a strong supporter of Darwin and made important contributions to the spread of evolutionary theory in continental Europe. But rather than following the laws identified by Von Baer, he formulated his own '*biogenetic law*', which is often summarised as follows: 'ontogeny recapitulates

Table 2.1: Quotation from Ernst Haeckel, *Generelle Morphologie der Organismen. Erster Band: Allgemeine Anatomie der Organismen* (1866), p. 197

Wir wissen, dass jeder Organismus während seiner Ontogenie eine Stufenfolge von niederen zu höheren Formen durchläuft, welche der Phylogenie seines Stammes in Ganzen parallel läuft, und wir können also von ersten Stadien der embryologischen auf der ersten Stadien der palaeontologischen Entwickelung durch Deduction zurückschliessen.	We know that each organism, during its ontogeny, goes through a succession from primitive towards higher stages, which is largely parallel to the phylogeny of its stem group, and therefore, through deduction, we can infer the first stages of the palaeontological development from the first stages of the embryological development.

phylogeny'. In other words: during the embryonic development, an embryo very rapidly passes through the stages of all its evolutionary ancestors. A human being is first a fish, then amphibian, reptile, and to finally become a mammal. Haeckel even argued that phylogeny is the mechanical cause of the succession of ontogenetic stages: evolution adds new stages to those already existing and so the older stages get compressed in early embryological development. Table 2.1 provides a quote from Haeckel's book, clearly indicating his point of view. Many people are familiar with the illustration he created, showing the embryos of fish, salamander, turtle, chicken, cow, rabbits and humans next to each other, stressing the similarities, while the adult animals largely differ. Haeckel's depiction has always been heavily criticised because of his showing the embryos in extremely idealised forms in order to make his point.

Von Baer did not like Haeckel's recapitulation theory. He criticised it fiercely in his fourth law of development. Still, Haeckel's views became immensely popular in the beginning of the twentieth century. Nowadays, however, no evolutionary biologist will accept the biogenetic law in the form in which Haeckel formulated it. But the fact that the resemblance between the embryos of a phylum has an evolutionary background is not questioned by anyone. The similarity between a human embryo and a fish larva is not caused by the fact that humans recapitulate a fish stage in their development, but by the fact that humans and fish share a common ancestor.

The relationship between development and evolution is also evident from the fact that many evolutionary innovations find their origin in a delay or acceleration of a part of the development process. In the evolution of a new body plan, some parts 'get stuck' in a juvenile stage of the old body plan, while other parts continue their development. This is what we call *heterochrony*: evolution by asynchrony in the development of different organ systems, especially reproductive development relative to somatic development. There are different forms of heterochrony, as is set out in table. 2.2.

Table 2.2: Terms used to indicate various aspects of evolutionary change by adjusted developmental timing

Term	Definition
Heterochrony	General term for changes in the timing of development as a mechanism for the formation of new body-plans and new species
Paedomorphosis	The retention of juvenile characteristics in the adult stage, by means of neoteny, by ceasing somatic development prematurely, or by starting somatic development later, while allowing reproductive development to proceed unchanged
Neoteny	A form of paedomorphosis in which the rate of somatic development is delayed in each stage
Peramorphosis	The appearance of new traits in the adult stage through accelerated development, early start of the development, or postponed completion of the development; opposite of paedomorphosis

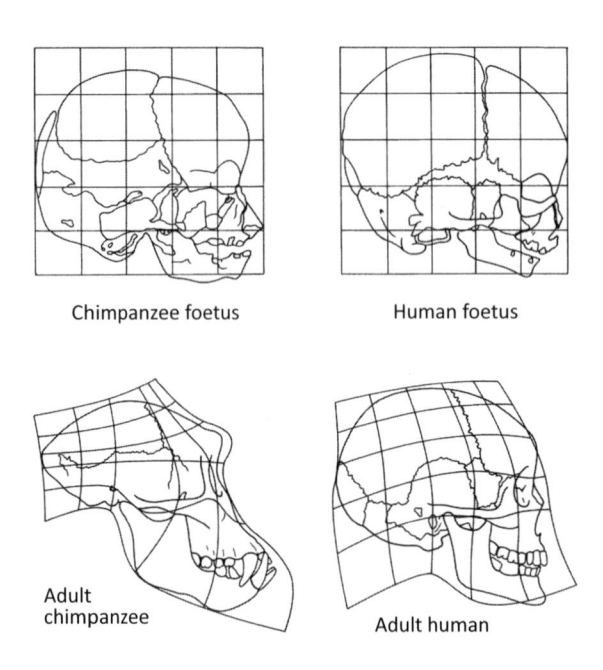

Fig. 2.2: The resemblance between skulls of humans and chimpanzee is much larger in the foetal stage than in the adult stage. The morphological changes of the skull during chimpanzee adult growth are lacking in humans: humans retain the juvenile shape.

Heterochrony has been very important in human evolution. Our brains, for instance, have many juvenile features, which allow us to learn new things during our whole life, while for a lot of animals this is limited to a short juvenile period. Another example is the shape of the human skull, when compared with that of the chimpanzee. The typical prognathic snout of an adult chimpanzee only emerges when the animal becomes adult (fig. 2.2). The baby chimpanzee has a flat (orthognathic) face, just as we have.

Heterochrony, and neoteny in particular (table 2.2), is a crucial aspect of the hominin evolutionary lineage. What's more, it seems as if the delay in juvenile development is a process that started with the first hominins, and proceeded continuously throughout the evolution, up to *H. sapiens* today. As a result, a human being can be regarded a baby chimpanzee that has become sexually mature. In chapter 1 we observed that *H. floresiensis* can be considered a neotenic form of *H. erectus*. The same applies to *Autralopithecus* relative to the great apes and to *H. sapiens* in relation to *H. heidelbergensis*. Neotenic characteristics of modern humans are: a flat face, small eyebrow, small nose, small teeth, large eyes, and a body with little hair. The degree of neoteny is stronger for females than for males, which explains some of the sexually dimorphic features of the human body. We will touch upon this topic again in chapter 3 when we discuss human 'nakedness'.

Still, human beings cannot be regarded as a neotenic chimpanzee in all respects. Some parts of the human body rather show the opposite. For instance, the pelvis of a human baby is more similar to an adult chimpanzee than an adult human. It seems that the evolution of the pelvis, with regard to upright walking, has been subject to accelerated development. This is indicated with the term *'peramorphosis'* (table 2.2).

The evolution of body plans, the modern variant of Haeckel's law, and character evolution through heterochrony, all illustrate that development and evolution are tightly linked together. In the remainder of this chapter, therefore, we will first discuss some key principles of developmental biology.

Cleavages and germ layers

Although each animal commences its development as a fertilised egg (a *zygote*), the first embryonic stages soon diverge. However, there are a number of general principles that apply to the entire animal kingdom; these will be discussed here, before detailing human embryology.

The first cleavages (divisions) of the zygote lead to a clump of cells, the so-called *morula*. The morula still has the same size as the fertilised ovum; the only thing that happens is that one large cell is subdivided into a large number of small cells. Afterwards the cells settle at the outside, creating a cavity in the centre. The morula becomes a vesicle, the *blastula*. The cavity in

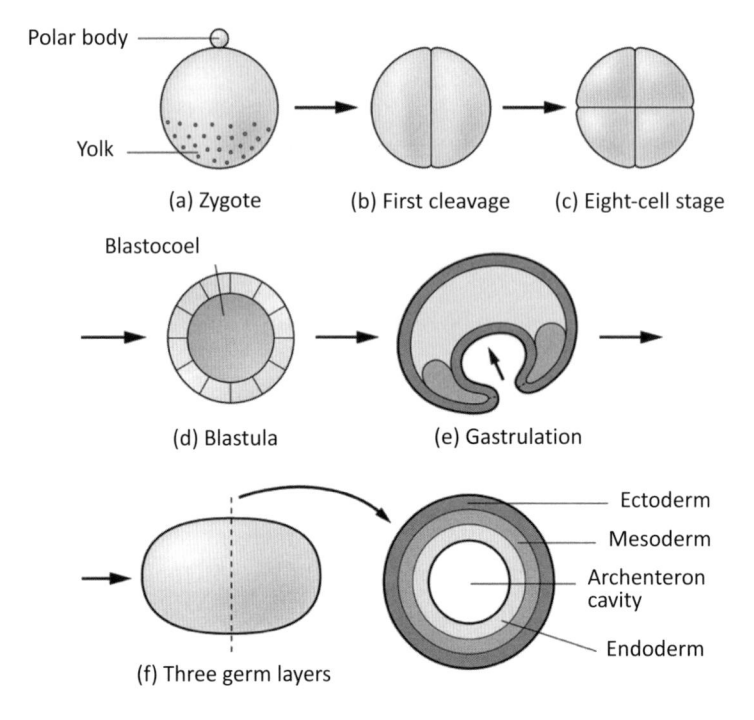

Fig. 2.3: Development of a fertilised egg (zygote) into a gastrula. During gastrulation, the blastocoel is compressed and the archenteron ('ancient intestine') is formed, opening outward via the blastopore. The mesoderm develops between the ectoderm and the endoderm. The three germ layers form the starting point for organogenesis.

the blastula is called the *blastocoel* (coel, to be pronounced as 'seel', derives from the Greek word '*koilia*', meaning cavity) (fig. 2.3).

Subsequently, the blastula undergoes an intensive process during which the cells at the surface rapidly multiply and flow inwards at one location (fig. 2.3). An invagination develops that eventually fills up the entire cavity. The blastocoel is compressed and makes way for a new cavity, the *archenteron* (primitive gut). The opening of the archenteron outwards is called the *blastopore* (which is a confusing term as the blastopore is not the opening of the blastocoel!).

Due to the inward flowing of cells an embryo with two cell layers is created; the outside is called *ectoderm* and the inside is called *endoderm*. In addition, a third layer, the *mesoderm*, is soon formed between those two layers. The three cell layers, also called three germ layers, are the starting point for further development. All organs of an animal body are derived from one of the three germ layers. The invagination of cells and the formation of three germ layers is known as *gastrulation*. The primitive embryo, formed

FROM OVUM TO HUMAN   43

Table 2.3:   Some embryological terms used to indicate features of the early animal embryo

Term	Definition
Morula	Solid clump of cells, resulting from a large number of cell divisions of the zygote
Blastula	Vesicle, structure in the early embryological development that has cells on the outside, leaving a cavity in the centre
Blastocoel	Cavity in the blastula (pronounced as 'blas-tuh-seel')
Invagination	Local inward migration of cells from the surface, as part of the gastrulation
Archenteron	Primitive gut, resulting from gastrulation
Blastopore	Opening of the archenteron; develops into anus with the vertebrates
Gastrulation	Formation of a three-layered embryonic stage from a blastula, including inward migration of cells
Gastrula	Early embryonic stage with three germ layers, including a primitive gut (archenteron)
Ectoderm	Outer germ layer of the gastrula; forms skin and the nervous system
Mesoderm	One of the germ layers of the gastrula, positioned between the endoderm and ectoderm; forms connective tissue, bone, muscles, heart, kidney and reproductive organs
Endoderm	Inner germ layer of the gastrula; forms stomach, intestine, liver, pancreas, lungs, bladder and prostate

as such, is called a *gastrula*, an embryo with an intestine. An overview of all these embryonic terms is shown in table 2.3.

In the mesoderm another cavity forms, the secondary body cavity or *coelom* (pronounced 'see-luhm'). It is called the secondary body cavity because the first one is the gut itself. The way in which the coelom is formed differs quite substantially between animal groups. We restrict ourselves here to the situation in vertebrates, where the mesoderm and the coelom are derived from the gut and develop as lateral pouches that grow to surround the intestinal tube. The coelom is recognised in the adult human body as the chest cavity (containing the heart and the lungs) and the abdominal cavity (containing the gut and its digestive glands). The kidneys and the reproductive organs have their own pouches of the coelom (called nephrocoel and gonocoel, respectively). The mesoderm walls of the coelom cover all internal organs with a membrane (in the human body called pleura and peritoneum in the chest and abdominal cavity, respectively).

Thus far we have described the development of an idealised animal in common terms. Indeed, this is what happens for some animals, for instance, model animals such as sea urchins, zebra fish and African clawed frogs, but a lot of animals significantly deviate from this common scheme. This

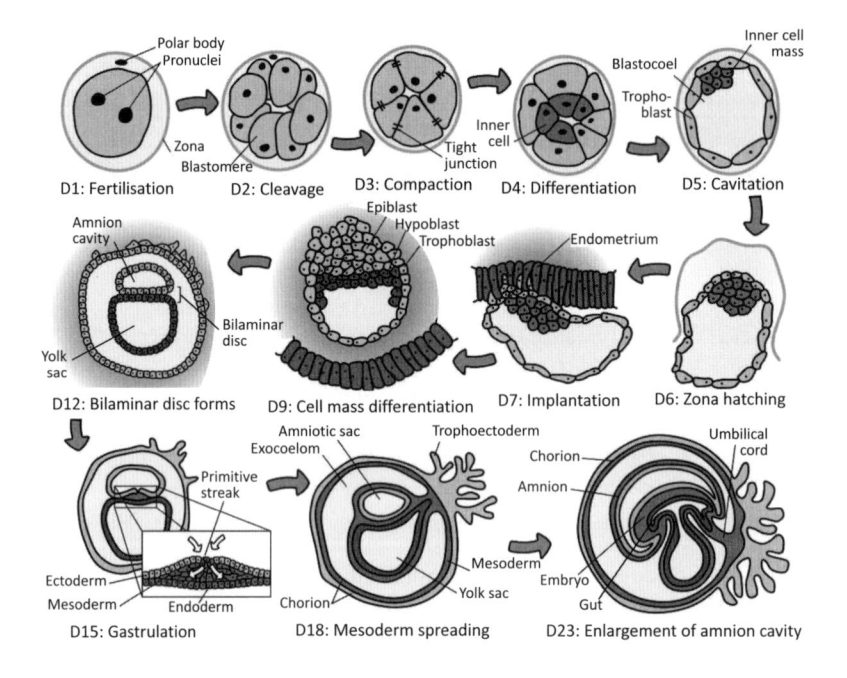

Fig. 2.4: Showing the various stages during early embryonic development in a typical mammal, such as mouse or human. On day 6 (D6) the bastula (blastocyst) 'hatches' from the glass membrane, derived from the zona pellucida. On day 7 implantation into the uterine epithelium (endometrium) begins, and this is completed on day 9. Primitive streak formation and gastrulation start on day 15; this process is enlarged to show the inward migration of cells. After gastrulation the amniotic cavity grows around the embryo, which thus becomes surrounded by two membranes, chorion and amnion. The yolk sac is included in the gut cavity while the gut grows into the umbilic cord.

is predominantly because of the quantity of yolk material in the ovum. Species of which the embryo is soon able to swim around and gather food have relatively little yolk, but species that are fully dependent on the reserves imparted by the mother have a lot of yolk. The eggs most rich in yolk are found in birds; due to their confinement in an eggshell, everything the chick requires must be already present in the egg. In the case of human beings, just as with all other placental mammals, not much yolk is needed, as the embryo, after implanting in the uterine wall, is nourished by the mother.

The amount of yolk in the egg also has consequences for the cleavages. Cell division requires migration of the chromosomes through the cell and this is difficult in cells with a lot of yolk. This is what causes the diversity in the early embryonic stages that we saw at the basis of the hourglass in fig. 2.1. In *holoblastic* cleavages (human, zebra fish, African clawed frog and similar) the entire egg is involved in the cleavage. But if the amount of yolk is substantial, the cell divisions are limited to the upper side of the ovum. This is called *meroblastic* cleavage. In the extreme situation, such as with birds, a germ disc forms (from which the embryo develops), which lies on an enormous amount of yolk. The lower part of the egg, not involved in the cleavages, factually remains one single cell, the yolk sac. The embryo subsequently incorporates the yolk sac during gastrulation. This is possible as the cells of the embryo spread around the yolk sac and the yolk starts functioning as a protrusion of the intestine. While the embryo is growing, the yolk is digested and reduced. For many fish, the larva hatches out of the egg while it still has a yolk sac, allowing the larva to survive the first days of its swimming life while depleting the yolk store.

In placental mammals the embryonic development after the blastula stage differs substantially from the scheme shown by lower chordates, which of course is a consequence of implantation of the embryo in the uterus and the formation of a placenta. The human blastula (usually called *blastocyst*) is asymmetric; having a number of small cells that surround the blastocoel, and a clump of larger cells (called the '*inner cell mass*') on one side (fig. 2.4). The side with the inner cell mass is called the embryonic pole and this is the part that makes contact with the endometrium of the uterus first. In the meantime the inner cell mass has differentiated into two layers called *epiblast* and *hypoblast*. A layer of epiblast cells subsequently expands towards the embryonic pole and forms a cavity beneath it, the amnionic cavity. The *amnion*, surrounded by its amniotic membrane, expands greatly and during later development grows around the whole embryo, protecting the foetus. The presence of an amnion distinguishes Amniota (reptiles, birds and mammals) from the lower vertebrates (amphibians and fish). It is one of the crucial structures that allowed life on land, since the amniotic egg does not need to develop surrounded by water.

The formation of the amnion and the implantation in the endometrium are completed by the ninth day of pregnancy in humans. It is important to note that amnion formation takes place before gastrulation. This is what makes the development of placental mammals so complicated, because the implanted structure that undergoes gastrulation is very different from a blastula proper.

Gastrulation starts by epiblast and hypoblast cells forming a so-called '*bilaminar disk*'. The epiblast cells subsequently form a '*primitive streak*', a longitudinal ridge along the anterior-posterior axis of the embryo that initiates gastrulation. Starting at the posterior side, epiblast cells migrate inside to define the three germ layers, ectoderm, mesoderm and endoderm. The proceeding wave of migration of epiblast cells is described as a 'polonaise'. Meanwhile the hypoblast cells differentiate to surround the (small) yolk sac, which is continuous with the embryonic gut to nourish the embryo, later to be included in the umbilical cord, together with trophoblast cells.

The trophoblast interacts with the endometrium to form a *placenta*. The formation of the placenta has been considered an evolutionary miracle, suddenly appearing, out of the blue, in mammals. This is, however, not correct: many viviparous animals, be it invertebrates or fish, have a structure that connects the foetus with the reproductive tract of the mother. Such structures are under extremely strong selective pressure and evolve very quickly. In mammals four different types of placentas have evolved which can be classified to the degree of invasiveness of the trophoblast. The human placenta is of the *'hemichordial'* type, most likely the ancestral type within the mammals, which has been simplified repeatedly in other lineages. Placenta-forming trophoblast cells lose their cell membranes and form a *syncytium*, a multinucleate tissue. This syncytium also dissolves the cells forming the walls of the maternal blood vessels. In this manner the maternal blood is in direct contact with the foetal trophoblast tissue, which of course aids in efficient transport of nutrients and waste products from mother to foetus and vice versa. Interestingly, in primates the genes involved in the formation of syncytia (syncytins) have been 'co-opted' from retroviruses. The evolution of innovative body plan features by deploying ancient genes and structures is discussed further in chapter 3.

Returning to the general trends of embryonic development, we have noted above that the gut is the first internal organ; it is formed during gastrulation, which literally means intestine formation. In an evolutionary sense, the intestine is the first, because even the most primitive animals (polyps, flatworms) also have a gut. But shortly after the intestine, at an early stage of development, the nervous system is being formed. The third organ is the heart. Gut, nervous system and heart develop in interaction with each other and require signals from each other's development, emphasising the 'wholeness' of this early embryonic stage.

The complex process of formation of a nervous system is called *neurulation*. We describe only the neurulation of vertebrates here, which was extensively studied in fish and amphibians and in principle also applies to human development. At the dorsal side of the embryo, right under the skin, lies a cartilage-like rod, the notochord, consisting of tightly packed, highly vacuolated cells. The *notochord* ('chorda') is characteristic for the phylum Chordata, to which humans also belong (*cf.* table 1.1). Interestingly, the notochord of chordates has a precursor in invertebrates. In annelids a longitudinal medial muscle, called *axochord*, is located near the ventral nervous system. The cells of this muscle show a gene expression profile similar to the chordate notochord.

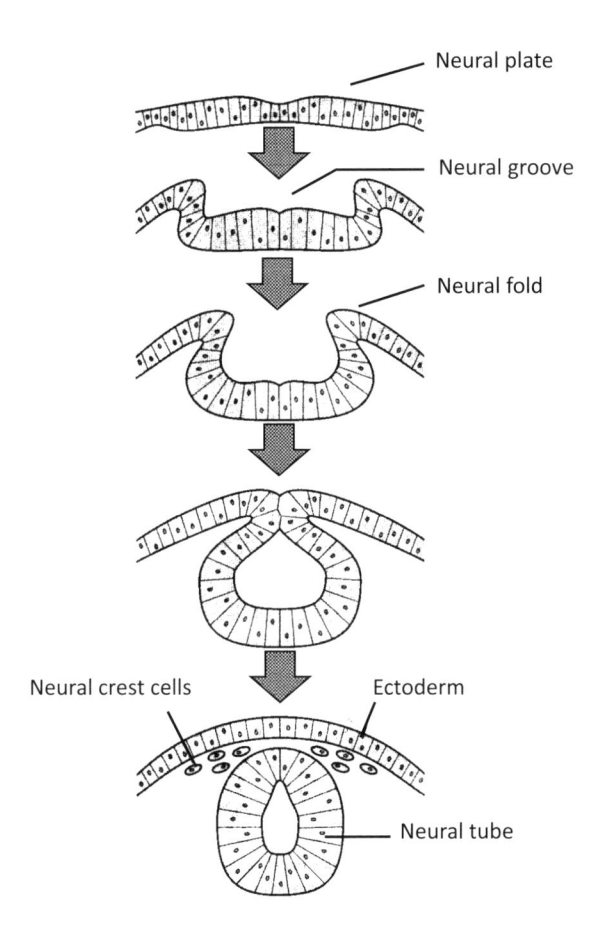

Fig. 2.5: Formation of the neural tube at the dorsal side of a vertebrate embryo (neurulation). Upon closure of the tube, neural crest cells are released, migrating through the embryo and contributing to the development of various organs.

Presumably, the muscle has lost its contractile properties, the cells became vacuolated and the whole structure came to be lying on the dorsal side after the protostome-deuterostome conversion (to be discussed below).

The notochord is of crucial developmental importance because it triggers the formation of the nervous system. Later on, the notochord vanishes: it is incorporated in the vertebral bodies of the spinal column and also contributes to the intervertebral disks.

Directly above the notochord there is a thickened cellular layer in the ectoderm, the neural plate, which stretches from front to back (fig. 2.5). The centre of this plate sinks downwards and curls, creating a tube, the *neural tube*. The edges to the left and right of the neural plate are called *neural crest*.

The neural tube closes from front to back, as a result of which it becomes detached from the skin. If this doesn't happen in a proper way, the child will have a so-called spina bifida, a serious disorder, which often leads to disability.

As the neural tube closes, groups of cells to the lateral sides are constricted from the neural crest: the *neural crest cells*. Also these cells have a precursor outside the vertebrates: cephalic melanocytes of sea squirts (a urochordate) can be mutated to behave like neural crest cells. In a vertebrate, the cells migrate through the body and contribute to the formation of the heart, the skull and the renal cortex, among others. It is fascinating to realise that the formation of some internal organs is so crucially dependent on cells that detach from the nervous system. It illustrates how strongly the various embryonic processes are interdependent and able to influence each other. We will meet the neural crest cells again in chapter 3, when we discuss the evolution and development of the heart.

Axes to provide direction

All bilaterally symmetric animals have a front (*anterior*) and hind (*posterior*) side, as well as a back (*dorsal*) and belly (*ventral*) side. In addition, a limb or leg has a *distal* side (the end most distant from the body) and a *proximal* side (the side closest to the body). The construction of axes, mostly by means of gradients of signal proteins, is typical for the entire embryonic development. Axes provide the embryo with a coordinate system allowing organs to be developed at certain locations. Without a coordinate system, it would be impossible to generate a body plan.

Even before fertilisation the egg has an axis, meaning it has an upper side and a lower side. This can be taken literally, as usually the eggs orientate in the earth's gravitational field and, for instance, in the case of zebra fish, the topside is darker than the lower side. The topside is called the *animal pole*, whereas the lower side is the *vegetal pole*. Generally, the mother imposes the animal-vegetal axis of the egg. As an example, the concentration of the signal protein VG1 is higher at the vegetal than at the animal pole (fig. 2.6). This is a result of the mother depositing *Vg-1* mRNA at the yolk-rich side of the ovum. The cells at the vegetal pole that are formed after fertilisation take up this mRNA and convert it into VG1 protein. The VG1 protein subsequently functions as a signal molecule, in turn eliciting other reactions. Also mRNAs of other maternal genes, like *Xwnt-11* and *VegT*, have a clear gradient in the unfertilised egg.

It often so happens that mRNA is provided with the ovum. It is an example of a *'maternal effect'*: a non-genetic influence of the mother on her offspring. Through maternal effects, environmental conditions may influence the phenotype of the next generation during the mother's life, clearly a non-Darwinian adaptation mechanism. It often happens with insects. For example, eggs laid in summer develop in a normal way, but in autumn, under the influence of shortening days, the mother lays eggs that go into diapause and only hatch after the winter. The maternal influence on development of offspring is not be underestimated. Research on insects shows that the majority of mRNA molecules in a developing egg derive from the mother and not from the egg itself. These mRNAs are activated and translated after fertilisation; the first cleavages are almost completely managed by maternal proteins. Only during gastrulation does the embryo start to produce its own mRNA. Maternal effects are also important for humans, as will be explained in chapter 4.

The first embryonic axis, the dorso-ventral axis, is determined by the location of fertilisation in relation to the maternally induced animal-vegetal axis. As soon as a sperm cell penetrates the oocyte, the cell membrane and the outer layer of the cytoplasm rotate approximately 30 degrees into the direction of the location where the sperm cell made contact, the so-called *cortical rotation* (fig. 2.7). The inner cytoplasm will thereby remain in place.

The spot where the sperm cell penetrated subsequently becomes the ventral side. Directly opposite that point, the dorsal side is established. This takes place

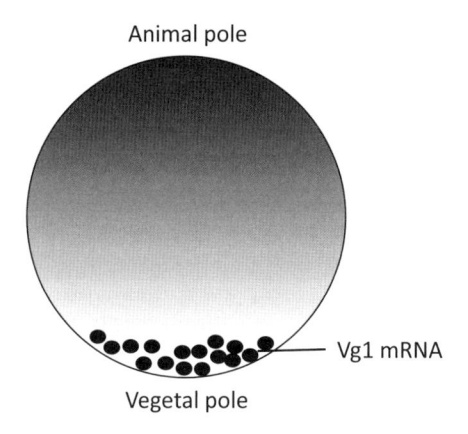

Fig. 2.6: At the vegetal pole of an unfertilised egg of the African clawed frog (toad), *Xenopus laevis*, there is a high concentration of mRNA of the maternal gene *Vg-1*. The cells forming in that location after fertilisation will take up this mRNA, transform it into protein and start to secrete VG1. Due to the gradient of *Vg-1*, the egg is provided with an animal-vegetal axis.

under the influence of so-called dorsalising factors, such as the VG1 protein and a signal-transduction system known as the *Wnt* pathway. The dorsalising signals also induce the so-called *Spemann centre*, which is the location where gastrulation starts and later on the blastopore is created (fig. 2.7).

The first cleavage after the cortical rotation runs parallel to the animal-vegetal axis, through the point where the sperm cell penetrated. The second cleavage also runs parallel to the animal-vegetal axis, perpendicularly to the first. The third cleavage is equatorial, perpendicular to the first two. In this way the eight-cell stage of the embryo already has a coordinate system: both up-down direction and front-back direction are fixed.

The scheme described above is derived from classical embryological work with the African clawed frog, *Xenopus laevis*. For other animals different factors are involved. In the case of the chicken, gravity plays an important role in establishing body axes. A mouse egg initially doesn't seem to have any polarity and the location where the sperm cell penetrates and the spot where the second polar body is secreted determine the animal-vegetal axis. (The *polar body* is one of two small haploid cells that emerge during the meiotic divisions of the oocyte; the second one is usually only shed after fertilisation.) In the case of other mammals, cells outside the blastula play a role when establishing the anterior-posterior axis.

The way in which the first body axes are established therefore appears to vary for different animals. Maternal influences, division of the yolk, secretion of the polar body, the spot of sperm entry, gravity and cellular influences from outside the embryo, all these factors can play different roles for different animals. Why evolution has produced such a diversity of mechanisms for such an important process, the establishment of embryonic body axes, remains a mystery.

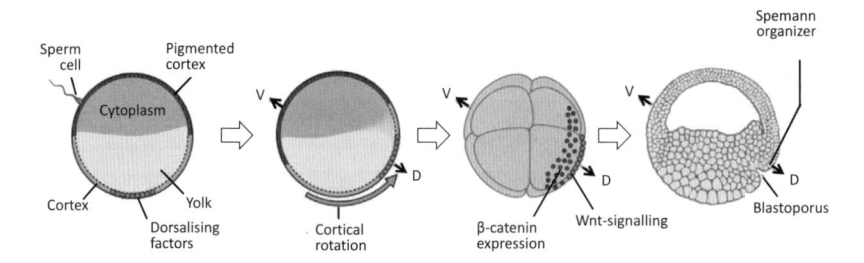

Fig. 2.7: Establishment of the dorso-ventral axis during early development of the *Xenopus* egg. Directly after fertilisation, the outer layer of the egg is subject to a rotation (cortical rotation) relative to the cytoplasm. As a result of this, the dorsalising centre moves from the vegetal pole to what becomes the dorsal side. Subsequently, under the influence of *Wnt* signalling, the dorso-ventral axis is established. The Spemann centre then initiates gastrulation.

Although during the early embryonic development the anterior-posterior axis is set in terms of direction, the front and back of the animal aren't yet established. It appears that in many cases this is done only after gastrulation. As previously discussed, gastrulation initiates the formation of the blastopore, the opening of the archenteron. For most invertebrates this opening becomes the mouth. However, in vertebrates the blastopore develops into the anus, while the mouth opens at the other end. This distinction is so fundamental that the entire animal kingdom, as far as they are bilaterally symmetrical, is divided into two main groups, *Protostomia* ('mouth first') and *Deuterostomia* ('mouth second'). Several features of body plan development are correlated with this distinction, such as coelom formation, the orientation of the division plane for successive cleavages, the degree of fate determination of blastomers, and the position of the nervous system in relation to the intestine (see table 2.4).

The majority of animals belong to the protostomes. For instance, arthropods, (*e.g.* insects, crustaceans and spiders), molluscs (*e.g.* snails, shellfish and squids) and worms are included in the protostomes. Deuterostomes include chordates (*e.g.* vertebrates) and echinoderms (such as starfish, sea urchins, sea cucumbers and sea lilies). The first bilaterally symmetrical animals are likely to have been protostome; these are called 'Urbilateria'. These ancient animals around 550 Ma BP gave rise to a new lineage with

Table 2.4: Differences between Protostomia and Deuterostomia

	Protostomes	*Deuterostomes*
Blastopore becomes	Mouth	Anus
Fate of cells after cleavages largely	Determined	Undetermined
Orientation of cells in successive cleavages	Spiral-shaped, leading to a twisted arrangement of cells in the morula	Parallel and perpendicular to the division plane, leading to radial stacking of cells
Mesoderm formed from	One fixed cell in the 25-cell stage (4d)	Outgrowths from the gut
Coelom formation by	Fission within the mesoderm	Pouches developing from the gut vacity
Position of the nervous system	Ventral	Dorsal
Position of the heart	Dorsal	Ventral
Representative large phyla	Arthropoda, Mollusca, Annelida	Echinodermata, Chordata
Important research models	*Nematostella, C. elegans, Drosophila, Tribolium*	*Strongylocentrotus*, zebra fish, clawed frog, chicken, mouse

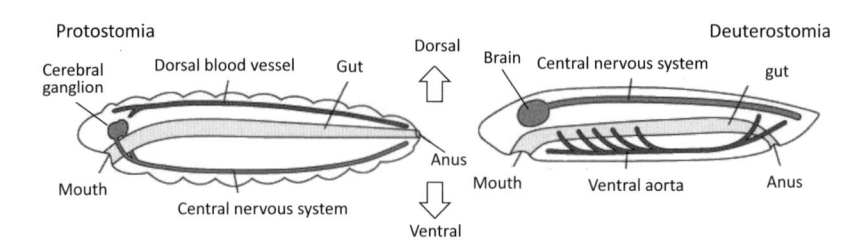

Fig. 2.8: Position of the gut, the body artery and the central nervous system relative to each other, for protostomes and deuterostomes.

reversed body axes: underside became top and front became back. This fundamental reversal can still be read from the position of the nervous system: in all protostomes, the nervous system is positioned ventrally, below the intestine; for all deuterostomes the nervous system is positioned dorsally, above the intestine. We saw above that the homology between the protostome axochord and the deuterostome notochord fits this reversal. Still, there are some animals (brachiopods) that are classified as protostomes but have a deuterostome development. The crucial issue is the posterior or anterior signal that emanates from cells near the blastopore. The axis orientation is actually more basic than the development of a mouth or an anus.

The protostome-deuterostome conversion can also be derived from the expression of developmental genes. For *Drosophila*, the gene *Decapentaplegic* (dpp) determines the dorsal identity; the homologous human gene, *Bone morphogenetic protein 4 (BMP4)*, determines the ventral identity. The gene *Sog* of *Drosophila* contributes to ventral development; the homologous human gene, *Chordin*, determines dorsal development. Embryonically we are strongly related to invertebrates, such as *Drosophila* and *C. elegans*, but first we need to reverse both of our body axes!

Model animals in developmental biology

Our knowledge of developmental mechanisms is based on a relatively small number of model animals. Experimental research on human embryos is, for obvious reasons, not only impossible, but also undesirable and therefore we rely on animals that serve as models for human beings. Adequate model organisms share important biochemical and physiological processes with humans, processes that were conserved throughout evolution. Fortunately the genetic regulation of development is largely similar across the entire animal kingdom, allowing us indeed to obtain essential information about human development, especially in the early stages, with the help of model animals.

Traditionally, six model animals are distinguished in developmental biology: the nematode worm, *Caenorhabditis elegans*; fruit fly, *Drosophila melanogaster*; zebra fish, *Danio rerio*; African clawed frog, *Xenopus laevis*; chicken, *Gallus gallus*; and mouse, *Mus musculus*. Over the past years a number of species have been added, such as the sea anemone, *Nematostella vectensis*, the flour beetle, *Tribolium castaneum* and the sea urchin, *Strongylocentrotus purpuratus*. The largest part of our knowledge of developmental biology is still based on the six classical species. Fig. 2.9 pictures the evolutionary relationships between humans and these six model animals.

The fruit fly, *Drosophila melanogaster*, and the nematode, *Caenorhabditis elegans*, are the best-known invertebrates in developmental biology. These two animals were the first for which the entire genome was sequenced, which gave them a huge advantage. Both *C. elegans* and *Drosophila* are easy to grow; they can be genetically manipulated; they have a short generation time; and they produce many offspring.

C. elegans belongs to the phylum Nematoda, a species-rich group of small vermiform animals that live in all kinds of habitats, particularly in the soil. Many species parasitise plants, but nematodes are best known for the many parasites living on animals and the human body. The pinworm, *Enterobius vermicularis*, a gut parasite that many children contract at some point, is also a nematode. *C. elegans*, however, is a free-living, bacteriovore representative of the phylum, which is found in rich soil and compost heaps. The usefulness of *C. elegans* as an experimental model was discovered by the South African-American molecular biologist Sydney Brenner in 1965.

Over 40% of the protein-coding genes of *C. elegans* have a homologous gene in human beings. All genes in *C. elegans* can be knocked down by use of RNA interference. To achieve this, bacteria expressing a double-stranded RNA are fed to the worms, the RNA being designed to inactivate a specific mRNA thus preventing translation of the gene of interest. The study of phenotypes after gene knockdown helps to elucidate gene function.

The body of *C. elegans* consists of 954 cells. It is known exactly how these cells develop from a fertilised egg. The fate of each cell is determined from the beginning and the whole development can be depicted as a cellular fate map. *C. elegans* is a classical example of embryonic development through determinate cleavages, a system appearing in strong or weak form in all kinds of protostomes, but not in deuterostomes (see table 2.4). In fish and amphibians, the cleavages before gastrulation are still indeterminate, which means that the fate of the daughter cells is not fixed after cell division.

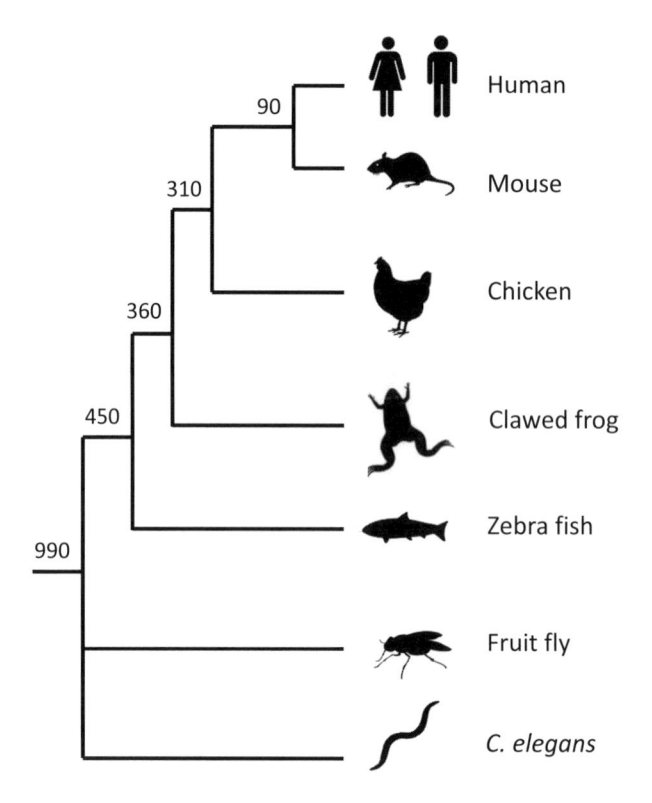

Fig. 2.9: Evolutionary relationships among the six animal model species of developmental biology, in relation to humans. The numbers in the tree indicate the divergence times in millions of years before present.

Also, the inner cell mass of the human blastocyst does not have a fixed fate, which is proven by the fact that a blastocyst may sometimes spontaneously divide into two, whereby both embryos, after implantation in the uterus, may develop into full-fledged babies: a monozygotic twin.

The fruit fly, *Drosophila melanogaster*, is one of the oldest models of developmental biology. The insect has been used for experimental genetic research for more than one hundred years and became especially famous thanks to the work of the American geneticist Thomas Hunt Morgan in the beginning of the twentieth century. Interestingly, human development is much more similar to that of flies than is often assumed. Many genes that regulate fruit fly development are homologous to the genes involved in the development of vertebrates. Therefore, it is no surprise that some of the most important principles of developmental genetics were derived from research with fruit flies. A very important group of animal developmental genes, the *Hox* genes, were first discovered in *Drosophila*.

The human genome contains approximately 1.4 times the number of genes in *Drosophila*, but the evolutionary conservation of these genes is strong enough to identify homologs (about 60% of the genes). This also holds for genes that cause human diseases, which can thus be examined in the fruit fly. Such studies may provide essential information about gene function, which may in turn contribute to the development of drugs.

Another big advantage of *D. melanogaster* and *C. elegans*, the two model invertebrates, is that experiments with these animals, from an ethical point of view, are more acceptable than experiments with vertebrates. They do not fall under Animal Experiments Acts and experiments do not require complicated and time-consuming licence applications.

Apart from the many advantages, the use of *Drosophila* also has disadvantages, caused by its ecology. The larvae of *Drosophila* develop in rotting fruit, a habitat that is but shortly available. The larval development is adapted to these conditions and is highly accelerated, compared with other insects. Also, the genome is rather 'trimmed'; many genes that do appear in other insects have disappeared in the fruit fly genome. For this reason, the flour beetle, *Tribolium*, which has recently been used more and more, is an animal that, in terms of development, can be regarded as much more typical for insects than *Drosophila*.

Ever since the 1930s the zebra fish, *Danio rerio*, has been used as a vertebrate model for embryogenesis and development. Mutants are well visible and manipulation/transplantation experiments can be performed easily. Later, in the 1980s, different genetic techniques were developed, such as cloning, mutagenesis and transgenesis, which made the zebra fish even more valuable for biomedical research. Large-scale genetic screening of mutants has provided essential information about the embryogenesis of vertebrates. Although the genome is rather large (approximately 2 billion base pairs), its sequence was recently fully elucidated. This genome information also shows the high degree of relatedness with human beings.

From the zebra fish different genetic mutant lines were developed for research on different forms of cancer. There are, for instance, mutants that develop leukaemia and skin cancer, allowing us to get more detailed insight in these diseases. In addition, models have been developed for studies of behaviour. The development of the zebra fish, from zygote to larva, is shown in fig. 2.10.

A model that is even closer to humans than the zebra fish is the African clawed frog, *Xenopus laevis*. A related species, *X. tropicalis*, is also often used. A practical reason for choosing *Xenopus* as a model is that oocytes can easily be isolated and manipulated (injection, transplantation). In the

1920s it became known that the body axes of vertebrates are established by signals from a group of cells that already have this positional information. Such tissues are called *morphogen*. In the 1960s the Dutch embryologist Pieter Nieuwkoop managed to transplant a part of the morphogenic cells to the other side of the blastula. A second body axis emerged from this, with associated segments, resulting in an embryo with two head structures and two spinal columns. Consequently the morphogenic centre is called the *Nieuwkoop centre*. Fig. 2.11 provides a schematic view of the experiment by Peter Nieuwkoop.

An additional advantage of *Xenopus* as a model for biomedical research is the high degree of conservation of biochemical and physiological processes, also expressed in the homology of *Xenopus* genes in relation to human genes. By comparing the zebra fish and *Xenopus* with the human genome, it became clear that important gene families, such as certain transcription factors, are indeed present in the frog and human genomes, but not in zebra fish. This is an additional indication that *Xenopus* shows more resemblance to human beings than the zebra fish.

Xenopus is also famous for the urine-based pregnancy test. Around 1930, the British zoologist Lancelot Hogben accidently discovered that female frogs, injected with the urine of pregnant women were induced to lay eggs.

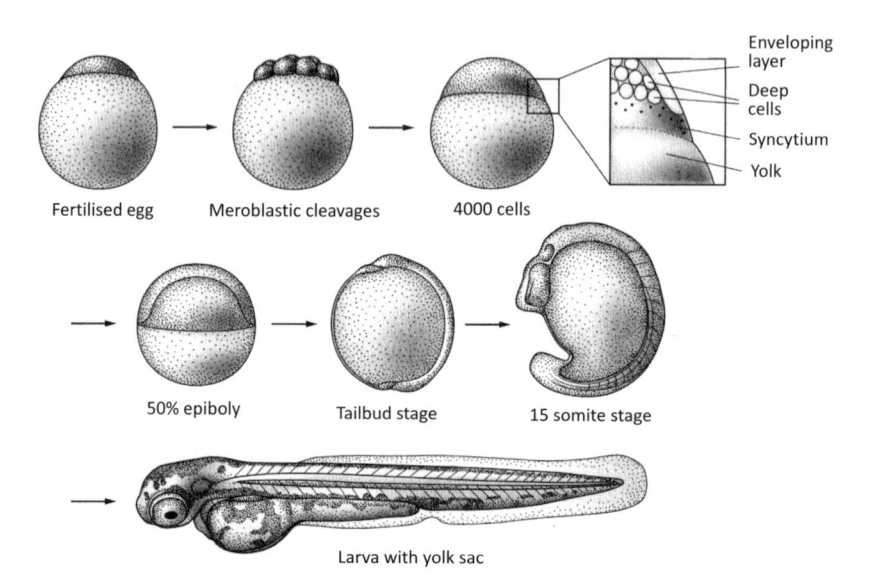

Fig. 2.10: Embryonic development of zebra fish. The cleavages are limited to the animal pole, while the yolk remains undivided and, during development, is enclosed by the embryo as a result of proliferating cells growing around the yolk sac (epiboly).

This resulted in the first reliable pregnancy test, which was to be used for many years. Later it became evident that the effect is caused by a peptide called *'human chorionic gonadotropin'* (hCG), a hormone excreted by the placenta which acts on the maternal corpus luteum in the ovarium, due to which the production of progesterone is sustained and subsequent ovulations are suppressed. The frog experiment to demonstrate pregnancy has now been replaced by a direct measurement of hCG.

The chicken, *Gallus,* is mainly used for research on *organogenesis* (formation of organs). The advantage is that a chicken egg is easily accessible; the entire development can be monitored without much surgical intervention. The chicken egg is also suitable for transplantations and visualisation techniques. As such, research involving the chicken has made a major contribution to the knowledge about the migration routes of the neural crest cells. An obvious disadvantage of the chicken is that bird development includes a range of specialisations that are not typical for human beings, for instance, due to the very large amount of yolk in the bird egg.

The house mouse, *Mus musculus,* is by far the most important model organism for biomedical research. The fact that mice are easy to breed, have a relatively short generation time, and show molecular, biochemical, physiological and even behavioural processes that are all fairly similar to those in humans, is what makes this animal so essential for developmental biological research. In addition, the genome of the mouse has been fully assembled, which has generated a huge amount of information about gene functions. Another significant advantage is that every gene in the mouse genome can be switched off or replaced by a homologous copy, for instance,

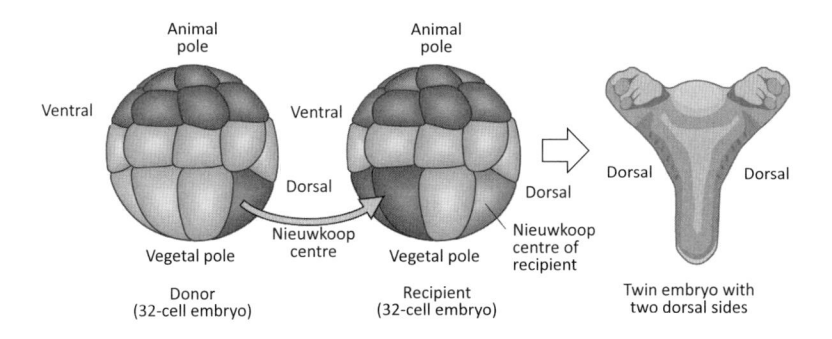

Fig. 2.11: Schematic view of a classical experiment conducted by Peter Nieuwkoop in 1969, with eggs of the African clawed frog. By transplanting cells from the vegetal pole to a donor, an embryo was created with two dorsal sides, proving that the transplanted cells contain a dorsalising factor.

a human gene, by use of homologous recombination. Many centres for laboratory animals that breed mice are able to supply transgenic mutant mice for biomedical research. Furthermore, it is possible to genetically manipulate embryos. A large disadvantage of the mouse is of course the fact that it is a highly developed animal, which raises ethical concerns against experimenting with them. Research on mice in many countries is very strictly regulated in Animal Experiments Acts, and an Ethical Committee is installed to evaluate the plans for any experiment.

The development of the mouse is depicted in fig. 2.12. As we have seen above, the embryo interacts with the endometrium in the blastocyst stage. After implantation, between the seventh and ninth week, the embryo rotates around its longitudinal axis (fig. 2.12). From that moment on it is surrounded by two membranes (the *chorion* stemming from the outside of the blastocyst and the endometrium, and the *amnion*, formed directly after implantation); these two membranes protect the embryo, continue to surround the foetus as it grows in the uterus and are broken only at delivery.

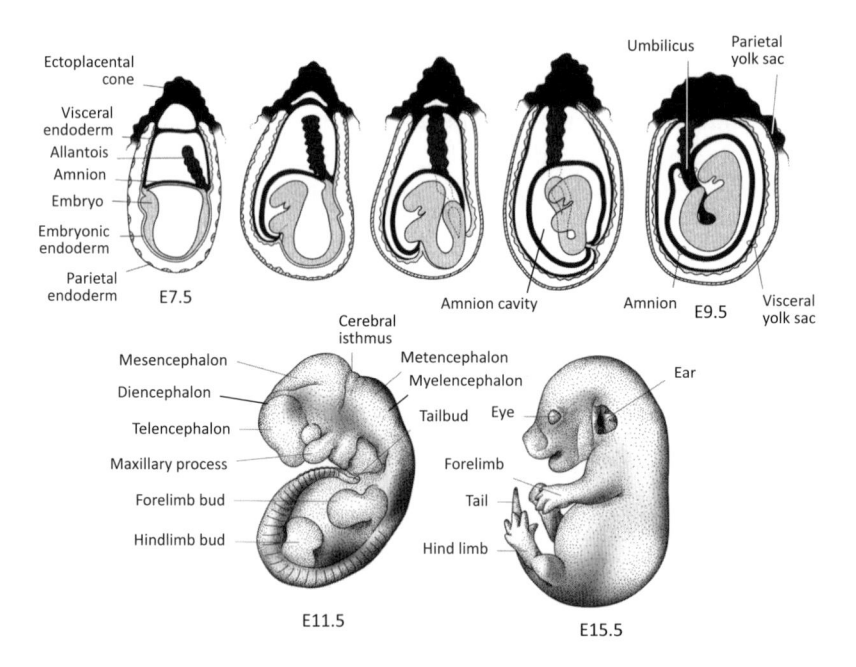

Fig. 2.12: Embryonic development of the mouse. The stages are indicated as 'embryo-days' (E7.5, etc.). Between E7.5 and E9.5 the embryo rotates along its horizontal axis, whereupon the intestine closes, the umbilical cord is being formed and the embryo is positioned into the amniotic cavity. Afterwards, other organs are formed, firstly the eyes and brain.

The molecular toolkit for development

The development of body plans is regulated by specific gene families that are strongly conserved within the animal kingdom. One important group of these genes are homeotic genes, also referred to as *Homeobox* genes, or abbreviated, *Hox* genes. *Hox* genes all encode transcription factors that have a specific DNA-binding domain, a so-called homeodomain. They owe the name 'homeotic gene' to the fact that mutations in these genes usually lead to *homeosis*: transformations of a body part into another body part, for example, an antenna becomes a leg, a halter becomes a wing, *etc*. Please note that only genes that regulate the anterior-posterior axis and the associated structures are called *Hox* genes. Some other developmental genes have a DNA-binding homeodomain, but are not *Hox* genes. *Hox* genes function in close cooperation with genes involved with cellular signalling, mitosis, cell adhesion and programmed cell death (apoptosis), in order to accurately regulate the development of a body part in the right place and at the right time.

The *Hox* genes underwent duplications in the different lineages of the animal kingdom. These duplications were probably essential for the development of new morphological structures in the body plan. We know that two structures are *homologous* if they originate from the same ancestor. Despite homology, the function of the structure may change (the forefeet of hoofed animals, for example, are homologous with those of bats). Also, genes can be homologous, but a complication arises in the case of gene duplication. In that case we need to distinguish two types of homology: *orthology* and *paralogy*. The *Hox* genes form a classical example by which the difference between orthologs and paralogs can be illustrated (fig. 2.13).

Two genes are orthologous if their sequence similarity is due to common descent. The orthologous genes of a group of species show the speciation of that particular group. Two genes are paralogous if their similarity is due to a duplication within an ancestral species. Paralogous genes in the same genome generally have a slightly different function; for example, in the case of enzymes they degrade different substrates or are expressed in different tissues. The *Hox* genes and other large gene families in the human genome have both orthologs and paralogs, sometimes many of them, reflecting a complicated pattern of duplication and ancestral transmission.

Hox genes were first discovered during large-scale mutation-screening studies with *Drosophila* at the end of the 1970s. Edward Lewis, Christiane Nüsslein-Volhard and Eric Wieschaus were later awarded the Nobel Prize for their discovery.

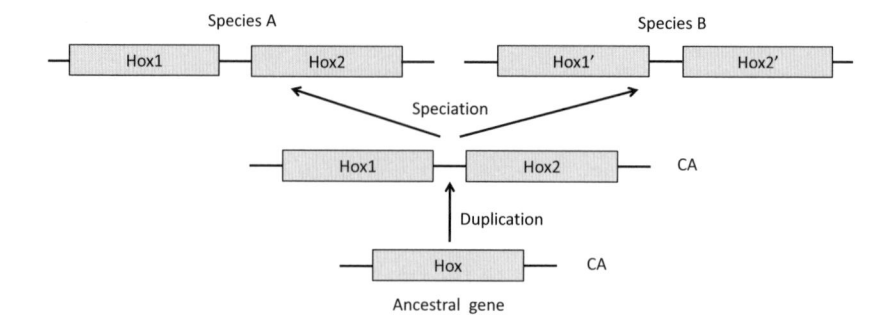

Fig. 2.13: After gene duplication and speciation, two types of homology can be distinguished. Hox1 and Hox2 are paralogs, just like Hox1' and Hox2', whereas Hox1 and Hox1', and also Hox2 and Hox2', are each other's orthologs.

Ultrabithorax (*Ubx*) was the first *Hox* gene ever to be isolated. Switching off this gene, in the case of *Drosophila*, turned out to provide an extra pair of wings at the location where normally halters are developing. *Drosophila* belongs to the insect order Diptera (two-winged insects, such as flies and mosquitoes), a group, which in contrast to most insects has two wings instead of four. Instead of the second pair of wings, dipterans have halters, rod-shaped structures on the third thoracic segment that contribute to stability during flight. The mutation studies showed that *Ubx* suppresses the wing-specific gene expression in the third thoracic segment; when switching off *Ubx*, the gene expression is no longer suppressed, allowing the formation of a normal wing. Thus the halter can be regarded as a non-developed wing.

Other *Hox* genes do not inhibit but activate the formation of certain body parts. The *Hox* gene *Antennapedia* (*Antp*) regulates the formation of legs at the first thoracic segment of the fruit fly. If this gene is switched off as a consequence of a mutation, antennae develop at the location where legs should have been formed. Spectacular results have been achieved by overexpression of *Antp*, causing *Antp* protein to be expressed in the wrong place at the wrong time during development. The resulting mutant fruit flies turned out to develop legs on their head instead of antennas. This also explains the name of the gene *Antennapedia* (fig. 2.14).

A remarkable property of *Hox* genes is their position in the genome: they are organised in tandem formation, in a cluster. Also, the order of *Hox* genes in the cluster is unique. *Hox* genes that regulate the development of body parts in the front (anterior) segments also appear to be positioned at the 'front', meaning, the 3' side, of the gene cluster. Genes that regulate the body parts in the rear (posterior) segments of the body are likewise positioned at the 'back' (5') of the gene cluster (remember that RNA polymerase reads

DNA in the 3' to 5' direction). Both *Ubx* and *Antp* are at the centre of the *Hox* gene cluster, because they regulate the development of body parts (legs and wings) in the thorax of the *Drosophila* body plan.

The genomic sequence of *Hox* genes appears to be similar across worms, flies, fish and humans. The position of a gene in the cluster has hardly changed over a period of 400 million years of evolution! Undoubtedly, this is related to the strict necessity to express the right genes in the right order. In addition, the increase in complexity of body structure from invertebrates to vertebrates was accompanied by multiplication of the *Hox* gene cluster, from one cluster in invertebrates to four clusters (A-D) in the genomes of vertebrates, such as fish and humans.

Fig. 2.15 shows the ortholog and paralog relationships between the *Hox* genes of *Drosophila* and those of humans. A hypothetic ancestor is shown from which the *Hox* clusters of both fruit fly and humans are assumed to be derived. In the lineage towards humans, the *Hox* cluster was duplicated twice, due to which *H. sapiens* has four copies of a gene of which the ortholog occurs only once in *Drosophila*. The four human *Hox* clusters are positioned on different chromosomes. Most likely the two rounds of duplication occurred at the origin of the vertebrates. This is concluded from the fact that the invertebrate chordates (lancelets and sea squirts) lack quadruple *Hox*

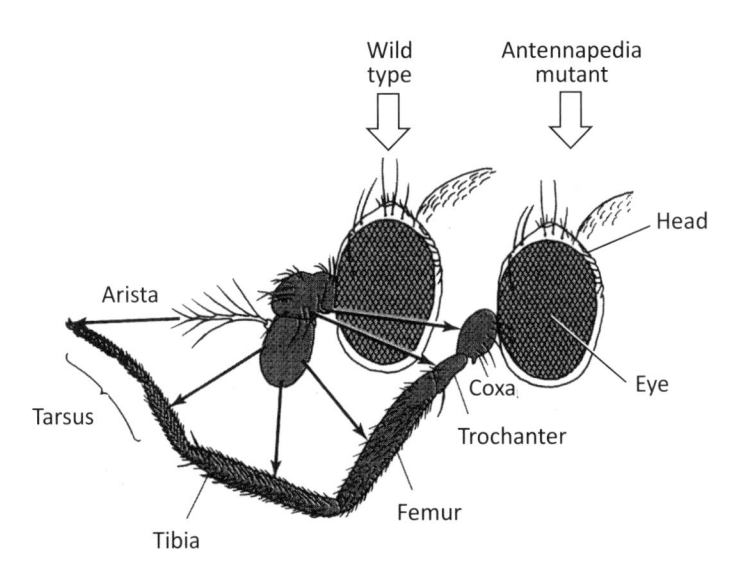

Fig. 2.14: The antennapedia mutant of *Drosophila* (right) has a leg on its head, where normally there would be an antenna (left).

clusters. After the duplications a number of genes were lost, due to which some *Hox* genes in the human genome (for example, *HoxA7*) have only one paralog (*HoxB7*) instead of three (*HoxC7* and *HoxD7* do not exist).

The arrangement of *Hox* genes in clusters was presented as a classical case of gene order conservation over species for a long time. In genomics this is called *co-linearity* or *synteny*. It was assumed that conservation of synteny of *Hox* genes was imposed by the necessity to precisely regulation gene expression in space and time. However, now that we know more genomes, it turns out that there are many exceptions. The synteny is actually mostly restricted to vertebrates. Looking at the animal kingdom as a whole, four different patterns can be distinguished: (1) organised (vertebrates), (2) disorganised (sea urchin), (3) split (*Drosophila*) and (4) atomised (*Oikopleura*). How the temporal and spatial regulation of *Hox* gene expression is regulated in the case of disorganised or atomised arrangements of the genes is unknown at the moment.

The positional expression of *Hox* genes was first investigated in detail in the posterior part of *Drosophila*'s *Hox* gene complex, better known as the *Bithorax* complex. In this part of the cluster there are three *Hox* genes, *Ultrabithorax*, *abdominal A* and *abdominal B* (fig. 2.15). If all three genes are knocked out by means of mutation, a larva will emerge that merely

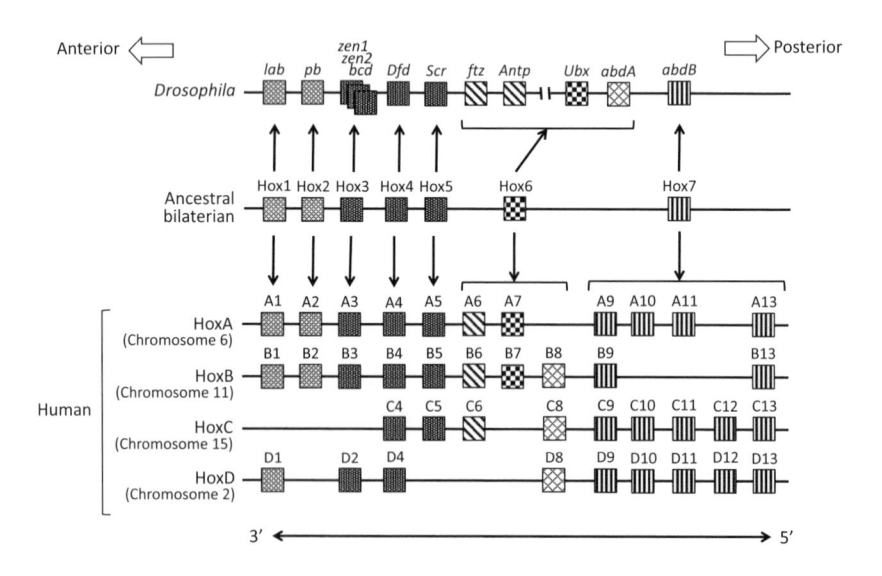

Fig. 2.15: Overview of *Hox* clusters in the genomes of *Drosophila* and human, and a hypothetical initial situation for the ancestor of all bilaterally symmetric animals. Genes positioned beneath one another can be considered orthologs. In the vertebrates the *Hox* cluster was duplicated twice, due to which humans have four clusters against one for *Drosophila*. The expression domains along the anterior-posterior axis have been retained.

consists of the three front segments (head and thoracic segments 1 and 2) and nine segments which have the same identity as thoracic segment 2. If subsequently *Ubx* protein is injected into these mutants, a thoracic segment 3 turns out to be formed, but all abdominal segments have the same identity as abdominal segment 1. Only after the injection of *Abd A* and then *Abd B* mRNA will the differential development of the thoracic and abdominal segments fully recover.

Performing similar experiments it was discovered that the information for the development of adult *Drosophila* body parts is already present in special groups of cells in the larva. These precursor cells are called *imaginal discs*. During pupation virtually all larval tissues are degraded and the adult animal (the imago) develops from imaginal discs that the larva carried in its body all the time. If you transplant cells of the imaginal discs for wings to an imaginal disc for a leg, the new tissue will grow a wing in addition to a leg. Apparently the cells in each imaginal disc contain the genetic information to form a specific body part, even during the early stages of the development.

The development of wings for fruit flies has been fully resolved in terms of genetics and serves as a textbook example. Expression of the transcription factor *Engrailed* at the posterior side activates expression of the signalling gene *Hedgehog*. The *Hedgehog* signal in turn activates expression of *Decapentaplegic* in the cells at the border between anterior and posterior, thus defining the front and rear sides. When this process is successful the dorsal-ventral axis is created by means of a similar cascade, but this time it is a combination of the transcription factors *Apterous* and the cellular signal genes *Wingless* and *Serrata* (part of the so-called *Notch* signalling pathway). These axes coordinate the further process of wing development, by expressing, for example, *Vestigial* (*Vg*) and *Sal* later, which determine wing morphology and the construction of wing veins, respectively.

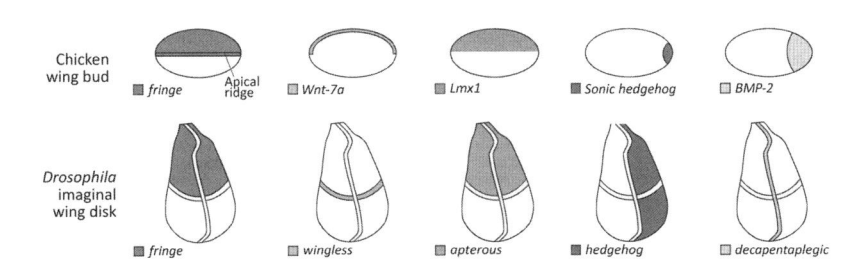

Fig. 2.16: Expression domains of five homologous development genes (some of which with different names) involved in the growth of wings of *Drosophila* and the wings of chicken are very much alike.

The work on *Drosophila* teaches us that the development of a body part always consists of interaction between *Hox* genes and cellular signalling genes at a certain position in the body, at a certain moment. This pattern is generic and shows striking similarities among invertebrates and between invertebrates and vertebrates. Fig. 2.16 shows that the expression domains of the developmental genes *Fringe*, *Wnt-7a*, *Lmx1*, *Sonic Hedgehog* and *BMP-2* in the development of chicken wings strongly resemble the expression of the homologous genes in the development of *Drosophila* wings. The wing of a chicken and the wing of a fruit fly are of course two totally different structures. Still, on the molecular level there is homology: two entirely different tissues are influenced by the same molecular machinery to generate structures with similar functions. This pattern (analogy at the morphological level, homology at the molecular level) is also observed in the formation of the lens eye of vertebrates and the compound eye of insects.

Fig. 2.16 also illustrates that the toolbox genes have different names for different animals, although they are orthologs. There are historical reasons for this. The biologists who studied mutant *Drosophila*s often named a genetic locus after the phenotype of the mutant. In those days, the gene had not been precisely characterised; it was sequenced only much later. This is how, for example, the name *Hedgehog* came about: a gene that, if mutated, stimulated the formation of spines on the body, as a result of which the larva looked like a hedgehog. When the ortholog of *Hedgehog* was discovered in the human genome, the British geneticist Robert Riddle, at the Harvard lab of Cliff Tabin, jokingly named this gene *Sonic Hedgehog*, after the well-known video game *Sonic the Hedgehog*, which he had just given to his six-year-old daughter. This name (abbreviated to *SHH*) was maintained. *SHH* turned out to be a gene of paramount importance, not only in early development, but also in the development of the brain. Mutations in *SHH* cause a number of serious disease profiles.

New axes for limbs

During the outgrowth of a limb, such as the front leg and the hind leg of a vertebrate, three new axes must be constructed at a specific location. A leg not only has an anterior-posterior axis and a ventral-dorsal axis, but also a proximal and distal side. The anterior-posterior axis is constructed first; it is determined by *Sonic Hedgehog* (*SHH*), which is activated at the posterior side. This activation area is designated *zone of polarising activity* (ZPA). *SHH* then starts a cascade of signals, leading to expression of the signalling

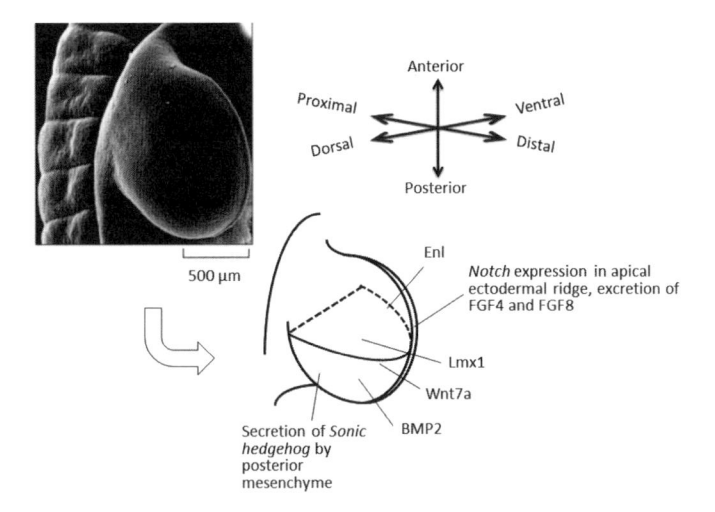

Fig. 2.17: Signals associated with limb development. The limb results from a bud (left). At its posterior side mesenchyme cells form a 'zone of polarising activity' (ZPA) (bottom right). At the apical end, subsequently the 'apical ectodermal ridge' (AER) develops. Both centres secrete signal molecules (Enl, Wnt7a, Lmx1, BMP2), which eventually establish the three-dimensional identity of the limb and its outgrowth.

gene *Fibroblast Growth Factor 8* (*FGF8*) in the epidermal layer, at the knob at the apical side (more or less diagonally across the ZPA). This zone is called apical ectodermal ridge (AER). The AER maintains distal outgrowth of the limb by promoting cell divisions in the underlying cell layers (the 'progress zone'). Subsequently, *SHH* activates *Bone Morphogenetic Protein 2* (*BMP2*) at the posterior side, determining the front and rear side (fig. 2.17). Vertebrate *BMP2* has the same function as *Decapentaplegic* in *Drosophila*, which defines the posterior side of the wing.

Finally, the dorso-ventral axis is determined by activation of *Engrailed* at the ventral side, again under the influence of *SHH* signals. *Engrailed* in turn activates *Wingless 7* (*Wnt-7*) and *Lmx1*. *Wnt-7* and *Lmx1* are homologous to *Wingless* and *Apterous* of *Drosophila*, respectively, which, in the case of fruit flies, are also responsible for the development of a dorso-ventral axis. This shows that the development of limbs in vertebrates and invertebrates uses the same molecular toolbox as is deployed for the wings of flies. More than that, the formation of limbs for vertebrates shows great similarity, at the molecular level, with the formation of parapodia in worms and tube feet in sea urchins. Apparently, the same toolbox is used to create any outgrowth from the ectoderm that must serve the purpose of locomotion.

Finally, during the full development of a limb *apoptosis*, programmed cell death, plays an important role. That process, for example, is important for the formation of single fingers and toes. Apoptosis is activated in places where *Bone Morphogenic Protein 4* (*BMP4*) is active, between the phalanxes of fingers and toes.

The fact that the developmental genetics of the human body plan largely matches that of models like mice and chicken, is illustrated most dramatically by patients with genetic defects in development genes. For example, in the case of *Holt Oram syndrome*, the arms are not or not fully developed. Usually, merely a hand structure is formed, without upper arms or forearms. This syndrome turns out to be caused by a mutation in the T-box transcription factor *Tbx5*, a gene that is essential for the development of front limbs. Furthermore, in 75% of the cases of Holt Oram syndrome a heart defect exists. This suggests that *Tbx5* is also involved in the development of the heart.

A syndrome that is caused by poor functioning of the apical ectodermal ridge is congenital (inborn) *hip dysplasia*. The phenotype consists of a deformation of the hip joint, which is caused by incorrect development of the proximo-distal axis in the AER (fig. 2.18). Problems with the expression of *SHH* in the ZPA also lead to congenital defects. A familiar example is *synpolydactyly*, in which case defects occur in the number of fingers and/or toes. This syndrome is associated with a shortage of *SHH* transcription in the ZPA, possibly in combination with mutations in *Hox13* genes. Finally, we can state that an increasing number of cancers turn out to be associated with defects in the developmental programme. For instance, incorrect expression of *Hox10* genes appears to be associated with breast cancer, leukaemia and

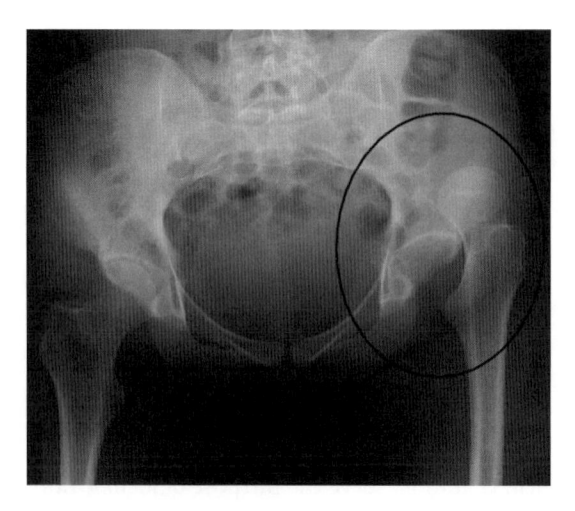

Fig. 2.18: X-ray diagram of the pelvis of a patient with serious hip dysplasia, caused by poor functioning of the apical ectodermal ridge during the embryonic development of the limbs.

ovarian cancer. This comes as no surprise, as *Hox* genes control mitosis, differentiation and programmed cell death directly.

In this chapter we have discussed and compared the genetics and evolution of invertebrate and vertebrate body plans. Using this, a general pattern can be described, consisting of five steps. The first is the development of body axes to fix directions and positions over time. A next step of specialisation is the creation of body parts, such as organs and limbs (organogenesis). This is usually accompanied by redefining of axes in primordial cell groups at specific locations in the body. After these new axes have been constructed, a unique genetic programme is deployed, eventually leading to the full development of the organs. After complete differentiation of the body plan, the role of the *Hox* genes is completed; they are switched off and will no longer be expressed for the rest of the life of the individual. If *Hox* genes are not properly deactivated, this usually leads to serious disease profiles, such as different forms of cancer.

3. Our tinkered body

Just as with all other animals, the human body carries the reminders of an evolutionary past. In our body, typical fish features can be identified that we share with all vertebrate animals. In addition, many traits can be identified that we do not share with fish, but only with reptiles, and obviously we also have the characteristics of a mammal. These similarities prove that our body is the result of a long process of evolutionary change. Over and over again, development has been adjusted by mutations in the molecular toolbox and the changes that proved to be beneficial have been retained. The human body is, just as with *Australopithecus* (table 1.4), a mosaic of ancestral and derived features.

This chapter discusses a number of organ systems of the human body and we will learn that many organs have awkward morphologies that are hard to understand and often inconvenient. The human body evidently wasn't designed 'on the drawing board' by an engineer. In that case the appendix, the tailbone (coccyx), the embryonic gill arches, the curious course of the larynx nerve, *etc.*, wouldn't have existed. We can only understand these structures if we involve their evolutionary background.

Tinkers, watchmakers and a Boeing 747

In a famous article of 1977, the French biologist and Nobel Prize winner François Jacob (1920-2013) described the action of evolution as that of a *'bricoleur'*, a *'tinker'*. A tinker was a travelling craftsman who repaired pots, pans and various kitchen utensils (fig. 3.1). In the bazaar of Middle Eastern towns like Isfahan, a place in Iran that is known for its brassware, you can still find them, sitting down on a stool and rhythmically hitting a copper kettle with a hammer, in order to properly shape it. A travelling tinker used metal objects that he happened to have with him, for example, a rivet to repair a leaking pan. He didn't make anything from scratch; he tinkered with what was at hand, as long as it took to make it good enough. This is how evolution works, too, said Jacob. Evolution is a *'jeu des possibles'*, a game of possibilities.

Another way to express the 'tinkering' aspect of evolution is implicated in the term *'exaptation'*, introduced by Stephen J. Gould and Elisabeth S. Vrba in 1982. They reasoned as follows. There are numerous characteristics in the human body that date back to an ancestor, yet in that ancestor had a

Fig. 3.1: Engraving of a tinker, who used to travel around and carried out all kind of repairs on metal utensils. The way in which the tinker was working with the tools and ironware he happened to have with him, according to François Jacob, is comparable to the action of evolution.

function different from the one it has now. Take, for instance, the protein *alpha-lactalbumin*, that is part of the enzyme complex lactose synthase. Mammals use lactose synthase in the mammary gland to produce lactose from glucose and galactose. But alpha-lactalbumin has not, as such, evolved to serve this function. It descends, via a relatively small number of changes, from a much older protein, *lysozyme*. Lysozyme acts against bacterial cell walls. In human beings it is excreted in sweat, tears and saliva, as well as through breast milk.

So, the function of alpha-lactalbumin in the synthesis of lactose, according to Gould, cannot be viewed as an adaptation in the Darwinian sense; the protein was not shaped under the continuous influence of natural selection for the function of lactose production. Therefore Gould introduced a new term: 'exaptation': a feature that was subjected to natural selection in an ancestor, evolved to serve a specific purpose in that ancestor, but that was 'co-opted' for a new function in the evolutionary descendant. Alternatively the term '*pre-adaptation*' is used: lysozyme is a pre-adaptation for alpha-lactalbumin.

This book will show us that *co-optation* (choosing from existing structures) occurs quite often in evolution; almost all parts of the human body can be seen as co-opted features, as exaptations, however, Gould's term 'exaptation' never really became commonplace. Also the term 'pre-adaptation' has fallen out of favour, due to the suggestion that something can evolve in view of a non-existing, future function.

The image of evolution as a tinker is at variance with the argument of '*intelligent design*', an idea that dates back to a book published 1802 by William Paley, called *Natural Theology: or, Evidences of the Existence and Attributes of the Deity*. In this book, Paley discusses all kinds of examples from nature and human society that, according to him, are the result of a higher power, a divinity, beyond doubt. He starts his book with the famous watch analogy (table 3.1): the existence of the watch is the very proof for an intelligent watchmaker. The English biologist Richard Dawkins played upon this analogy in his famous book *The Blind Watchmaker* (1986). Dawkins demonstrated that complex biological structures, even when they are more complex than a watch, could evolve via a process of mutation and natural selection. Starting from random mutations, and selecting the best variant every time, a structure ultimately can be changed radically. In this way one can – to use Dawkins' words – even climb *Mount Improbable*. The crucial point is that you do not start all over again every time, but you make use of what you have already, you continue with a structure that has seen millions of years of adaptations. Evolution is a cumulative process.

Table 3.1: Excerpt from the first chapter of William Paley, *Natural Theology* (1802)

In crossing a heath, suppose I pitched my foot against a stone and were asked how the stone came to be there, I might possibly answer that for anything I knew to the contrary it had lain there forever ... But suppose I found a *watch* upon the ground, and it should be inquired how the watch happened to be in that place, I should hardly think of the answer which I had given, that for anything I knew the watch might have always been there ... the inference we think is inevitable, that the watch must have had a maker – that there must have existed, at some time and at some place or other, an artificer or artificers who formed it for the purpose which we find it actually to answer, who comprehended its construction and designed its use.

The astrophysicist Fred Hoyle (1915-2001) made an argument that, in terms of scope, is comparable with that of Paley. Hoyle attempted to estimate statistically what the chance would be for the emergence of life from non-living components. He stated that: 'The chance that higher life forms might have emerged in this way is comparable to the chance that a tornado sweeping through a junkyard might assemble a Boeing 747 from the materials therein'. Due to the vivid image that Hoyle created, his argument received a lot of attention, but of course it is not correct, because evolutionary theory does not suggest that complex organisms emerged in one blow.

In the 1990s, the theory of 'intelligent design' again drew attention, due to the publication of the book *Darwin's Black Box* by Michael J. Behe (1996). Behe pointed out the enormous complexity at the cellular and molecular level and introduced the concept of *'irreducible complexity'*. A system of which the complexity cannot be reduced consists of many different parts, each contributing to the overall function. Such an irreducibly complex system cannot work if one of the components fails, and thus cannot evolve either via random assembly.

Behe's best-known example is the bacterial flagellum: the rotating structure, spinning like a whip and driving the bacterium forward. The entire flagellum consists of more than fifty proteins: a filament, a hook, a series of proteins that anchor the flagellum in the outer membrane and the cell membrane, and an intracellular basal body with a propulsion system. Each of the proteins is essential for the function, making the system, said Behe, irreducibly complex, so it cannot have emerged in accordance with a Darwinian process.

However, in the meantime it has been shown that many bacteria have basic precursors of the bacterial flagellum, so-called type-III secretion systems (fig. 3.2). This system consists of a 'needle' which pathogenic bacteria use to detect the proximity of a host, and secrete proteins that play a role in the infection. The core of the system is formed by a collection of 24 proteins

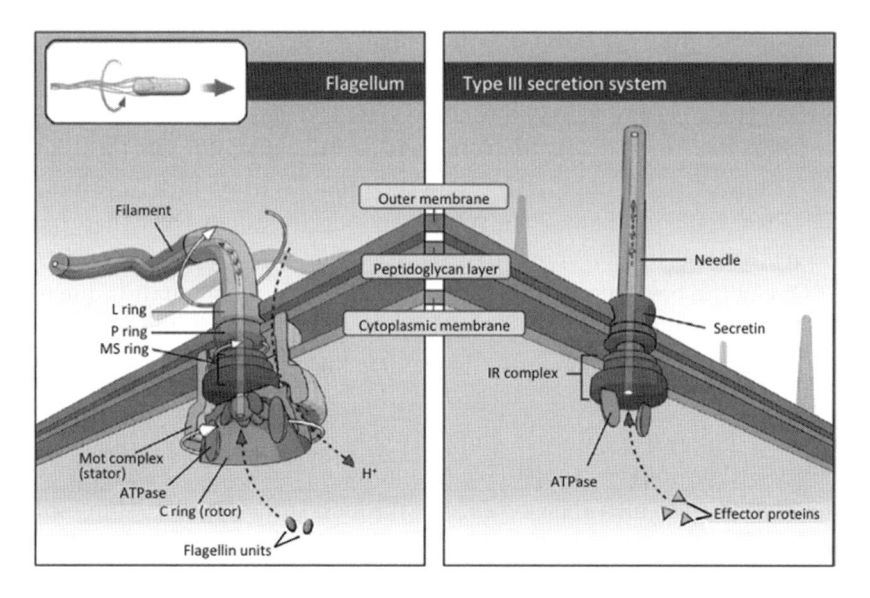

Fig. 3.2: Homologies between the proteins that are part of a type-III secretion system and proteins that are part of a bacterial flagellum. This shows that complex structures can evolve following a step-by-step process, making use of gene amplification, the addition of new and existing proteins that take on new functions.

to which genes have been added through duplications. So, after all it is possible to obtain a complex structure by gradual addition of increasingly more elements, whereby each subsequent stage is an evolutionary improvement. And, while accumulating more and more complexity, the function may change radically.

Creationists still quote the arguments of Paley, Hoyle and Behe in detail, whenever they want to prove that evolutionary theory is not correct and that nature itself is proof for the existence of God. But evolutionary biologists such as John Maynard Smith and Richard Dawkins have shown time after time that the reasoning of Paley, Hoyle and Behe is false, as evolution never creates complex structures all at once. Evolution operates through tiny steps, each of which is an improvement of the previous. Due to the strong objections, the theory of 'intelligent design' has now been pushed to the background entirely.

The naked human

One of the most externally obvious human characteristics is the absence of body hair. It is for this reason that Desmond Morris, in his famous book from 1967, called the human being a 'naked ape'. All the more remarkable is that we know so little of this aspect of the human body. When did humans begin losing their body hair? Was it all at once or gradually? Were *Australopithecus*, *H. erectus* and Neanderthal still hairy? Why did the head, the armpits and the pubic area retain their hair? Why do men have more hair than women? What was the actual benefit of little body hair? There has been much speculation about these questions among numerous scientists.

One of the best-known theories is that the lack of body hair is due to sexual selection. With little body hair, the form of the human body is better expressed, which stimulates sexual arousal in the other sex. Individuals who were less hairy than average, would have higher fitness, because the other sex preferred them. In this way, humans would gradually have become 'more naked'. According to a comparable reasoning, nakedness leads to stronger bonding between men and women. As nakedness stimulates sexual arousal, men and women prefer to be in each other's proximity. Also copulating while facing each other, something quite rare anywhere else in the animal kingdom may be connected to this. Nakedness as well as facing each other during sexual arousal stimulates pair bonding, which is beneficial for the

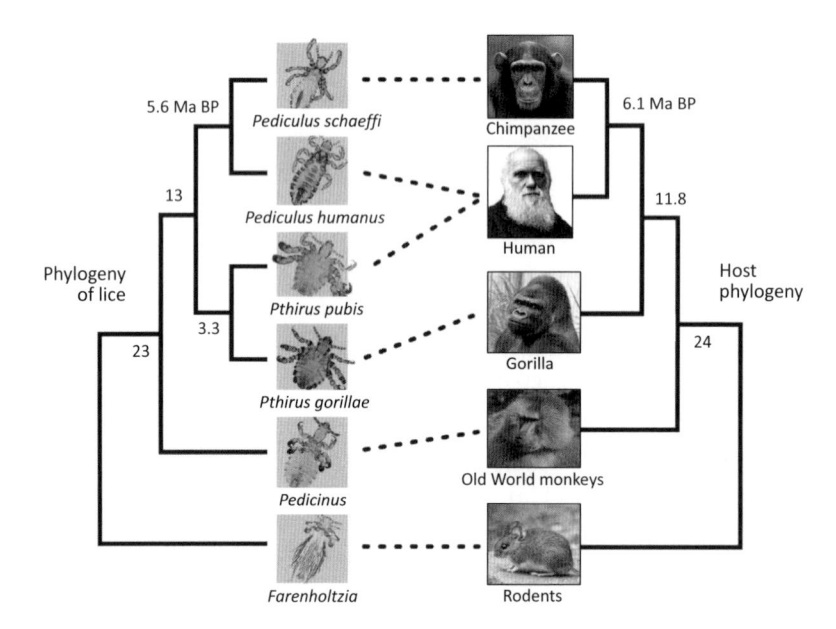

Fig. 3.3: Evolution of lice, compared with the evolution of their hosts (primates). The human head louse (*Pediculus humanus*) probably evolved with the hominins (5.6 million years ago), but the human pubic louse (*Pthirus pubis*) diverged from the gorilla body louse only 3.3 million years ago, and must therefore have 'jumped over' at the time of *Australopithecus*.

successful raising of children, whom, in the case of humans, need a long period of care.

A second theory stresses the role of parasites. On a hairy skin parasites like fleas and lice can easily find a hiding place; loss of body hair reduces the parasitic burden, which is beneficial. It is interesting to note that the human head louse (*Pediculus humanus*) is related to the body louse of the chimpanzee (*P. schaeffi*); the divergence time between the two species is estimated at 5.6 million years, a period that would correspond with the emergence of the hominins (see chapter 1). The human pubic louse, however, is another species (*Pthirus pubis*), which is most related to the body louse of the gorilla (*P. gorillae*); for these lice the split is no older than 3.3 million years (fig. 3.3). Thus, the pubic louse must have jumped onto hominins while these already existed. The human pubic hair may have emerged after the loss of body hair, because the great apes have little hair in the pubic area, compared to the rest of the body. The evolution of lice can also be used to estimate the origin of clothing. The split between the human head louse and the human body louse (two subspecies of *P. humanus*), is dated between 72-42 ka BP. Since the body louse can only survive on a naked body that is clothed, the habitual use of clothing may date back to around that date, which would fit with the time *H. sapiens* was venturing into colder areas outside Africa.

The third theory is that of the '*aquatic ape*'. This idea has arisen in different forms ever since the beginning of the twentieth century. It is usually attributed to the British marine biologist Alister Hardy (1929). The theory was promoted much by the Welsh writer and evolutionary anthropologist Elaine Morgan and the Flemish doctor Marc Verhaegen. According to the aquatic ape theory, the hominins, after they started walking upright, did not evolve any further on the savannah, but made their living near the water, spending much time in the water, like real aquatic animals. Many terrestrial mammals that secondarily adopted an aquatic living have a body with little hair, for instance, whales, dugongs, hippopotamuses and the like. According to the same theory, many other characteristics of the human body can be viewed as an adaptation to aquatic life. The greatest weakness of the theory is that no hominin fossils are known from sediments of coastal seas. All fossils of hominins are recovered from layers that were terrestrial during the fossilisation.

The fourth, most widely accepted theory about human nakedness states that little body hair contributes to a better heat regulation in a warm and sunny climate, such as the African savannah (*cf.* fig. 1.1). The nakedness would, as such, be a direct consequence of walking on two legs. Due to upright walking, the brain is exposed to intensive solar radiation – but primates do not have a system to properly cool off the brain. Ungulates have a so-called *rete mirabile*, a network of blood vessels just beneath the brain. Via a system of reverse flow, relatively cold blood coming from the veins in the nose cools off the warm blood in the carotid artery. In this way overheating of the brain is prevented. But hominins did not have such a system, so it was necessary to prevent overheating of the brain in a different way. This was achieved by intensively exposing the majority of the body to the ambient air, preventing warming via evaporation of sweat. In this theory, therefore, bipedalism, of which the advantages were discussed in chapter 1, and lack of body hair are directly connected.

It is remarkable that there has been much more speculation about the 'use' of little body hair than about the issue how it emerged and when. A logical point of view is that mutations in keratin genes underpin the nakedness of hominins. Keratin is a protein with high concentrations in epidermis and hair. There are various genes that encode a keratin; jointly they are called KRTAP genes (*keratin-associated proteins*). They are subdivided into two main groups: keratins rich in cysteine and keratins rich on glycine and tyrosine (fig. 3.4).

Recently, the evolution of all known KRTAP genes in mammals was elucidated. It is a gene family in which many genes have become *pseudogenes*.

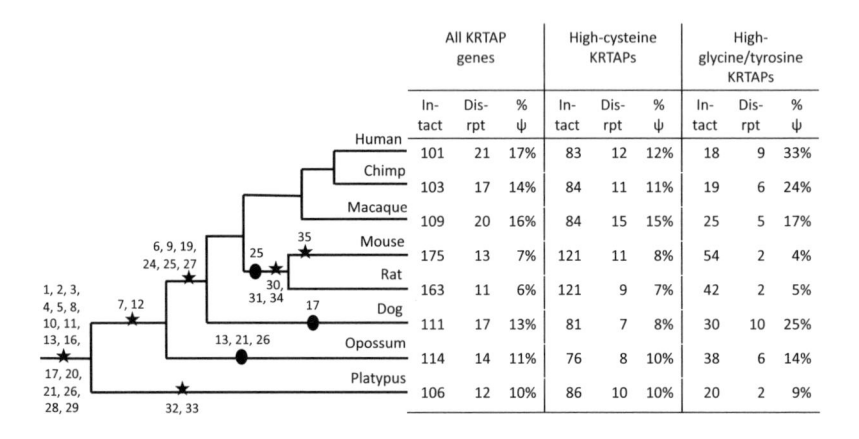

	All KRTAP genes			High-cysteine KRTAPs			High-glycine/tyrosine KRTAPs		
	In-tact	Dis-rpt	%ψ	In-tact	Dis-rpt	%ψ	In-tact	Dis-rpt	%ψ
Human	101	21	17%	83	12	12%	18	9	33%
Chimp	103	17	14%	84	11	11%	19	6	24%
Macaque	109	20	16%	84	15	15%	25	5	17%
Mouse	175	13	7%	121	11	8%	54	2	4%
Rat	163	11	6%	121	9	7%	42	2	5%
Dog	111	17	13%	81	7	8%	30	10	25%
Opossum	114	14	11%	76	8	10%	38	6	14%
Platypus	106	12	10%	86	10	10%	20	2	9%

Fig. 3.4: Evolution of keratin genes (*keratin-associated proteins*, KRTAP) in a phylogeny of eight mammals. The KRTAP genes are divided into two main groups, cysteine-rich and glycine/tyrosine-rich. Within the groups 35 subfamilies are distinguished (indicated by numbers). The asterisks indicate emergence of a subfamily; the circles indicate loss of a subfamily. Disrpt = disrupted, %ψ = percentage of genes with disrupted open reading frame (pseudogenes).

That means that a mutation has interrupted the open reading frame and a premature stop-codon has emerged. Some pseudogenes are partly transcribed but no functional protein is being produced. This is particularly the case if the mutation is fairly recent. In the case of old mutations, a pseudogene is hard to recognise and sometimes barely differs from a non-coding DNA sequence. However, the loss of keratin genes is relatively recent and many pseudogenisations are specific to human beings. For example, the human *HaA* gene (*hHaA*) is a pseudogene and is thus presented as *ψhHaA*, but the orthologous genes *cHaA* and *gHaA* in chimpanzee and gorilla, respectively, are functional. The mutation in human *HaA* is dated at 240 ka BP, which approximates the origin of *Homo sapiens*.

However, the complete KRTAP picture teaches us that there is little connection between the loss of keratin genes and body hair (fig. 3.4). In humans 17% of the KRTAP genes have turned into pseudogenes, a percentage that is barely higher than that of the chimpanzee (14%) and the rhesus macaque (16%). Due to the absence of a clear relationship between keratin genes and body hair not only do we know little about the function, but also about the cause of human nakedness.

Finally, another view on human nakedness connects it with developmental retardation, neoteny. Body hair typically emerges in the course of the juvenile period. In the womb, the foetus has a light hair covering, called *lanugo*. This hair vanishes just before birth, as a result of which the newborn baby has an almost smooth skin. The neoteny hypothesis implies that this phenotype has been retained, in the same way in which humans have retained a baby-like skull (fig. 2.2). Hair growth in the pubic area, in the armpits and on the male face is under the control of sex hormones and so would have escaped the developmental retardation acting upon body hair. This hypothesis corresponds with the observation that the degree of neoteny is stronger for women than for men, just as is the case for other characteristics (see chapter 2).

Adaptations to bipedalism in the locomotor apparatus

One of the most striking human characteristics is our walking on two legs. In chapter 1, we stressed that the evolution of bipedalism should be regarded as the greatest hominin innovation, allowing for a radical new way of life. We see that in the course of evolution the human body has been further adjusted in numerous places in order to optimise bipedal locomotion. In the skull, the *foramen magnum* was positioned at the centre of the skull basis, so as to be able to carry the skull straight on the spinal column. The shoulder joint was oriented to the side, making it possible to freely swing the arms next to the body. In relation with the broadening of the shoulder girdle, the chest became cylindrical instead of conical, like in chimpanzees. The spinal column obtained an S-shaped, curved structure (*kyphosis* in the upper back and *lordosis* in the lower back), allowing for better absorption of shocks as a result of walking.

The greatest changes, however, can be found in the *pelvis* (fig. 3.5). These adaptations started very early in the evolution of the hominins and were apparently essential for upright walking. The pelvis of the chimpanzee is

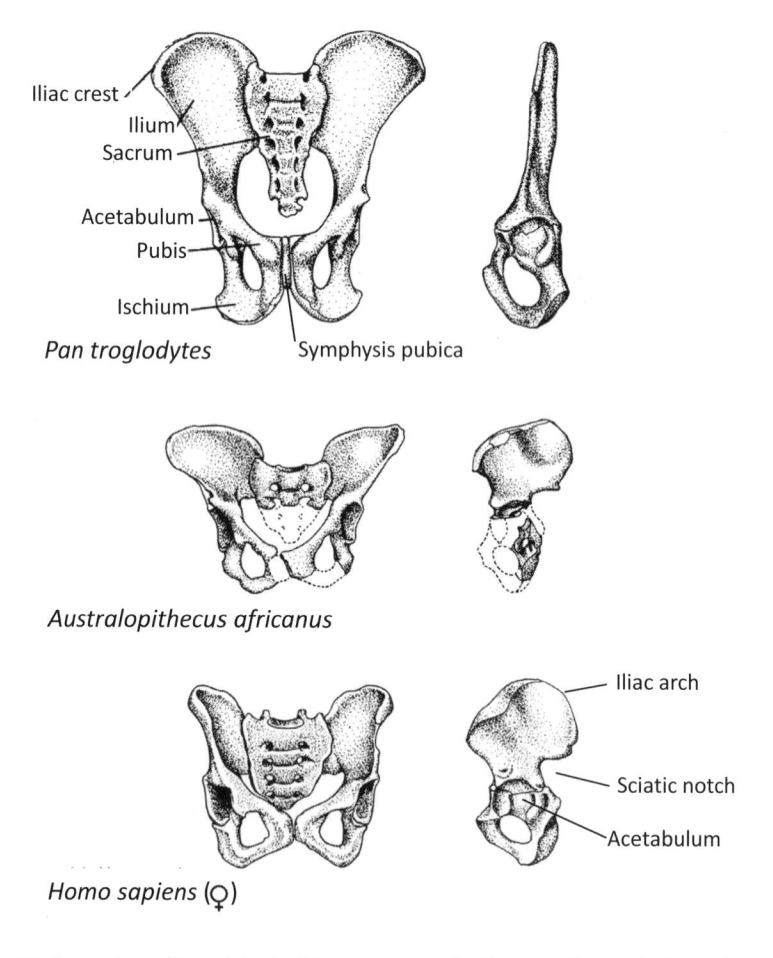

Fig. 3.5: Comparison of the pelvis of a chimpanzee, *Australopithecus* and human indicates the profound evolutionary changes involved with upright walking.

a flat structure, while the pelvis of the hominins became bowl-shaped, supporting the intestines. In addition, our pelvis is shortened and somewhat tilted in relation to the chimpanzee: the *ilium* is curved backwards (forming a '*sciatic notch*' below it), and the *pubis* protrudes forward.

The pelvis of *Australopithecus* clearly has the bipedalism-adapted human form, yet also shows its own typical traits (fig. 3.5). The widely flaring hipbones, very different from chimpanzees but also from humans, are particularly remarkable, a situation that persists up to *H. habilis*. Apparently the older hominins had a different way of upright walking than we do. It seems that the basic biomechanical requirements for bipedalism, including changes in the muscle attachments on the ilium that we see in *Ardipithecus* and *Australopithecus*, were realised early, but further refinements involving the lower parts of the pelvis were achieved much later, only in *H. erectus*.

In addition to the locomotor adaptations of the pelvis, the distance between the left and right joint socket of the femur (*acetabulum*) became larger. This latter change relates to the enlarged birth canal, which was necessary to give birth to children with larger brains in the later hominins. Broadening of the pelvis to some extent conflicts with our walking on two legs (which calls for a smaller width), and evolution has forged a compromise. The consequences of this compromise are seen in human childbirth. During delivery, the baby ideally conducts a 'corkscrew manoeuvre' and to prevent obstetric problems the assistance of a midwife is desirable. Inspection of female Neanderthal pelvises and infant heads shows that a difficult delivery was also common in Neanderthal. Related to the birth canal adaptation the human pelvis is sexually dimorphic. The lower angle between the left and right pubic bone is smaller than 90° for men, but larger than 90° for women. This is one of the most consistently sexually dimorphic characteristics of the human skeleton.

Ilium, ischium and pubis are connected to the spine by means of the *sacrum*, which consists of a number of fused vertebrae, the sacral vertebrae. Humans typically have five sacral vertebrae and five lumbar vertebrae, while chimpanzee has six sacral and four lumbar vertebrae. This shift may be associated with the necessity for more flexibility in the spinal column to ease upright walking. The number of sacral vertebrae in *A. afarensis* is reported to be four, however, the later species *A. sediba* has five, like *H. erectus* and *H. sapiens*. As it is unlikely that, starting with a chimpanzee-like ancestor, the number of sacral vertebrae would have become six, then four and subsequently five, the count in *A. afarensis* may be wrong. But given the fact that modern humans also show some variation (some people have six sacral vertebrae, without any medical consequences), the differences could also be the result of some degree of developmental instability in the definition of segments towards the end of the spinal column.

The changes in the pelvis were accompanied by changes in the function of muscles that move the legs. The *musculus gluteus maximus* of great apes is an *abductor* (it moves the leg sideways), but in humans it is an *extensor* (it moves the leg backwards). The human gluteus maximus has a better leverage on the backwardly curved ilium, which helps in bipedal locomotion. The abductor function in humans is performed by the m. gluteus medius and minimus. The m. sartorius, which connects the knee with the pelvis, was given a function in the forward extension.

Not only the adaptations in the pelvis, but also the changes in the femur emerged at an early stage (fig. 3.6). The distal part of the femur has an inward-oriented extension, the neck of the femur or *collum femoris*, ending in a round head (*caput femoris*) that pivots in the acetabulum of the pelvis. This is the largest and strongest joint of the human body, because it supports the entire body weight (with the exception of the legs). The human caput

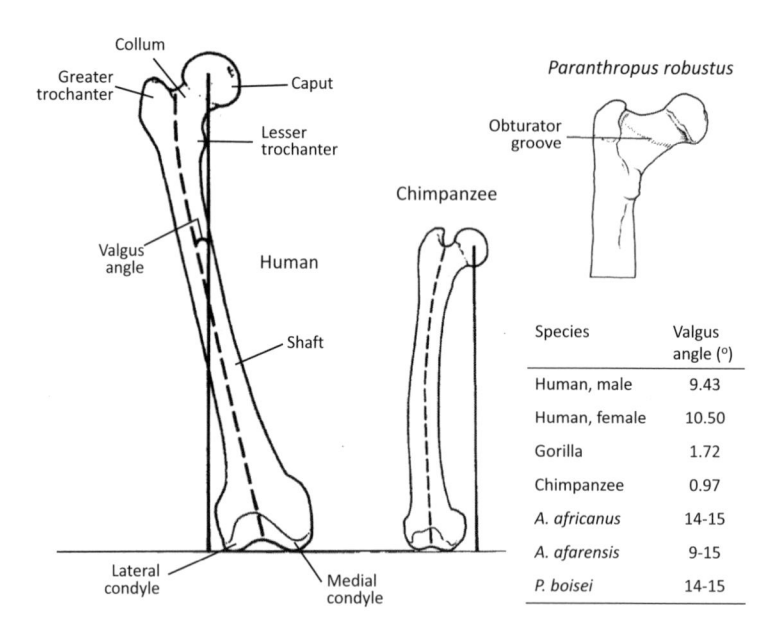

Species	Valgus angle (°)
Human, male	9.43
Human, female	10.50
Gorilla	1.72
Chimpanzee	0.97
A. africanus	14-15
A. afarensis	9-15
P. boisei	14-15

Fig. 3.6: Left: a human femur, seen from the front, with the condyles of the knee joint placed on a horizontal surface, as a result of which the inward-oriented position is apparent. The angle between the perpendicular and the longitudinal direction of the femur is the valgus angle or Q-angle. For the chimpanzee (centre) this is virtually zero degrees, for humans, it is approximately 10 degrees. In *A. afarensis* the angle is comparable with that of humans (table in the image). Right: the upper part of the thighbone of *Paranthropus rubustus*, with the obturator groove in the collum.

femoris is much larger than that of the chimpanzee, after correction for the difference in body weight.

The angle that the collum femoris makes with the longitudinal axis of the femur, is called the *inclination angle*. This is about 60 degrees for chimpanzee but in the case of humans it is (on average) 48 degrees. The smaller angle allows the force exerted on the joint to be better guided towards the thighbone. The inclination angle is, however, quite variable; it increases during life from 30 degrees in children to even 60 degrees in the elderly. The inclination angle is also often reported as a supplement, so 132 degrees rather than 48 degrees for the average human and 120 degrees in chimpanzee. Another interesting aspect is a transversal groove on the neck of the femur, caused by the tendon of one of the pelvic floor muscles, the *m. obturatorius internus* (fig. 3.6). This muscle runs from the inside of the pubic bone towards the thighbone. Due to upright walking, the tendon rubs the neck of the femur, causing a groove. The presence of such a groove in the femur neck of *Orrorin tugenensis* (an early hominin that we came across in chapter 1) was an important argument to classify *O. tugenensis* as an upright walking hominin.

If we go further down the human body, we meet the knee. In humans the lower leg makes a slight angle with the upper leg. This so-called *valgus angle* is measured by placing the femur on an even surface (fig. 3.6). In the case of the chimpanzee, the lower leg runs down in an almost straight line from the upper leg, with a valgus angle of practically zero degrees. For humans the femur is directed inward, causing the lower leg to move closer to the mid line, a clear adaptation to walking on two legs, as it prevents excessive shifting of the weight to the left and right while walking, which would cause a waddling gait. The average valgus angle for women is larger than that of men, due to the fact that women have a broader pelvis (fig. 3.6).

The foot also shows important adaptations to bipedalism. The foot of *H. sapiens* is a platform, while that of apes is a prehensile foot. That was also the case for the first hominins (for instance, the foot of *Ardipithecus ramidus*, fig. 1.6). The heel bone (*calcaneus*) is extended backwards in humans, favouring the leverage effect of the calf muscle, which is connected with the back of the heel bone via the Achilles tendon.

For *Australopithecus sediba*, based on the fossil foot bones, a biomechanical analysis was made of its locomotion (fig. 3.7). This hominin walked with a strong *overpronation*, meaning that the foot made a significant rotation around its longitudinal axis during walking. Generally, a certain degree of pronation is required to move the weight from the outside of the heel towards the inside of the forefoot and the big toe, but if pronation is too large, it may lead to all sorts of physical complaints. Overpronation may

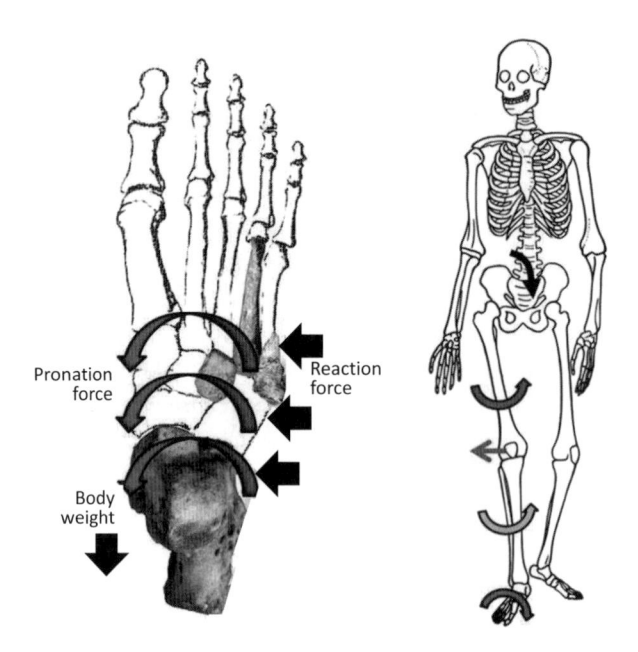

Fig. 3.7: Reconstruction of the locomotion of *Australopithecus sediba*. Based on the shape and articulation of the foot bones, *A. sediba* must have walked with strong overpronation. On the right, the forces on the lower leg, knee joint, upper leg and pelvis are shown, all necessary to compensate the pronation force (projected on the bone structure of humans).

be absorbed by compensatory forces in the lower leg, the knee and the upper leg (fig. 3.7). From the fossil bones of *A. sediba* we may conclude that the knee joint and the hip were adapted to this. Just as was shown for the pelvis of *A. afarensis* and *H. habilis*, this analysis illustrates that the early hominins did walk on two legs, but compared with modern humans, had a different way of walking.

Finally, we also see adaptations in the toe bones (*phalanges*). In humans the toe bones are stretched instead of curved and the big toe (*hallux*) is positioned next to the other toes, which decrease in size from median to lateral (a condition known as *entotaxony*). The angle of the big toe relative to the longitudinal direction of the foot is known as *hallux abduction*. This shows a gradual change throughout the hominins: from 28 degrees in *A. afarensis*, via 12 degrees in *H. erectus* towards 8 degrees for modern humans.

The hallux abduction values are derived from fossil footprints (fig. 3.8). The best known are those at Laetoli, a town in Tanzania where, 3.7 million years ago, two upright walking creatures, probably a parent with a child, left their footprints in fresh volcanic ash, which later became fossilised.

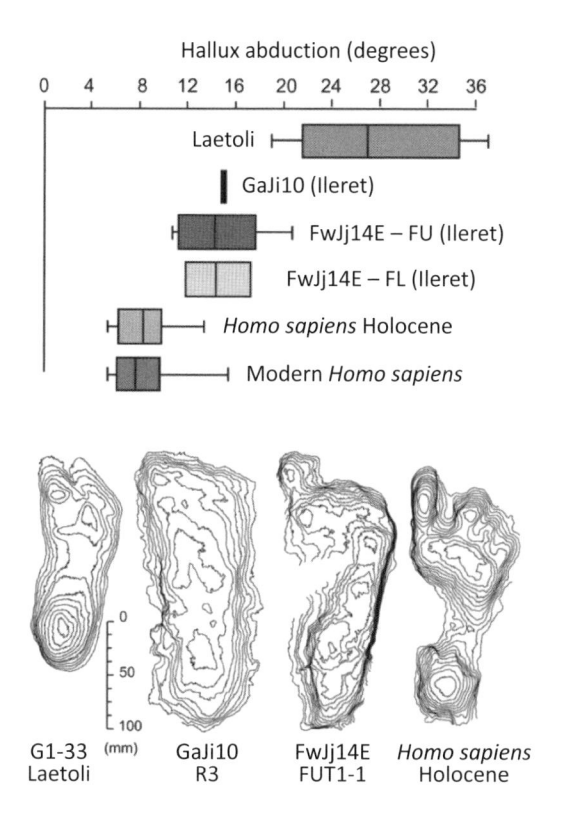

Fig. 3.8: A detailed study of archaic footprints makes it possible to estimate the hallux abduction of extinct hominins. G1-33 (Laetoli, 3.7 million years old) is of *Australopithecus*, GaJi10 and FwJj14 (Ileret, 1.5 million years old) are of *Homo erectus*. During the evolution of the hominins hallux abduction became gradually smaller.

The prints are attributed to *Australopithecus afarensis*. Such footprints are also known from *Homo erectus*, of 1.5 million years ago, near the town of Ileret nearby Lake Turkana in Kenya. Obviously the reduction of hallux abduction was a very gradual process, only reaching its modern value with *H. sapiens*, and it may be assumed that these changes are a consequence of directional natural selection associated with bipedal stride.

In summary we can state that radical changes were required for walking on two legs. Many of those changes, such as the pelvis and the femur, emerged at an early stage, as they were essential for upright walking. Other adaptations, such as the chest becoming more cylindrical, as well as the decrease of hallux abduction, gradually evolved and reached their modern state only with *H. sapiens*. Table 3.2 gives an overview of the various adaptations to bipedalism in the locomotor apparatus.

Table 3.2: Overview of changes in the skeleton associated with adaptations to upright walking

Skull	Foramen magnum in the centre of the basicranium, not at the back of the skull
Shoulder	Shoulder joint laterally oriented (towards the sides) instead of cranial (towards the skull)
Spinal column	S-shaped; on the average five lumbar vertebrae instead of four
Pelvis	Bowl-shaped instead of plate-shaped, shortened, tilted, pubis pointed forward, ilium curved backwards, offers attachment for m. gluteus maximus as an extensor instead of abductor
Femur	Inclination angle of femoral neck reduced, caput femoris enlarged
Knee	Valgus angle enlarged
Tibia	Elongated, relative to the femur
Foot	Calcaneus backwards extended, phalanges stretched instead of curved, entotaxony of toes, hallux abduction reduced

Gill slits, larynx and middle ear

In the course of evolution, the transition zone between head and thorax of the vertebrates was profoundly reorganised; in the embryonic development of humans we can still identify the remainders of this process. At the height of the neck the embryo develops cartilage arches that remind us of the gill arches of fish. These structures, developing in all chordates, are part of the phylotypic stage called pharyngula, and they date back to the first chordate animals, such as the lancelet *Amphioxus*. This invertebrate chordate has a series of clefts in its pharynx that open to the outside, through which the water that it swallows is forced out again. Between the slits there are reinforcements, the gill arches or *pharyngeal arches*, over which blood vessels and nerves are running. The structure as a whole is called a gill basket, but its original function is not that of a gill, but of a filtration device. Lancelets filter seawater in this way and feed on the plankton that is retained by the pharyngeal basket. Later, in the fish, the gill arches were used to attach the gills, an obvious case of co-optation. The gill basket lost its function as a filtration device when the fish developed jaws that allowed them processing larger food parts. The gill arches have vanished in adult reptiles, birds and mammals, although they are still present in the phylotypic stage (*cf.* fig. 2.1).

In the human embryo structures homologous to gill arches can also be identified. The first arch is known as *Meckel's cartilage*. The dorsal part of it soon vanishes into the skull base; a part of the os sphenoidale is homologised with the first pharyngeal arch. The rest of the first arch is included in the

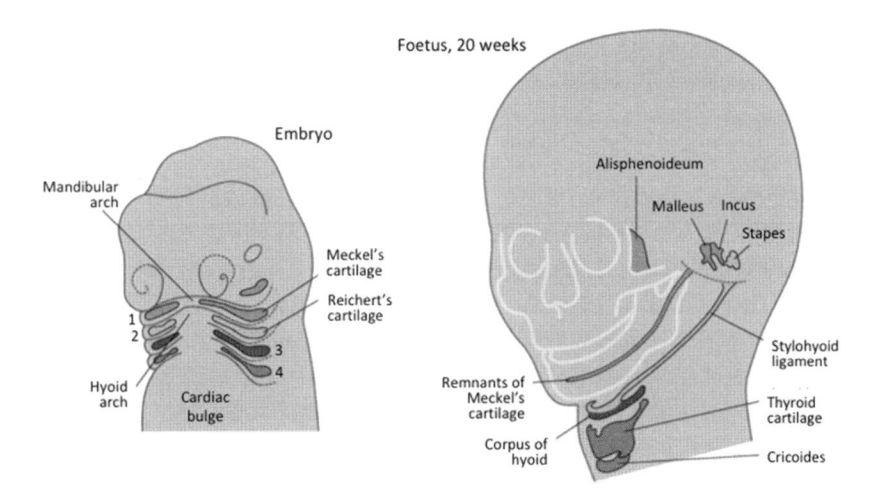

Fig. 3.9: Schematic diagram of reorganisations in the neck region during human embryonic and foetal development. Meckel's and Reichert's cartilages are homologous with the first and second gill arches of fish; they are transformed in favour of the lower jaw, the middle ear and a ligament that connects the hyoid bone with the skull. From the third gill arch the hyoid bone develops, while the larynx results from the fourth. On the left, the situation is shown for a five-week-old embryo, to the right, that of a foetus of twenty weeks old.

lower jaw, and from the rear part, two out the three middle ear ossicles are formed. The second pharyngeal arch, called *Reichert's cartilage*, forms a connective tissue strand that connects the *hyoid bone* (os hyoideum) with a protrusion of the temporal bone in the skull (the stylohyoid ligament); furthermore, this arch forms the third (actually the first, as we will find out below) middle ear ossicle, the *stapes*. The third pharyngeal arch forms the hyoid bone and the fourth arch contributes to the cartilage of the *larynx* (fig. 3.9).

As if this tinkering with the old fish structures weren't enough, there is also the strange aggregation of middle ear ossicles. As we know, we have three middle ear ossicles, *stapes*, *malleus* and *incus*. These fulfil an important role: they transfer the amplified vibrations of the eardrum to the ear. Our ancestors, the reptiles, however, have only one middle ear ossicle, homologous to our stapes. This bone is formed from the second pharyngeal arch (fig. 3.9). In the course of evolution, however, the jaw joint of the reptiles obtained an auditory function. In order to understand which changes that has brought about, we should study the bones that form the jaw joint in detail (table 3.3). The reptile jaw joint is formed by the *quadrate* in the skull and the *articular* in the lower jaw. In the evolutionary transformation, this jaw joint was shifted inwards, whereupon the bones obtained a new function

Table 3.3: Overview of the homologies between the bones of the jaw joint and the middle ear of reptiles and mammals; in reptiles, quadrate and articular form the jaw joint, while in mammals the joint is between squamosum and dental

	Reptiles		Mammals	
Skull	Squamosum		Squamosum	Skull
	Quadratum		Incus	Middle ear
Lower jaw	Articular		Malleus	
	Angular		Disappeared	Lower jaw
	Supra-angular		Disappeared	
	Dental		Dental	

in the middle ear. The articular in the mandibular was given a new life as the malleus; the quadrate in the skull was converted into the incus. In the lower jaw, two bones, angular and supra-angular, have disappeared. The result is that in mammals the joint is between the *squamosum* and the *dental* (table 3.3). The mammals developed a new jaw joint, because the original joint moved to the middle ear.

It is interesting that the formation of the mammalian middle ear, a quite remarkable case of developmental tinkering, is clearly proven both in human embryonic development and in the fossils. The German anatomist Karl B. Reichert described the process in 1836 (mind you, before Darwin's book appeared!). There are different extinct reptiles (belonging to Therapsida, the group from which the mammals emerged) that show beautiful transitory stages, and also a number of evolutionary 'try-outs' in the then rapidly radiating group of mammalian reptiles and early mammals, before the present situation settled. Also, in human embryonic development a shift takes place at the rear end of Meckel's cartilage, towards the middle ear. We have, as Stephen J. Gould called it in a famous essay from 1990, 'an ear full of jaw'. Gould thereby cited his favourite teacher at Harvard, John Burns, who was in the habit of triggering his students by reciting poems that included subtle puns. His poem 'Evolution of the Auditory Ossicles' runs as follows:

With malleus
Aforethought
Mammals
Got an earful
Of their ancestors'
Jaw

The intestines and the lung

The intestinal system of an animal usually shows a more tight association with its ecology, especially with its feeding strategy, than with its evolutionary background. Within many evolutionary lineages one may find a large diversity of digestive systems, suggesting a potential for rapid adaptation rather than evolutionary conservation, as seen in many of the other organ systems discussed in this chapter. Does the human digestive system fit into this general pattern?

The first observation when regarding the human intestine is that it is relatively short. Humans are often compared to pigs because they share many physiological characteristics with these animals, and pigs, as well as their wild relatives, are primarily omnivorous, like humans. However, the human gut is considerably smaller than the pig's (fig. 3.10). The midgut-to-hindgut ratio is comparable, though, and in both cases the hindgut (*colon*) is haustrated (subdivided in pouches).

Herbivores generally have much longer digestive tracts, often with specialised, sacculated, capacious or subdivided compartments for digesting cellulose with the aid of microbial communities. Various parts of the intestine (stomach, ileum, caecum, colon), can take on this function,

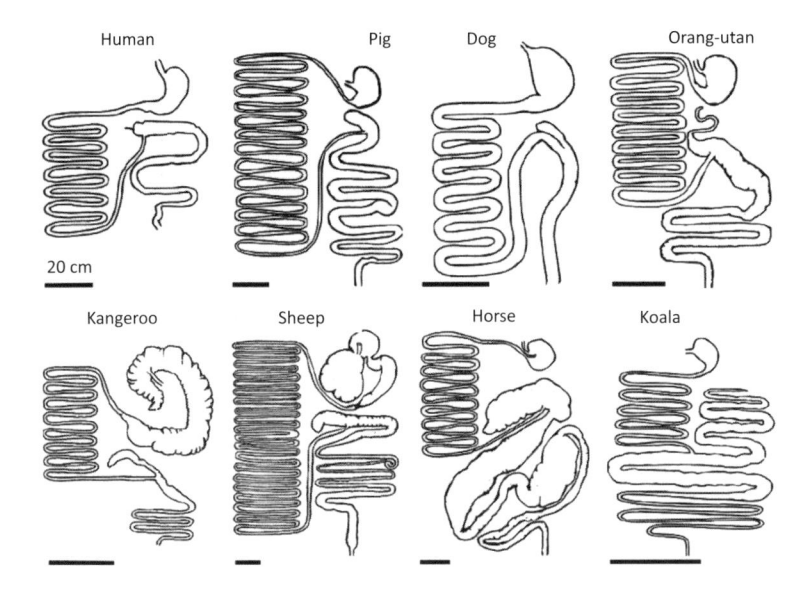

Fig. 3.10: Schematic pictures of the intestinal tracts of a number of mammalian species, illustrating the large morphological variation in association with feeding habits.

depending on whether the animal is a *foregut fermenter* (kangaroo, sheep) or a *hindgut fermenter* (horse, rat, koala). Carnivores, exemplified by the dog in fig. 3.10, generally have a short digestive tract.

The question may be asked: If the intestinal tract rapidly evolved in response to a feeding strategy, what could have been the selective forces shortening the human gut? The most common explanation is that humans started to cook their food. We have created our own, unique, feeding niche and in this way have influenced our own evolution. Calculations show that eating only raw materials cannot generate enough energy to sustain the human body with the gut we have. In the words of the British primatologist Richard Wrangham, cooking should be considered a biological trait of our species to which our guts have adapted.

It is not clear when humans started to process (cook, roast or grill) their food. If cooking is really necessary for the human lifestyle it could have been as old as the practice of hunting, so 2 million years, starting with *H. erectus* in Africa. We will see in chapter 6 that the evidence for butchering animals using stone tools, maybe in association with attempts to get access to bone marrow, goes back even further, to 3.3 Ma BP in Africa. There is evidence that *H. erectus* in Asia (Philippines) was butchering animals by 709 ka BP. However, as we have seen in chapter 1, the archaeological evidence for fireplaces and scorched bones, in Europe and the Levant, does not go back earlier than 350 ka BP. How these different lines of evidence can be matched is not clear at the moment.

Another line of reasoning is that the relative shortness of the human gut has something to do with our enlarged brain. This is the core of the *'expensive tissue hypothesis'* formulated by the American anthropologists Leslie Aiello and Peter Wheeler in 1995. These authors noted that the human brain is an organ that demands a large amount of energy; yet, while the brain expanded its volume by a factor of 4.5 during hominin evolution, basal metabolic rate did not increase proportionally. All organs of the human body have a volume in accordance with expectations from other mammals, except the brain (much larger) and the gut (much smaller). Aiello and Wheeler argued that expansion of the brain, being under strong positive selection, forced the gut to become smaller. This was possible when humans were adopting an omnivorous feeding strategy, including high-energy food items such as meat. A negative correlation between brain mass and gut volume is also observed in other animals, so it could reflect a fundamental biological phenomenon. The gut communicates with the brain through hormones and sensory neurons in the epithelium that inform the brain about the gut's nutrient status, however, a convincing mechanistic explanation for the relation between gut volume and brain size has not yet been given.

The human gut is also characterised by the absence of a big *caecum* and the presence of a *vermiform appendix*. The appendix is often portrayed as a typical *vestigial* (rudimentary) organ, and in textbooks is listed among the evidence for evolution in the human body, but it is doubtful whether the common image of the appendix as a rudimentary organ is correct.

First of all, the appendix is not without function in the human body. Two functions have been described; it is a lymphoid organ responsible for the production of immunoglobin A (IgA), the major factor contributing to regulation of gut microbes, and it is a 'hiding place' or reservoir for commensal bacteria that may recolonize the gut after diarrhoea or other factors causing loss of microbial communities from the colon. Among these two functions the immunological role may be subject to redundancy, which is suggested by the observation that patients undergoing appendectomy do not suffer from lower IgA; apparently, the production is taken over by other lymphoid tissues in the gut. However, the replenishment function for gut communities is not redundant as people that have undergone appendectomy have a higher risk of colorectal diseases.

Secondly, the appendix is not a typical rudimentary organ because it has evolved several times among mammals, and where it evolved it was retained, suggesting it serves a function. Unlike, for instance, the wisdom teeth, which have a tendency to vanish even among modern humans, the appendix does not show an indication for continuous decline.

Finally, we note that the human intestine has quite a strong acid compartment, the stomach. Such compartments occur in many animals and their main function is to aid in protein digestion using pepsinogen and

Table 3.4: Comparison of pH in acidic compartments of the intestinal tracts of birds and mammals with various feeding habits

Feeding group	Example of species	Stomach pH (mean ± standard error over several species)
Obligate scavengers	Turkey vulture	1.3 ± 0.08
Facultative scavengers	Red-tailed hawk	1.8 ± 0.27
Generalist carnivores	Ferret	2.2 ± 0.44
Omnivores*	Domesticated pig	2.9 ± 0.33
Specialist carnivore	Insectivorous bat	3.6 ± 0.51
Hindgut fermenter herbivores	African elephant	4.1 ± 0.38
Foregut fermenter herbivores	Colobus monkey	6.1 ± 0.31

The data are based on 68 species in total
*including humans (stomach pH = 1.5)

HCl. However, an additional function of high acidity may be to act as a barrier against pathogenic bacteria. Since bacteria on meat and especially on carcasses left in the environment present more problems than the bacterial load of plant material, it is expected that the pH of the stomach would be lowest among scavengers and highest among herbivores. There is indeed good evidence for this trend (table 3.4). Interestingly, the human stomach has a pH that is quite below the average for omnivores, being more in the range of scavengers. This may be taken as evidence that scavenging has been an important aspect of the ancient feeding strategies of humans.

In connection with the intestinal tract we discuss in this section also the evolution of the lung. The reason to do so is that the lungs are basically blind sacs (*diverticula*) of the upper digestive tract. In the human embryo, we see the lung developing from a ventral groove of the pharynx, in the area of the pharyngeal arches. From this groove, a left and right sac develop, whereby the right one is somewhat larger. On the ventral side, the blind sacs grow in caudal direction, along the oesophagus past the heart (fig. 3.11). From the unpaired groove the trachea develops, and the blind sacs become the left and right lung. As the lung is not functional in the foetus, the development of bronchi and lung vesicles is spread across the entire development and is only completed shortly before birth.

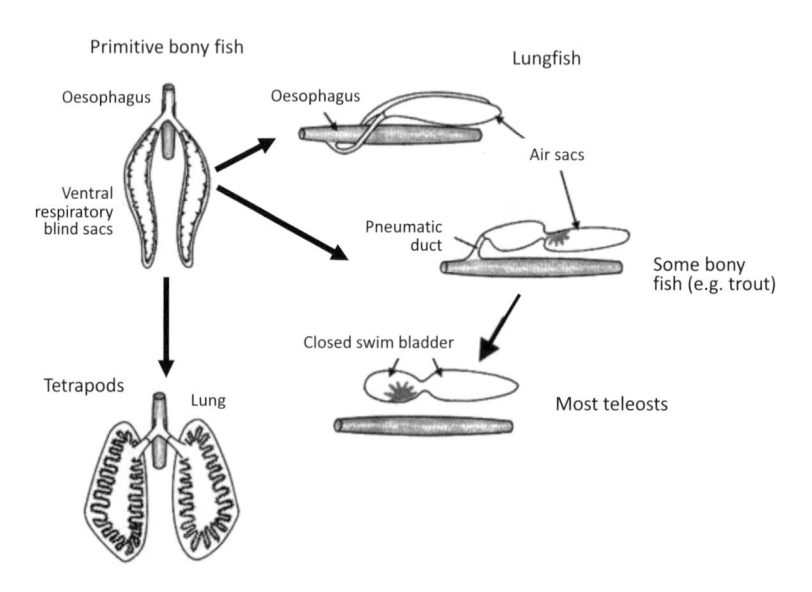

Fig. 3.11: Relation between the lungs of tetrapods and the swim bladder of bony fish. Both derive from a set of ventral diverticula of the oesophagus.

Comparative research has shown that the human lung is homologous with the swim bladder of fish (fig. 3.11). Also the swim bladder develops as an offshoot of the pharynx. In some fish the connection between the swim bladder and pharynx is still visible as the *pneumatic duct*. However, in the majority of higher fish (Teleostei) the swim bladder has become a separate organ that is sealed off from the gut. The swim bladder should not be regarded though as the predecessor of the lung; the situation is rather that both lung and swim bladder derive from the same pharyngeal blind sacs. In the case of fish these sacs grow to dorsal, where the left and right pouches merge into the swim bladder, while, in the case of the mammals, they remain ventral and paired. Lungfish, however, retained the original respiratory function, which allowed them to survive temporarily on land. This was the condition that was inherited by the first amphibians, from which reptiles, birds and mammals descend that all have fully developed lungs. The use of the pharyngeal blind sacs as a swim bladder thus represents a derived and not an ancestral condition! The swim bladder is another example of co-optation. Some fish show a transitional situation; lungfish, for example, have two lungs that lie dorsally to the pharynx, while the pneumatic ducts wrap around the pharynx and are connected at the ventral side. This all shows that there is a close relationship between the lung and the intestinal tract, both in embryology and in evolution.

Heart and urogenital system

The heart is one of the most complicated organs of the human body. Just like in all chordate animals, it is situated at the ventral side of the body, while for most invertebrates it is at the dorsal side. This situation refers to the reversal of body axes upon the evolution of deuterostomes from protostomes, as described in chapter 2 (fig. 2.8).

The development of the heart and the closed blood circulation are strongly connected with the emergence of the secondary body cavity, the coelom. This starts early in development. Directly after gastrulation, groups of *angioblasts* (blood vessel forming cells) constrict from the left and right mesoderm and migrate towards the ventral medial line, where they form a tube, which is part of the coelom; the coelom cavity around the heart tube is called the *pericardium* (fig. 3.12). In further development, also cells of the neural crest are involved (see fig. 2.5), that migrate from dorsal to the heart. These neural crest cells are of paramount importance; they form, for instance, the septum that separates the aorta from the pulmonary artery. If the neural crest cells

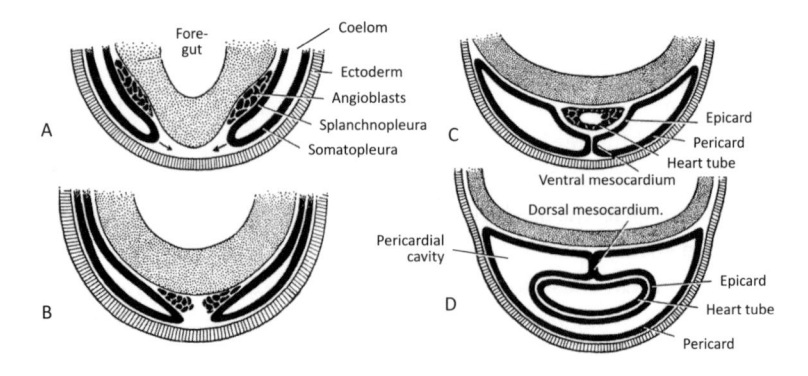

Fig. 3.12: Schematic view of the formation of the vertebrate heart. Angioblasts migrate medio-ventrally to form a tube that is surrounded by coelomic vesicles, which form the pericardial cavity. The heart is initially sustained by a ventral mesentery, later by a dorsal mesentery (mesocardium).

do not arrive in time, or in the wrong place, there is a risk of serious heart defects. Subsequently the heart tube curves in longitudinal direction. It loops in an S-shape, as a result of which the strongly contractile part, the ventricle, which was initially positioned at front, is now in a caudal position. Simultaneously, the heart descends in the chest.

The descent in the chest has a particular consequence: it causes a remarkable course of the nerve that innervates the larynx, the *laryngeal nerve*, a branch of the vagus nerve (fig. 3.13). What is happening here? The aorta is homologous with the artery that in fish runs over the fourth pharyngeal arch. We also find the larynx nerve there, while the arch itself forms the larynx, as we saw before. While the heart and the aorta descend in the chest, the larynx remains in the neck. As a consequence, the larynx nerve is looped: branching from the vagus it first runs down, loops inside the chest, around the aorta, and then runs up towards the larynx. This is the case for all mammals, even for, for instance, the giraffe! It is one of the many inconvenient 'tinker situations' in the human body that can only be understood from an evolutionary perspective.

Subsequently, *septa* (partitions) emerge in the heart, due to which the interior is divided into a left and right half. The septa first emerge in the atrium, and then in the ventricle. The lengthwise separation enables a complete double blood circulation. Also in evolution, we see that the double blood circulation appears stepwise: the amphibians still have an undivided heart with single blood circulation, while reptiles present a great diversity of hearts: various stages of a partly divided heart and partially double blood circulation (fig. 3.14). The right ventricle, which in most reptiles still serves the entire body, in mammals starts to focus entirely on the lung; the artery of the right sixth gill arch becomes the pulmonary artery. The left ventricle retains the connection with the artery of the left fourth gill arch; this becomes the aorta. The human left ventricle is homologous with the (undivided) fish heart, while the right ventricle was added during reptile evolution and so has a more recent evolutionary history. This could be a reason why so many congenital heart diseases concern the right ventricle and its connections to the great arteries.

The double blood circulation for human beings becomes functional only after birth. During foetal development in the womb, blood circulation is still single, which it should be, as long as the lung is not functional yet. The *foramen ovale*, an opening in the septum that separates the left and right atrium, makes it possible for the blood to flow from the right to the left atrium. In addition, there is second nexus, the *ductus Botalli* or *ductus arteriosus*, which directs any remaining blood from the pulmonary artery to the aorta. The ductus Botalli is homologous with the artery that runs over the left sixth gill arch in fish. So here an ancestral structure that has

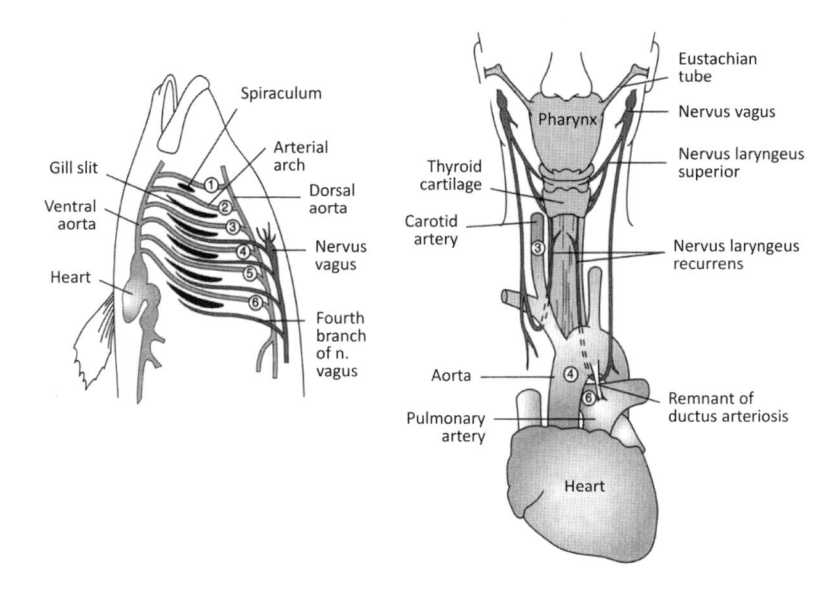

Fig. 3.13: In the course of evolution, the heart of mammals has descended inside the chest, while the heart of fish is still right behind the gills. The downward movement has forced the laryngeal nerve, a branch from the nervus vagus, to form a long loop around the aorta, with a recurrent branch going up to the larynx. It is a heritage of past evolution: the laryngeal nerve is associated with the fourth gill arch of fish, just like the aorta.

lost its function, is deployed temporarily in the embryo, another case of co-optation. We will see more examples of this principle below. Directly after birth the foramen ovale closes and the ductus Botalli atrophies within three days. If the foramen ovale does not close or if the ductus Botalli remains functional, heart disease emerges, which, depending on the size of the opening, can remain unnoticed or may lead to significant function loss (shortness of breath, tightness, growth retardation). In fig. 3.15 the complex blood circulation of the foetal heart is shown.

Just like the heart, the urogenital system is a fine example of a development process in which the evolutionary history can still be identified. It is no surprise that we call it *'urogenital system'*, which includes the kidneys ('uro'), the gonads ('genital') and their joint efferent ducts, because these organs are clearly interlinked, both in development and in evolution. In fish, the kidney is an organ that dorsally stretches to the left and right across the whole body. Later in evolution it becomes a more compact organ, limited to the abdominal cavity. In most fish and some amphibians we see that the testis uses the kidney's duct. This tube is called the *archinephric duct*. Urine and sperm are conducted via the same tube. In the case of the

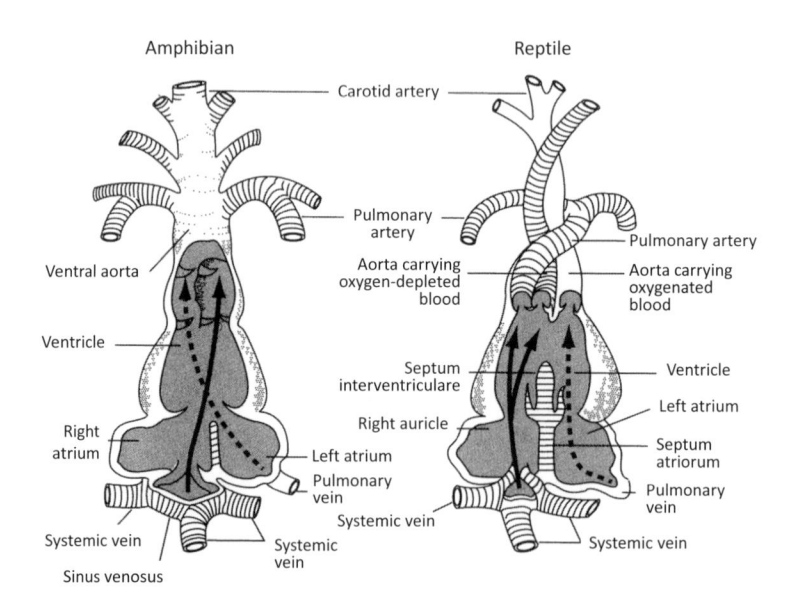

Fig. 3.14: The beginning of a double circulatory system can be identified in amphibians and reptiles. These groups show all kinds of transitional stages in the separation of oxygen-depleted and oxygen-rich blood, to a greater or lesser extent. For some species only the atrium is partitioned into a left and right part, for others also the ventricle.

reptiles a partition between the two functions appears: the archinephric duct specialises entirely on the testis (forming the *vas deferens*), while the kidney obtains its own, new, nephric duct, the *ureter*. This is also the situation in mammals.

This evolutionary trend is reflected in the organogenesis of the human urogenital system. The embryo develops three consecutive kidneys: the *pronephros* without duct, the *mesonephros* that passes over its duct (the archinephric duct) to the testis, and, later in the development, the *metanephros*, which has its own efferent tube, the ureter (fig. 3.16). The pronephros is in the neck region and not functional in humans. The mesonephros serves as an embryonic kidney up to the tenth week of pregnancy; in humans the archinephric duct is known as the *Wolffian duct*. From this tube, develop at a later stage, under the influence of the male hormone testosterone, the vas deferens, the seminal vesicles and the epididymis. In females, the Wolffian duct degenerates. In addition, along the tube of the mesonephros, a *paramesonephric* duct develops from the coelom in both sexes. This tube is known in humans as the *Müllerian duct*. It is, however, not in contact with the kidney function of the mesonephros and also is not continuous with the lumen of the gonad; the proximal end opens in the coelom. In women, the *oviduct*, the uterus and the upper part of the vagina develop from it.

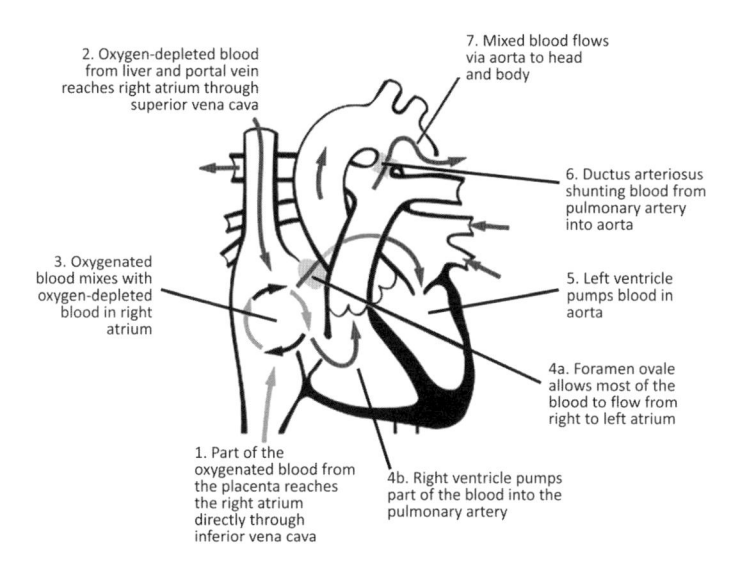

Fig. 3.15: Representation of blood circulation through the heart of an unborn child. There are two 'short circuits', the foramen ovale between the atria and the ductus arteriosus between the pulmonary artery and the aorta, which both prevent blood from being pumped to the lung.

The Müllerian duct is actively degraded in males, under the influence of the anti-Müller hormone. So the formation of the urogenital system is another fine example of how ancestral structures are co-opted and deployed to serve a new function: the male vas deferens was originally a kidney duct and functions as such in the embryo. The changes are summarised in table 3.5.

Another structure with an interesting evolutionary history appears in human embryology, the *urachus*. This is an evolutionary legacy of the *allantois* in the amniotic egg, which first appeared in the reptiles. We have seen in chapter 2 that in an early stage of egg development, all amniotes (reptiles, birds and mammals) develop an amniotic cavity. In birds and reptiles, a second cavity develops which is wrapped around the embryo, called *allantois*. It functions as a waste bag for the nitrogenous excretory products of embryonic metabolism. As the eggs of reptiles and birds are sealed off from the outside world, they have to store their waste products inside the egg, whereas placental mammals can pass them on to the mother. Having lost its storage function in mammals, the allantois serves as a connection between the foetal bladder and the umbilical cord (fig. 3.16). Via this connection, the urine of the foetus can be passed on via the umbilical cord and the placenta to the mother. After a while, when the bladder descends into the pelvis, the urachus reduces and after birth only a connective tissue strand between the bladder and the belly button is left. The urachus is a

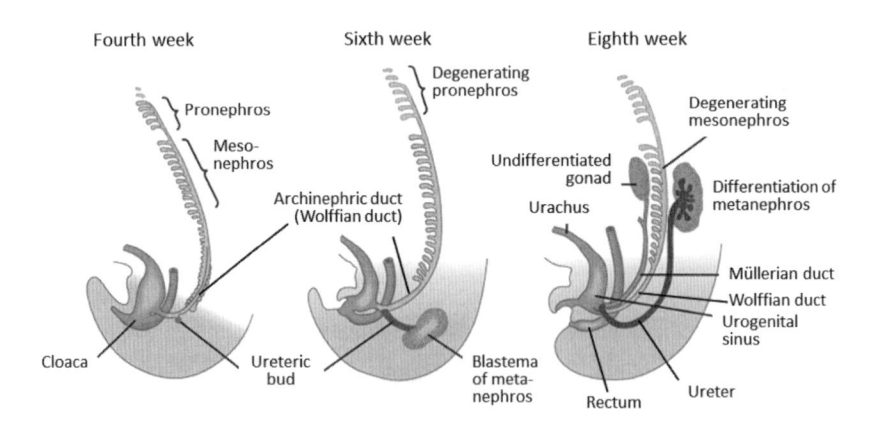

Fig. 3.16: During its development the human embryo has three renal systems that correspond to the kidneys of our evolutionary ancestors. First the pronephros is formed, which degenerates after six weeks. The mesonephros is active during the first stages of foetal development. From the eighth week on, a metanephros starts to develop, with its own duct, which becomes functional after 10 weeks. The metanephros is the kidney of the adult human. The duct of the mesonephros, also called the Wolffian duct, switches function to serve the male gonad as a vas deferens. For females, the Müllerian tube is going to function as the gonadal tract (oviduct and uterus).

Table 3.5: Organs of the urogenital system and their efferent ducts

Organ	Primary (embryological) duct	Secondary (final) duct
Pronephros	No duct	–
Mesonephros	Archinephric duct, up to week 10	–
Metanephros	Ureter, from week 10	Ureter
Male gonad (testis)	Archinephric duct (= Wolffian duct) from week 10	Vas deferens
Female gonad (ovarium)	Paramesonephric duct (= Müllerian duct)	Oviduct

typical example of reuse of an old evolutionary structure that served a different purpose in the ancestor (co-optation).

The last evolutionary development trend of the urogenital system to be discussed here, is the position of the testes. Adult males carry their testicles in the *scrotum* but in the unborn male child they are still in the vicinity of the kidneys, in the same location where the female ovaries are located. While the ovaries remain in the abdomen, the testes, shortly before birth, descend to the scrotum, via the *inguinal canal*. Infants are routinely checked for testicular descent, as failure could cause reduced fertility and increased risk of testis infections.

Many biologists have racked their brain about the 'use' of testicular descent (*descensus testiculorum*). There are several mammals without descended testicles. Even one of our closest relatives, the gorilla, carries his testes in the abdomen and not in a scrotum. Often it is claimed that the descent is required, because the formation of sperm cells is optimal at a temperature that is a few degrees lower than the body temperature. Also, the fact that the scrotum relaxes in warm conditions, while during cold, it is drawn towards the body by means of the m. cremaster, indicates the importance of a constant temperature of the testes. However, the presence of so many, normally fertile, mammals without descended testis shows that the lower optimal temperature is a consequence rather than a cause for testicular descent. Descensus testiculorum seems to be due to an internal, developmental factor in the first place, rather than being a response to some external factor.

Looking at the position of the testis from a comparative zoological perspective we observe that the descent fits into a trend that started in fish (fig. 3.17). In the evolution of the vertebrates the testes increasingly became a compact organ, instead of an oblong organ that stretched from the front to the back of the body. Moreover the testes gradually moved

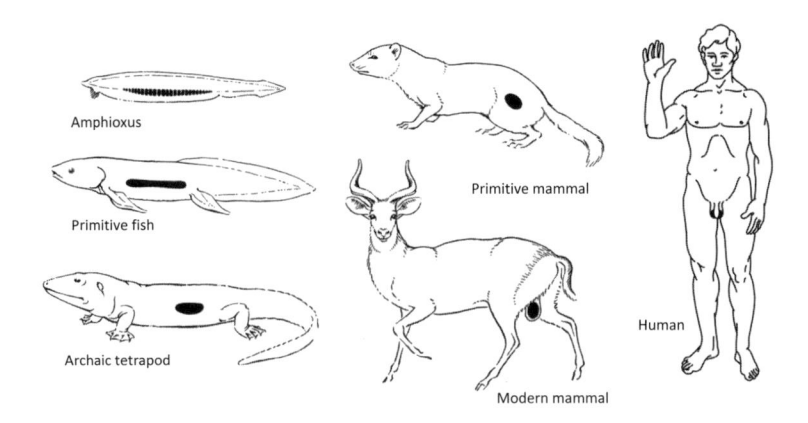

Fig. 3.17: Representation of the position of the testis for a number of vertebrates. The descent of testicles in a scrotum, occurring in several mammals, including humans, according to zoologist A. Portmann, can be seen as part of an evolutionary trend starting with the fish.

more caudal (downwards). Phylogenetic analysis suggests that *testicondy* (testicles descended but retained in the abdomen) is the ancestral condition in mammals and this situation has been retained in the first diverging placental mammals (Afrotheria and Xenarthra). In the lineage containing primates (Laurasiatheria and Euarchontoglires) testicondy evolved further to the fully scrotal condition. However, this process was not completed or was reversed in several species, hence the variation within mammals.

It is not very clear what would be the advantage for animals to carry their testicles in a scrotum. Many hypotheses have been proposed, including a contribution to sexual attractiveness or a demonstration of masculinity. The Swiss zoologist Adolf Portmann (1897-1982) argued that testicular descent resulted from an ongoing developmental emphasis on 'polarity' of the body. Moreover he described the visibly wearing of testicles as contributing to the '*Selbstdarstellung*' of the male physique. This concept can be traced back to the Canadian sociologist Erving Goffman, who, in a booklet published in 1956, argued that when a person engages socially with others, he tries to form himself an image of the situation, including his own role. Self-portrayal or expression of the self can be seen as a logical consequence of social interaction. Goffman's concept, designated *Selbstdarstellung* in the German translation of his book, was taken over by Portmann, who proposed that the behaviour of animals is also governed by this tendency. The philosophical concept of Portmann can, although intended much wider, be regarded as a form of sexual selection. In chapter 7 we will discuss the role of sexual selection in human evolution in more detail.

Evolution of the brain

Obviously our extremely large brain is the most distinctive feature to make us unique among all animals. During the evolution of the hominins, the volume of the brain has increased from approximately 400 cc in *Australopithecus* to 1,800 cc in Neanderthal. The progress was initially slow; the first species to be called *Homo* (*H. habilis*, 2 million years old) still had a small brain volume; the classification of *H. habilis* in the genus *Homo* is not based on its brain volume, but on the fact that this was the first species to use stone tools (see chapters 1 and 6). The large leap forward in the brain only started with *Homo erectus*, 1.5 million years ago.

The increase of brain volume within the hominins most likely was accompanied by an increase of cognitive abilities, such as toolmaking, habitat use and functioning in a social environment. Therefore, it is only logical to assume that each increase of the brain carried a direct advantage. Being able to function properly in a social context is often regarded as the most important driving force, but also the precision of eye-hand coordination and control over the locomotor apparatus, required for manipulation of tools and hunting, could have been selective factors. Recently investigators have pointed out that ecological intelligence (habitat use, hunting, culture) is a stronger driver for brain expansion than social intelligence (group behaviour, cooperation, competition) because social intelligence does not come with direct energetic gains to outweigh the high costs of a large brain, while ecological intelligence does.

Not only between species, but also within species an increase of brain volume has taken place. The most striking is the brain growth of Neanderthals. The oldest fossils have a volume of around 1,500 cc, while the later fossils reach values up to no less than 1,800 cc (bigger than that of humans). In *H. sapiens*, however, a modest decrease took place. Since Cro-magnon Man (40,000 years ago) the human brain volume has reduced from 1,500 to 1,400 cc.

We saw in chapter 2 that the nervous system is one of the first organs to develop in the embryo. Even before neurulation (formation of the neural tube) the neural plate at the dorsal front end of the embryo has formed primordia for three brain vesicles, which soon subdivide to form five secondary vesicles. These vesicles are connected to each other and continuous with the lumen of the neural tube (fig. 2.12). The two *cerebral hemispheres*, which in the adult human make out almost the complete brain, develop from the most frontal vesicle, the *telencephalon*. Together with the second vesicle, called *diencephalon*, this forms the forebrain. The third vesicle, *mesencephalon*, develops into the midbrain, while *metencephalon* and *myelencephalon* make up the hindbrain. A constriction between the midbrain and the hindbrain, called *cerebral isthmus*, is an important organising region for differentiation of the brain. The structures between the cerebral hemispheres and the brain stem are jointly called the *limbic system*. This includes nuclei derived from the telencephalon and diencephalon such as hippocampus, thalamus, hypothalamus, nucleus accumbens, amygdala, *etc*. From this developmental history, we realise that the brain is basically a hollow, fluid-filled organ and has realised its enormous expansion by elaborate infoldings of the walls of the system, the cortex. Both cortical hemispheres are divided into four lobes, the *frontal, parietal, temporal,* and *occipital* lobes.

The older parts of the brain are involved in emotion, aggression, sexual behaviour, motivation and memory, as well as in controlling all kinds of mostly unconscious physiological functions in the rest of the body. The limbic system is something we have in common with all vertebrate animals, while the newly developed cortex on top of it, the *neocortex*, is typical for mammals. Subsequently, in the case of primates, a part of the neocortex, the *prefrontal cortex*, was greatly enhanced. This part of the brain, which is directly behind the forehead, is involved in higher cognitive functions, such as attention, concentration and reasoning.

Going from chimpanzee to humans, the prefrontal cortex is no further enlarged, but it is striking that the connections between the prefrontal cortex and the limbic system have become much more extensive. These connections, which are typical for the human brain, are linked to the regulation of basal instincts (aggression, sexuality). Vice versa these very instincts, by integration with the prefrontal cortex, could have led to the typically human 'higher' cognitive features, such as creative expression, fantasy, musicality, *etc*. The psychologist Sigmund Freud called this the *'sublimation'* of our instincts: the conversion of basic urges into socially valued behaviours. The strong connections between the prefrontal cortex and the limbic system might be considered the neurological counterpart of Freud's sublimation.

Thus, just like other organs we have discussed in this chapter, the brain is an evolutionary mixture of ancestral and novel structures. The extremely large and complex neocortex is evidently a new development, but due to the intensive connections with the nuclei of the limbic system, ancestral structures are deployed for the purpose of our higher mental functions and typically human behaviours.

How large is our brain actually? If you want to compare the volume of the brain with that of other mammals, you must take into account the relationship between brain volume and body weight. Usually the volume or mass of an organ does not increase in direct proportion with the body weight. This is called *allometry*, a very common situation, as all small animals tend to have a shape that differs from that of large animals, even when they seem rather similar. There is *isometry* if the organ volume scales with a fixed factor relative to body weight. The study of scaling relationships in nature goes back to the great Scottish scientist D'Arcy Wentworth Thompson, who in 1917 published his highly influential book *On Growth and Form*.

In the case of the brain it appears that large animals have relatively small brains. If the brain weight (H) of all mammals is compared with the body weight (W) a curve emerges, which can be described by the formula $H = a W^b$. Taking the logarithm to the left and right this may be written as: $log\ H = log\ a + b\ log\ W$. This is an equation for a straight line with slope b if $log\ H$ is plotted on the y-axis against $log\ W$ on the x-axis. The value of b, estimated from regression over for all mammals turns out to equal 0.67. In the case of isometry a slope of 1 is expected. The actual slope being virtually equal to 2/3 suggests that the size of the brain is not related to the volume of an animal, but to its surface. This matches the basic organisation of the brain as a hollow vesiculated tube, as described above.

Taking allometry into account, you can compare the brain size of animals with different body weights using the encephalisation quotient (*EQ*) that is defined as:

$$EQ = \frac{H}{0.12\ W^{0.67}}$$

This equation measures the actual brain weight (H) relative to the allometry-derived brain weight that is expected for a mammal with body weight W (both weights measured in grams). Humans have an *EQ* of 8, which means that the brain volume is 8 times larger than you would expect for a mammal with the body weight of a human. Apes have an *EQ* of 2 to 3 – twice as large as that of an average mammal. Table 3.6 shows several values for extinct hominins and apes, illustrating again the enormous expansion of the brain in humans.

Despite the emphasis on brain volume as a factor that contributes to our high cognitive capacity, in modern humans there is just a weak relationship between intelligence and brain volume. Evidently, our mental abilities do not depend on brain volume, but predominantly on brain functions, actually on the complete circuitry between the different brain areas. Neuroimaging studies show that people with larger brains show spatial expansion of activity in the frontoparietal regions of the brain, more than activities in the sensory, motor and limbic systems. This fits in the evolutionary trend

Table 3.6: Encephalisation quotients (EQ) for a number of representatives of the Hominidae

Species	EQ*
Homo sapiens	8.07
Homo erectus	4.40
Homo habilis	4.31
Paranthropus boisei	3.22
Australopithecus afarensis	2.44
Pan troglodytes	3.01
Gorilla gorilla	1.61

*Calculated using Jerison's formula

towards increasing emphasis on brain areas and activities associated with higher cognition.

Male brains are slightly larger than female. Also, some parts of the brain show a certain degree of sexual dimorphism. The best-known sexually dimorphic brain area is the *'sexually dimorphic nucleus of the pre-optic area'* (SDN-POA), a part of the hypothalamus that is twice as large in men as it is in women. Another dimorphic area is the *nucleus suprachiasmaticus*, also a part of the hypothalamus. This nucleus is directly above the chiasma opticum (crossing of the optic nerves); it controls our biological rhythms and fulfils an important role in the regulation of sexual behaviour. Research in the 1990s by Michel Hofman and Dick Swaab of the Netherlands Institute for Neuroscience showed that the shape of this nucleus differs between the sexes and is considerably larger in homosexual men than in heterosexuals. Their discovery that gays differ from straights in their brains triggered a lot of controversy at the time.

Other cerebral nuclei are sexually dimorphic in a functional sense, but are morphologically identical. An example of this is the *hippocampus*, a paired tube-shaped structure that runs backwards from the underside of the brain and then bends forward to end in the temporal lobe. The hippocampus is involved in the retention of episodic memories, remembering smells and spatial orientation. The remarkable difference between men and women in dealing with spatial information is related to a different functioning of the hippocampus. Another functionally dimorphic structure is the *anteroventral periventricular nucleus*, a part of the hypothalamus with neurons that affect maternal instincts and secretion of the hormone oxytocin.

To better understand the evolution of our brain a big research effort is devoted presently to identifying genetic correlates of brain volume and

Table 3.7: Summary of human-lineage specific genetic changes associated with evolution of brain volume and function

Gene (abbreviated name)	Gene function	Type of change in human lineage	Relevance to brain expansion or brain function
Coding region mutations			
FOXP2	Forkhead protein P2, protein involved in brain development	Two non-synonymous substitutions, contrasting with great conservation in mammals	Mutations in this gene cause speech problems in humans, human FOXP2 in mouse alters behaviour and vocalizations
MYH16	Myosin heavy chain 16, palate- and jaw-muscle-specific myosin	Frameshift mutation, causing incomplete transcript, no protein	Musculus temporalis weaker, allowing growth of skull
HAR1F	Gene encoding a regulatory RNA involved in neuron specification and migration	Part of gene showing great sequence divergence in human lineage (HAR1)	Increased connectivity and neuronal layering
AUTS2	Protein involved in transcriptional regulation during neural development	Several coding and intronic regions showing sequence difference with Neanderthal	Neural development and enhanced brain growth
Duplications and deletions			
DUF1220 (NBPF)	Domain of unknown function 1220 in Neuroblastoma breakpoint family genes	Hyper-amplification of protein domain	Correlation between number of CON2-type DUF1220 domains and cognitive capacity
SRGAP2	Signalling protein highly expressed in brain development	Partial duplication	Partial protein from paralog inhibits original protein to extend dendritic maturation
NOTCH2NL	Protein involved in Notch signalling pathway involved in timing of development	Regain of function by ectopic gene conversion of a great ape pseudogene	Delayed maturation of neuronal progenitor cells, enhanced cortical neurogenesis
ARHGAP11	Regulatory gene preferentially expressed in radial glia	Human-specific duplication, deletion of 55 nucleotides in ARHGAP11B causing change of function	Enhanced proliferation of neural progenitor cells
BOLA2	Gene involved in iron metabolism during embryonic development	Part of a segmental chromosomal duplication showing signature of positive selection	Unclear functional significance at the moment
TBC1D3	Gene involved in vesicle transport	Multiple copies in human genome	Enhanced basal neural progenitor cells and cortical folding

Regulatory mutations			
HARE5 (FZD8)	Human accelerated region enhancer 5, acting upon FZD8, receptor protein in canonical *Wnt* signalling pathway	Mutations in regulating sequence	Faster neural progenitor cell proliferation and larger cortex
HAR21 (NPAS3)	Human accelerated region 21, acting upon NPAS3, brain-enriched transcription factor	Mutations in regulating sequence	Functional significance not clear yet
HAR246 (CUX1)	Human accelerated region 246, acting upon of CUX1, regulator of synaptic spines	Mutations in regulating sequence	Changes in cortical neural morphology

cognition. In 2006 bioinformaticians from the University of California, Santa Cruz, announced that they had identified 50 unique DNA sequences called '*human accelerated regions*', HARs, which were proclaimed sequences 'that made us human'. Their argument was that sequences showing little change within the vertebrates but an unexpected high substitution rate between chimpanzee and human must be related to some typical human features, such as bipedalism, diet opportunism and increased brain volume. More HARs were discovered later, making up a list of some 2,500, usually short, non-coding and presumably regulatory. However, in some cases HARs turned out to be part of an open reading frame and at least in one case a HAR appeared to be part of a gene encoding a regulatory RNA. Not surprisingly, a significant part of the HARs turned out to have a function in brain development. In addition, cognitive and neural disorders such as genetically determined autism susceptibility and schizophrenia, turned out to be related to mutations in HARs.

The suggestion from the HAR discoveries was in remarkable correspondence with earlier thoughts by Marie-Claire King and Allen C. Wilson who argued in 1975 that the similarity between humans and chimpanzee was so great that the differences in behaviour and cognition must be due to regulatory differences rather than structural differences in protein sequences. Later research has proven this argument to be correct. Evidence of altered transcriptional regulation comes from transcriptome analysis of brain samples. In fig. 3.18 expression differences of ten different transcription factors of a number of primates are compared. As shown in chapter 2, transcription factors active in development are normally quite conservative, and this is also evident from the constancy of their expression in nonhuman primates (fig. 3.18). However, several brain-specific transcription factors have quite a different expression in humans, both higher and lower. Since each transcription factor regulates the expression of many other genes, this explains why relatively few differences in DNA sequence between chimpanzees and humans can still cause large differences in functional capacity.

An important target for regulatory change in the developing brain concerns the *radial glia cells* of the embryonic brain. These cells serve as primary progenitors for neurons, astrocytes and oligodendrocytes and their proliferation determines to a great extent the number of cells in the adult brain. Regulatory changes underlying brain expansion include an earlier start of their proliferation, a higher activity and a delay in their maturation. Table 3.7 provides an overview of some of the most important regulatory changes as well as some other well-founded genetic correlates of brain expansion and brain function. We discuss some examples below.

One of the most striking changes concerns *DUF1220*, a protein domain (part of a gene) that was only recently discovered, because it was located in a section of chromosome 1 that was difficult to sequence and was lacking in the original assembly of the human genome published in 2004. The chromosomal section in which it is situated (indicated as 1q21.1) is unstable and subject to many genetic changes. Various brain diseases, such as microcephaly,

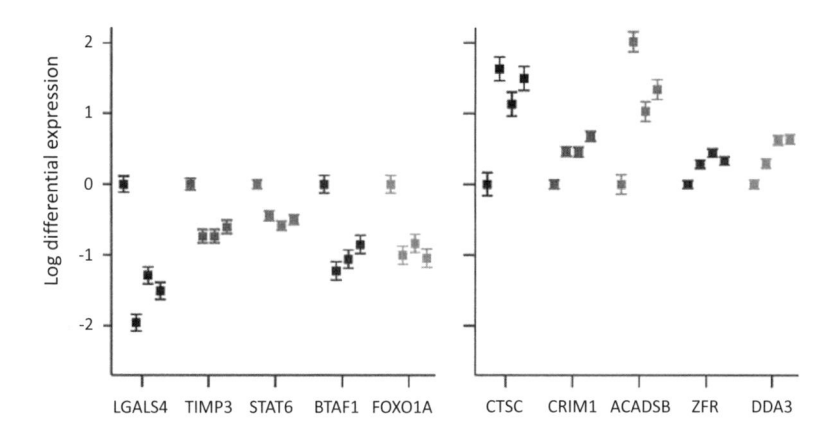

Fig. 3.18: Differences in expression between humans and primates for ten different genes encoding transcription factors active in the brain. The expression ratio per gene is given (on a log scale) for humans (set to log (1) = 0), chimpanzee, orang-utan and rhesus macaque. In the left graph five genes are shown for which expression in human is higher than in the three primates, the right graph shows five genes with lower expression, compared with the other primates.

schizophrenia and mental retardation, are associated with mutations in the 1q21.1 area. In the human lineage, the region rich in DUF1220 domains was relocated to the other arm of chromosome 1, by means of a large pericentric inversion, which was followed by gene loss and several segmental duplications. The number of DUF1220 domains was extremely amplified, up to more than 260 times. This happened in the form of triplets (fig. 3.19). The triplet of domains emerged before the split between humans and chimpanzees, but the subsequent triplet amplification and the inversion are human-specific.

Further analysis showed that DUF1220 emerged 300 million years ago, as a partial duplication of a gene called PDE4DIP (*phosphodiesterase 4D interacting protein*), a protein that binds the enzyme phosphodiesterase 4D to the Golgi apparatus. Subsequently, approximately 120 million years ago, this sequence came under control of a CM (*conserved mammalian*)

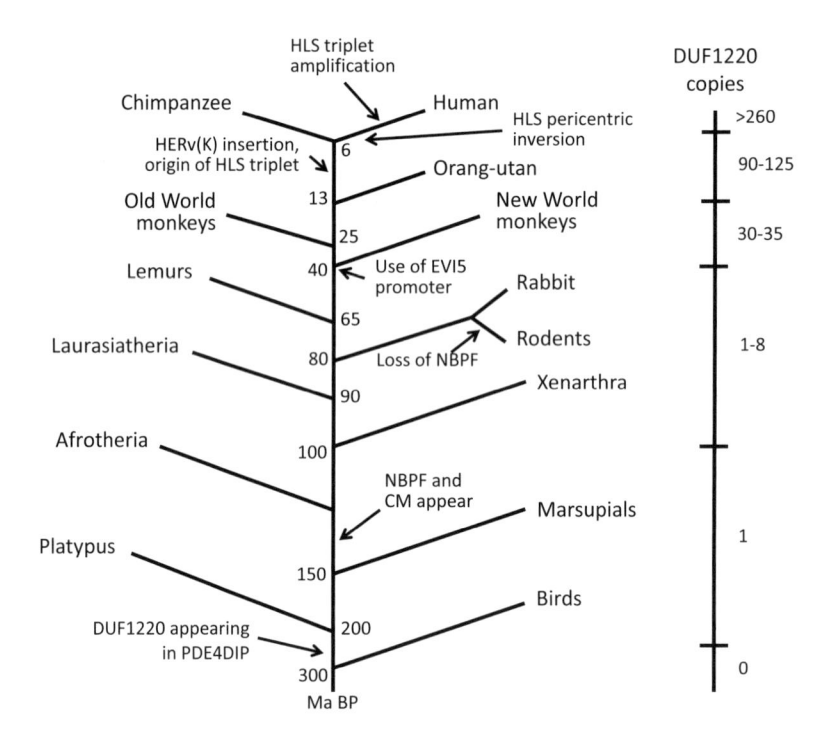

Fig. 3.19: The protein domain DUF1220, playing a role in brain development, first appeared in primitive mammalian species after a partial duplication from PDE4DIP. The new domain became part of the NBPF gene family and, due to various rearrangements, came under control of new promoters (CM and EVI5). In the course of evolution, DUF1220 was duplicated several times, to an extreme degree in the human lineage, after the formation of a domain triplet. The increase of DUF1220 copies runs parallel with the volume increase of the brain. HLS = human-lineage specific.

promoter, after which domain amplifications occurred up to eight times early in mammalian development. The resulting gene family is called NBPF (*neuroblastoma breakpoint family*) because it was discovered that a member of the family was mutated in a neuroblastoma patient (suffering from Hutchinson's disease). Amplifications proceeded in the lineage of primates, up to more than a hundred times in the apes, after which even more copies were added in the human lineage (fig. 3.19). The extreme speed of the whole process suggests that the genes with amplified DUF1220 domains were under strong positive selection all the time.

It is unknown how DUF1220 works exactly (DUF stands for *domain of unknown function*), except for the fact that it is expressed strongly in the brain. The number of DUF1220 domains also varies between individuals. There are six different versions of the domain, one of which is rather strongly correlated with the IQ score of children, with boys' mathematical capacity, in particular.

Another more recent fascinating and in some ways similar example of genetic change underlying human brain evolution is due to a gene called *NOTCH2NL*, which, situated in the 1q21.1 locus like the NBPF genes with their DUF1220 domains, was also affected by the human-lineage-specific rearrangements in chromosome 1. NOTCH2NL originated from a partial duplication of NOTCH2, encoding a protein in the NOTCH signalling pathway, controlling many aspects of the timing of development. The partial duplication occurred before the split between gorilla and chimpanzee, and resulted in non-functional pseudogenes in both species. However, in the human lineage, NOTCH2NL regained its function by an ectopic gene conversion, in which an intron and a promoter were added to the open reading frame. This gene then duplicated twice, producing three nearly identical functional NOTCH2NL genes in the present human genome, with a highly stimulating effect on radial glia cells.

It is interesting to note that the evolution of our highly evolved cognitive abilities, our 'becoming human' if you like, has been dependent on such queer 'genetic accidents' in chromosome 1, which, in the spirit of François Jacob, can be called a case of haphazard molecular tinkering *par excellence*.

4. There must be differences

When in 1859 Charles Darwin published his ground-breaking book *On the Origin of Species by Means of Natural Selection, or the Preservation of Favoured Races in the Struggle for Life*, biologists didn't have the faintest clue as to the mechanism of heredity. It is even more surprising that Darwin's assumption, that natural selection acts upon spontaneously and randomly emerging hereditary changes, turned out to be largely correct, even when his principle of natural selection was consolidated with an adequate theory of genetics only in the course of the twentieth century. In this chapter we will review the now well-established theory of population genetics, expanded with the principles of neutral evolution and some non-Darwinian phenomena like epigenetics, and apply it to human evolution.

Giant leaps, neutral fluctuations or gradual adaptation?

In retrospect, genetics started with the experiments of the Augustinian monk Gregor Mendel, which he conducted in the garden of St. Thomas's Abbey in Brno, in the current Czech Republic. Mendel published his findings in 1865; he even sent his publication to Darwin, but it remained unnoticed. Forty years later, around 1900, his works were rediscovered by the Dutchman Hugo de Vries, the German Carl Correns and the Austrian Erich von Tschermak.

The experiments with plants conducted by these biologists at first seemed to be in conflict with the Darwinian idea that evolution works through natural selection in response to small differences between individuals. The Dutch biologist Hugo de Vries (1848-1935) (fig. 4.1) was a professor at the University of Amsterdam. He conducted, for instance, experiments with the large-flowered evening primrose (*Oenothera erythrosepala*) in the greenhouses and the garden of the Hortus Botanicus in Amsterdam. Just like Mendel, he assumed that the hereditary material consisted of discrete units that corresponded with certain external features. He observed that evening primrose, when crossbred, could undergo substantial phenotypic changes, which he attributed to '*mutations*'. This interpretation was – in hindsight – not correct, because the phenotypic changes were in fact caused by a deviating system of chromosome pairing and recombination, specific to evening primrose. De Vries' views about the discrete nature of the hereditary material and about the role of mutations in evolution, however, were evidently correct.

Fig. 4.1: Portrait of Dutch botanist Hugo de Vries (1848-1935), one of the European scientists who rediscovered Mendel's laws in the beginning of the twentieth century. On the basis of his experiments with evening primrose, De Vries developed his 'mutation theory' for evolutionary change.

Based on his observations, De Vries stated that evolution proceeds in leaps. He described the emergence of new plant species as a consequence of mutations, in his book *Mutationstheorie*, written in German, in 1901. This view was not in line with Darwin's, who had claimed that evolution works through natural selection on small differences within the species. Thus, a difference of opinion arose between biologists who claimed that mutations were the driving force behind evolution (*mutationists*) and people who stated that natural selection was the driving force (*selectionists*). The difference between the two modes of evolutionary change is illustrated in fig. 4.2. Evolution via *macro-mutations* (mutations with large phenotypic effects) is also called '*saltation*'.

The discrepancy between mutationism and selectionism was solved between 1920 and 1940, with the '*Modern Synthesis*', a theory of evolution that combined heredity and natural selection, based on the works of American and English evolutionary biologists like Thomas Hunt Morgan, Ronald

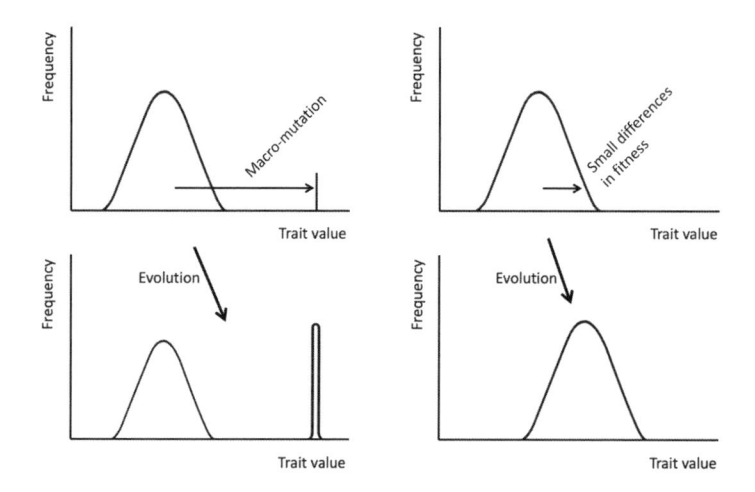

Fig. 4.2: Illustration of the mechanism of evolution by macro-mutations (left) and the classical model of natural selection acting upon small differences in phenotypic traits (right).

Fisher and Sewall Wright. The term 'modern synthesis' is due to a book published 1942 by Julian Huxley, in which he summarised the main findings of evolution in the first half of the twentieth century.

The basic insight of the Modern Synthesis was that both mutation and selection are required for evolution. Still, the movement that continued considering macro-mutations as the main driving forces, never really disappeared. A later representative of this is Richard Goldschmidt, an American of German origin who formulated his idea of the *'Hopeful Monster'* in 1940. New species with a highly divergent body plan would emerge through macro-mutations with large phenotypic effects. It was no surprise that Goldschmidt proposed this, because he was very interested in the integration of genetics and development biology. The theory of the hopeful monsters was heavily criticised at the time (how will a macro-mutant, which occurs as a rare event, reproduce itself if there are no other macro-mutants?), and is now totally pushed to the background.

The contrast between saltation and gradualism revived when in the 1970s Niles Eldredge and Stephen J. Gould introduced the concept of *'punctuated equilibria'*. These authors argued that major changes of body form occur within short evolutionary time windows and that such punctuational change, alternated by long periods of *stasis* or undirected fluctuation, dominates the history of life. Phyletic gradualism, *i.e.* continuous small changes within an evolutionary lineage guided by directional selection, would be rare. The concept of punctuated equilibria does not necessarily imply evolution by macro-mutations, it just acknowledges the fact that the intensity of selection, and consequently the rate of evolution changes over time. The idea was inspired by the fossil record, which indeed does show sudden bursts of diversity. It may also be applicable to human evolution, as the appearance of new hominins shows evidence of cladogenesis rather than anagenesis (see chapter 1). In fact, the period of 2.5-1.5 million years ago may be considered a punctuation in human evolution (*cf.* fig.1.4).

In developmental biology, mutations with large phenotypic effects are widely known. In chapter 2, we saw that mutations in the *Hox* genes of *Drosophila* can change an antenna into a leg (fig. 2.14). Research on sticklebacks has shown that a mutation in a single gene (*Pitx1*, also a member of the homeobox family) can suppress the presence of spines on the pelvic girdle, causing a new ecotype to evolve that can live on the bottom of a lake instead of only at the surface. Another example is the macro-mutation that underpins the emergence of maize from its ancestor, teosinte (fig. 4.3). In teosinte a sexual transmutation occurred that changed the male flowers on the side branches into female flowers, while the side branches were significantly shortened and the original female flowers on them disappeared; only at the top of the plant was the male flower retained. In this way, a plant emerged that could be

Fig. 4.3: Maize (right) is derived from teosinte (left) through a sexual transmutation. The male flowers at the top of the side-branches transformed to female, while the side-branches themselves were significantly shortened, as a result of which the cobs came to be positioned along the main stem. The male flower at the top of the plant remained unaffected.

regarded a monster, but after further selection by the first human agricultural societies in Mesoamerica turned out to be of immense value (chapter 6).

A modern form of mutationism finds inspiration in molecular genetics. It appears that the genome is characterised by profound dynamic changes that barely affect the phenotype. Richard Lewontin, an American geneticist and theoretical biologist, had proposed in the 1960s, inspired by his work on allozyme polymorphisms, that most of the variation on the molecular level is neutral. Mathematically oriented evolutionary biologists like Masatoshi Nei, Eugene Koonin and Michael Lynch showed that, depending on population size, neutral mutations can become fixed by random fluctuations of allele frequencies. Using mathematical models they demonstrated that even mutations with a small negative effect can settle in a (small) population by chance. Based on this, in 2007 Masatoshi Nei formulated his 'new mutation theory of phenotypic evolution'. In his book *Mutation-Driven Evolution* (2013), he further elaborated this theory. He stated that natural selection should not be regarded as the cause of evolution, but as a consequence, and that mutation as a transformation process is more important than selection. Mutations, he stated, are largely fixed via neutral processes. Later in this chapter, we will address the neutral theory.

The position taken by Nei, that neutral evolution is the most important mechanism of evolutionary change, represents in fact a stance against '*adaptationism*', the tendency of some evolutionary biologists to see adaptation everywhere in nature and to assume that natural selection is always in force, is unlimited and acts upon all phenotypic traits. Adaptationism was ridiculed in a famous article by Stephen J. Gould and Richard Lewontin from 1979, perhaps the most widely discussed article in the entire field of evolutionary biology. The authors compared adaptationists with Dr. Pangloss, a character from the satirical book *Candide, ou l'optimisme* by the French playwright and philosopher Voltaire. Dr. Pangloss has an optimistic answer to everything, no matter how preposterous. He even manages to defend the Lisbon earthquake of 1755 (causing 50,000 deaths) as a demonstration of the good, since the earthquake was 'for the best in the best of all possible worlds'.

Gould and Lewontin pointed out that we should not reason like Pangloss when explaining phenotypic variation. Many parts of the body plan of an organism are realised as a by-product of something else. They compared this, in a famous metaphor, with the spandrels of the San Marco Cathedral in Venice. Spandrels are triangular spaces between the dome and the arches that the dome rests on. These spaces do not serve a constructional purpose; they are simply a result of the way in which the cathedral was built. Often these spandrels, filled with cement, are decorated with beautiful frescoes.

But to claim that the purpose of the spandrels is to show beautiful pictures is the pitfall of the adaptationist. In the same spirit, the American geneticists Michael Lynch and Eugene Koonin have called for pluralism in the explanation of evolutionary phenomena rather than overemphasising 'the awesome power of natural selection' that 'has come at the expense of reference to any other mechanisms', as Lynch qualified it. However, the 'adaptationist school' of evolutionary biology has strong defenders in the British biologist Richard Dawkins and the American philosopher Daniel Dennett, and many others. In favour of adaptation as a far-reaching driving force, the American evolutionary geneticist Laurence Hurst showed that seemingly neutral mutations in the genome may still be under selection for all kinds of unanticipated reasons. Whether the reach of selection includes every detail of the genome or is limited to eliminating phenotypic monstrosities is still an open question to date. No doubt the answer will be somewhere in the middle.

The emergence of variation

Evolution makes use of genetic variation, and variation emerges through mutation. Mutations are spontaneous changes in the DNA, caused by errors during DNA replication, defects during meiotic recombination, unequal crossing-over, gene conversion, the action of mobile elements, imperfections during repair of strand breaks, and environmental influences, such as UV radiation and reactive chemicals. The resulting changes in the DNA can take many forms. In table 4.1 an overview of the various types of mutation is provided.

A special role is played by the mutating action of viruses and mobile elements. In the human genome 44.7% of the sequence is due to transposable elements:
– DNA transposons: 2.9%
– Long terminal repeat (LTR) retrotransposons: 8.3 %
– Short interspersed nuclear elements (SINEs): 13.1%
– Long interspersed elements (LINEs): 20.4%

This situation as such is not uniquely human, as mammalian genomes on the average contain 50% transposons and remnants of transposons, while plants contain even much more (up to 90% of a genome). The human genome is particularly rich (1.1 million copies!) in so called *Alu*-elements, which are SINEs. The mutational effects of retrotransposons are due to their mechanism of propagation, which involves an RNA intermediate,

Table 4.1: Overview of the many different types of mutations and the resulting polymorphisms

Type of mutation	Explanation
Point mutations, leading to SNPs (single nucleotide polymorphisms)	Substitution of one nucleotide for another
Microsatellite, VNTR (variable number tandem repeat)	A site with sequentially repeated copies of a core sequence changes by increasing or decreasing the number of repeats
Insertion and deletion (indel)	As a result of errors in DNA replication, or due to recombination, unequal crossing-over, or as a result of translocation by a virus, a new sequence is inserted or deleted from the existing sequence
Duplication and gene loss	As a result of unequal crossing-over during meiosis one of the descendants gets two consecutive copies of the parental DNA, or an entire gene vanishes
Gene conversion	During mitosis or meiosis, double-strand breaks in one chromatid are repaired using the other chromatid as a template, resulting in a chromosomal segment being copied to the homologous chromosome creating allelic diversity in gene families
Inversion	A part of a chromosome is positioned in reverse direction; this could be pericentric (around the centromere) of paracentric (in one of the chromosome arms)
Translocation	A DNA segment of part of a chromosome is transferred to a different chromosome, or two DNA-segments exchange places
Synonymous mutation, non-synonymous mutation	Mutation in a gene which does, or does not, lead to a different amino acid in the encoded protein, depending on which base has changed in the triplet
Micro-mutation, macro-mutation	Mutation with a small or large phenotypic effect
Structural mutation	Mutation in an open reading frame which leads to modification of the (primary) structure of a protein
Regulatory mutation	Mutation in a non-coding sequence which leads to a modified expression of a gene

which by *reverse transcription* forms a DNA segment that may integrate in any place in the genome. The mutations caused by transposition are not always destructive; they are often neutral and sometimes may be a source of novelty. We saw in chapter 3 that some features of the human body plan (placenta formation, expansion of the brain) are due to co-optation of genes and promoters from mobile elements.

Mutation rate strongly varies per section of the DNA. Some spots in the genome (loci) mutate frequently, up to 10^{-3} times per generation, while other elements hardly ever change (for example, 10^{-7} to 10^{-9} times per generation).

The first type of locus is called *hypervariable*. These are pieces of DNA that mutate so quickly that everybody has his or her own variant. Considering a number of them together, a designated combination of alleles appears in only one individual in the world population. In forensic science these markers are used to associate DNA traces with a suspect. Right across are loci that barely change over many generations (conservative genes). The difference is in the evolutionary pressure on the locus: a mutation that causes loss of function in a gene that is of crucial importance for survival is quickly eliminated, for example, because the foetus is not viable. The consequence is that such mutations will not lead to permanent substitutions and we cannot identify them in the population. Mutation frequency is also related to the activity of the DNA element where the gene is located. For example, by means of chromatin condensation, a part of the DNA can be inactivated and isolated from cell metabolism thus becoming less susceptible to the mutating influences of viruses and mobile DNA elements.

If mutations are unidirectional, that is, if an allele A mutates into an allele a without ever mutating back, then, given enough time, the original allele A will disappear from the population. However, often we will have to consider the backwards mutation rate as well (from a to A). Then if we designate the forward mutation rate by u and the backwards mutation rate by v, the frequencies of the two alleles will reach an equilibrium ('*mutational equilibrium*'), which is given by:

$$q_{eq} = \frac{u}{u+v}$$

where q_{eq} is the equilibrium frequency of allele a. This shows that when the two mutation rates are equal, the equilibrium frequencies of a and A will both become 0.5, and when v is small compared to u, q_{eq} will approach unity. However, in the absence of other processes, the rate at which allele frequencies reach mutational equilibrium is usually quite slow; it may take several thousands of generations. Therefore, mutational equilibrium is usually not considered an important factor in human populations.

By far the most common mutation in the human genome is a substitution of one nucleotide; this leads to '*single nucleotide polymorphisms*' (SNPs, pronounced as 'snips'). The human genome has an estimated 10 million of such variable positions, or approximately 1 in every 300 base pairs. By far most SNPs are in noncoding stretches of the genome and are selectively neutral. If they are in the open reading frame of a functional gene, phenotypic consequences may occur, especially in the case of a *non-synonymous substitution* (table 4.1). Synonymous substitutions are changes (usually in the third base of a triplet) that do not cause changes in the encoded protein. This is a result of *redundancy* in the genetic code: multiple triplets encode the same amino acid.

As SNPs can be found throughout the genome and the genomic environment of most SNPs is known, they are excellently suitable in genetic mapping studies: Where are certain phenotypic features, for example, diseases, encoded in the DNA? This question is addressed in so-called *genome-wide association studies* (GWAS), whereby the SNPs are correlated with disease profiles and are used as markers for the genomic position. In this way the gene responsible for the disease is ultimately identified.

An example of a medically relevant SNP is a polymorphism in the *LDL receptor*: a protein that is anchored in the cell membrane and is able to bind lipoproteins from the blood, after which these are incorporated in the cell via endocytosis. These lipoproteins are carriers of cholesterol, a fatty substance that does not easily dissolve in blood and must therefore be transported with assistance of carrier molecules. There are two types of those carriers: proteins with high density (*high-density lipoproteins*, HDLs) and with low density (*low-density lipoproteins*, LDLs). The latter are bound by the LDL receptor. There are several SNPs in the gene encoding the LDL receptor; one of them is called the 'J.D. mutation', after the patient in which it was described for the first time. This mutation involves a triplet TAT that is mutated to TGT, as a result of which amino acid 828 of the protein has changed from tyrosine to cysteine (fig. 4.4). This substitution happens to be in the C-terminus of the protein that directs the receptor to a coated pit in

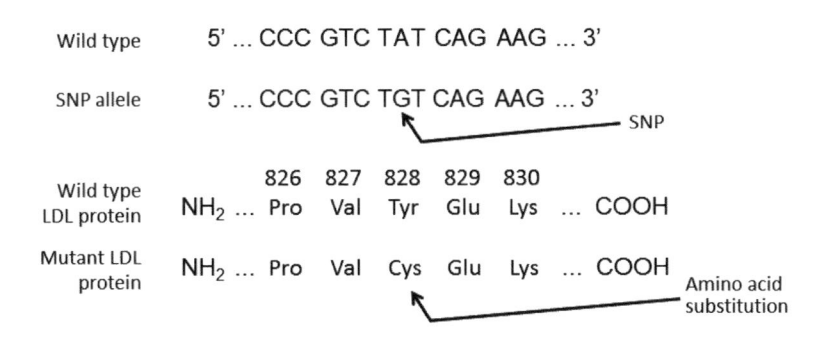

Wild type	5' ... CCC GTC TAT CAG AAG ... 3'						
SNP allele	5' ... CCC GTC TGT CAG AAG ... 3' — SNP						
		826	827	828	829	830	
Wild type LDL protein	NH_2 ...	Pro	Val	Tyr	Glu	Lys	... COOH
Mutant LDL protein	NH_2 ...	Pro	Val	Cys	Glu	Lys	... COOH — Amino acid substitution

Fig. 4.4: Representation of the DNA sequence (top) and the amino acid sequence (below) of a small part of the human LDL receptor, a membrane protein that recognises extracellular lipoproteins with low density (binding cholesterol), and initiates their uptake by the cell. A SNP in the second position of a triplet leads to a protein in which tyrosine at position 828 is replaced by cysteine, with significant loss of function.

the cell membrane. The mutated protein does not settle in the membrane, the cell is unable to bind LDL from the blood, the cholesterol content of the blood becomes chronically high, eventually leading to an increased risk of cardiovascular diseases. In 1985, Michael Brown and Joseph Goldstein received the Nobel Prize for the disentanglement of the mechanism of cholesterol uptake by receptor-mediated endocytosis.

A second important source of mutations is formed by changes in the number of repeats in *microsatellites* (table 4.1). The term 'microsatellite' dates back to the traditional purification technique for DNA that makes use of density centrifugation with cesium chloride, where strongly repetitive DNA concentrates as a separate band ('satellite DNA') in the gradient, due to the deviating GC-content. This mainly consists of DNA that stems from the areas around the centromeres. Nowadays, these pieces of DNA with repeats up to approximately one hundred bases are called *minisatellite* and repetitive DNA with even shorter repeats (a few bases) is indicated as 'microsatellite'. Microsatellites are also called STR (*Short Tandem Repeats*) or SSR (*Simple Sequence Repeats*). In a more general sense, they are referred to as VNTRs (*Variable Number Tandem Repeats*), which also includes minisatellites.

A microsatellite is characterised by a core sequence and a repeat number. In the case of a polymorphic microsatellite with core sequence CAGA, a mutation could, for instance, consist of a change from five to six repeats (from

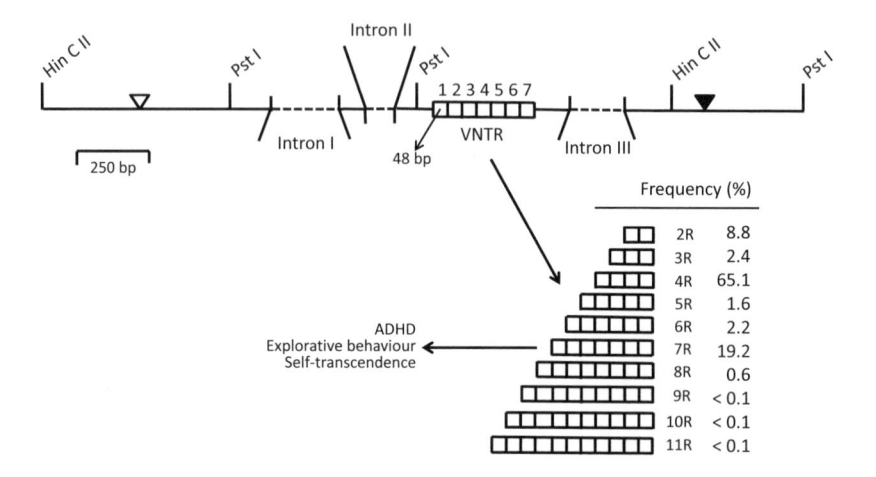

Fig. 4.5: Organisation of the dopamine receptor D4 gene, encoding a G protein-coupled receptor in the brain that responds to dopamine. The gene has three introns. Some of the restriction sites (Hin C II and Pst I) are marked. Exon III has a VNTR that consists of a core unit of 48 bp (16 amino acids) that is repeated two to eleven times. The frequencies of the different VNTR alleles are indicated. Allele 7R is associated with ADHD and some other behavioural phenotypes.

$(CAGA)_5$ to $(CAGA)_6$). The mutation is usually due to DNA polymerase losing contact with the DNA strand during DNA replication, and subsequently continuing in a 'wrong', meaning shifted, position. Microsatellites can reside in coding DNA sections or in non-coding DNA. If they are in coding DNA, it will always concern a core sequence of three base pairs, or a multiple of three.

An example of a VNTR in a coding DNA is the polymorphism in *dopamine receptor D4* (fig. 4.5). This protein is anchored in the cell membrane of a neuron and on the outside has a binding site for the neurohormone dopamine. Binding of dopamine leads to a conformational change of the protein, whereby in the cell a G protein is released, causing a cascade of reactions. DRD4 thus ensures that the neuron reacts to dopamine. One of the intracellular loops of the protein contains a VNTR that consists of a core sequence of 48 bases (16 amino acids). In most people this is repeated about four times (this is called the 4R allele), but a variety of repeats, from two to eleven times, occur, in varying frequencies. The 4R allele is considered the ancestral type, from which other alleles have emerged through mutation. The 7R allele is special: it is associated with certain behavioural phenotypes, such as explorative behaviour, ADHD and self-transcendence (the ability to 'rise above yourself' and undergo spiritual experiences). It is estimated that 7R emerged 30,000 to 50,000 years ago, a time period that coincides with the 'cultural explosion' in Europe (see chapter 6). As 7R occurs at a relatively high frequency, there must have been positive selection on it, which could be the case if the phenotype with increased explorative behaviour had an advantage.

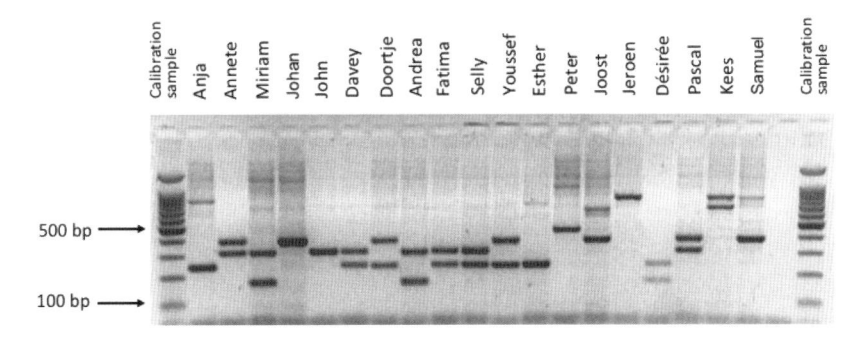

fig. 4.6: Example illustrating inter-individual variability due to a VNTR designated D16S5 in a group of children (IMC Weekend School Amsterdam-West 2006; the names have been altered). The image shows an electrophoresis gel on which DNA fragments have been loaded after a PCR that amplifies the locus as well as part of its genomic environment. The gel separates the amplicons according to size. The image shows that the locus has seven alleles in this group, that seven out of the nineteen children are homozygous, while twelve are heterozygous. The lanes to the left and right are calibration samples, which allow the size of DNA fragments to be estimated.

The VNTR of DRD4 is in a gene, but generally microsatellites are in non-coding sequences. An example of this is given in fig. 4.6. Genetic differences in this case were visualised using a PCR that amplifies a DNA segment, including the variable locus. Because of the polymorphism, the amplicon has a variable length, which after staining becomes visible as bands at different heights in an electrophoresis gel (fig. 4.6). It turned out that in a group of nineteen children, there were no less than twelve different heterozygotes for this locus. This illustrates the population genetic principle revisited below that for loci with many alleles the number of heterozygotes is always much larger than the number of homozygotes.

In some cases microsatellite mutations have significant medical consequences. An example is Machado-Joseph disease (MJD), a neurodegenerative disorder which causes adults to gradually lose control over their locomotor apparatus. The disease first appeared in the Middle Ages among a group of Sephardic Jews in Portugal, who then took it to the Azores in the fifteenth century, where it is expressed relatively often. The disorder originates in a microsatellite in the MJD gene that encodes a protein with a crucial role in the cytoskeleton of muscle cells. The microsatellite concerns a repeat of a CAG triplet (coding for glutamine); in the case of healthy people this triplet is repeated 12 to 44 times, but with MJD patients the number of repeats is 61 or more.

A third group of mutations to be discussed here, are *indels*: insertions and deletions. Such mutations could concern a short sequence (one or a few nucleotides) or a relatively long one (typically several hundreds of base pairs, yet increasing to 10 kbp). The short indels are caused by errors during DNA replication, the long ones due to recombination and unequal crossing-over during meiosis. Indels form an important source of differences between humans and chimpanzee. Roy Britton, a geneticist at the California Institute of Technology, demonstrated as early as 2003 that many pieces of DNA, whether in human beings or in chimpanzees, are inserted or deleted, resulting in a large number of differences. These differences are not included in the truism that humans and chimpanzee are 98.5% identical in terms of their DNA. According to Britton, this percentage should rather be 95% when all indels are taken into account. Upon the publication of the chimpanzee genome in 2005 this idea was confirmed, but nowadays it is taken with a pinch of salt, as the indels, especially the short ones, in many cases can be attributed to technical errors arising in genome sequencing.

Mutations that comprise larger parts of the DNA, often also with far-reaching phenotypic consequences, concern *duplications, inversions, translocations* and loss of genes. Gene duplication and gene loss are important mutation symptoms that are especially essential in the evolution of species. Quite often we find that evolutionary lineages after splitting from their ancestor are subject to expansions or reductions in the number of genes in a gene family, most likely as a result of adaptation to new conditions. For example, the *CYP* genes, which encode enzymes that can degrade potentially toxic substances, have been subject to much lineage-specific duplication. The human genome has 57 *CYP* genes, but *Drosophila* has 83 and the plant *Arabidopsis* has 244 *CYPs*. The homology relationships between the genes of different species are very difficult to establish in such large gene families, as each species has its own a set of paralogs (see chapter 2, fig. 2.13). To compare species the paralogs are sometimes clustered and the phylogenetic relationships derived using *clusters of orthologous groups* (COGs).

An example of a duplication that was of paramount importance for human evolution, is the wonderful story of SRGAP2. In 2012, a team of scientists led by Evan Eichler (University of Washington, USA) discovered that SRGAP2 (*Slit-Robo Rho-GTPase-activating protein 2*), involved in the development of the neocortex, has duplicated three times in humans, once 3.4 million years ago and then again 2.4 and 1 million years ago (fig. 4.7). Neanderthals and Denisova Man also have the four paralogs. The duplications are associated with the expansion of the neocortex and the increase of the number of connections between neurons at the transition from *Australopithecus* to *Homo*. The curious thing is that two of the three duplications have led to pseudogenes, while paralog SRGAP2C, that emerged 2.4 million years ago, encodes a part of the protein; it comprises the promoter and the first nine exons, including the functional domain. This incomplete copy is transcribed and then suppresses the original gene SRGAP2A because the peptide dimerizes with the active protein. As a consequence the maturation of dendritic spines (places where synapses are forming) is slowed down during the development of the brain, whereupon

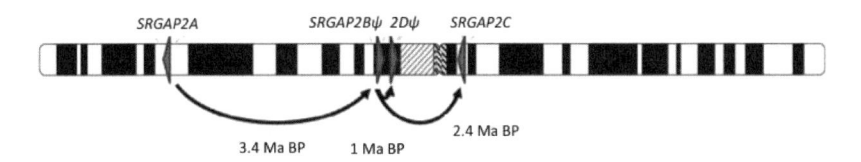

Fig. 4.7: Model for the evolution of SRGAP2, a protein involved with the migration of neurons and the formation of synapses during development of the brain. The gene, which is on chromosome 1, has been subject to three duplications (3.4, 2.4 and 1 million years ago), which produced four paralogs, two of which (SRGAP2B and D) are pseudogenes. SRGAP2C encodes a part of the protein and suppresses the original protein SRGAP2A by dimerisation.

the spines grow longer. The idea is that this micro-morphological change enables the pyramid cells to integrate more synaptic input. It can be regarded as a form of molecular neoteny, a crucial process in the evolution of the brain (*cf.* table 3.7).

Gene loss is also a mechanism of evolution. According to a recent analysis, humans have 19,537 pseudogenes, so almost as many as the number of functional genes. Of these, 67 are human-lineage specific, implying that they are still functional in our close relatives. In chapter 3 we discussed the many pseudogenes in the KRTAP family (keratin-associated proteins, fig. 3.4). Pseudogenisations abound in gene families with many paralogs. As a result, the phenotypic consequences are relatively limited, as other members of the family can (at least partially) take over the function. Loss of genes with just one copy in the genome usually does have an effect, although even that is not necessarily dramatic. The best-known example is the inactivation of GULO (*L-gulonolactonoxidase*), a crucial enzyme for the synthesis of vitamin C. In the human genome the open reading frame of this gene is disrupted by a mutation, which means that we cannot synthesize vitamin C and we need to take it in through the diet. The mutation goes back to the emergence of the haplorhines within the primates, which includes tarsiers, Old World monkeys and apes (*cf.* table 1.1). Also guinea pigs and bats have an affected GULO gene. It is believed that the pseudogenisation of GULO for the primates was left without fitness effects because monkeys and apes take in sufficient vitamin C by eating fruits.

Genes involved in chemoreception lead the list of human pseudogenes. An exceptional number of *olfactory receptors* have been lost: proteins in the olfactory epithelium and the tongue that react to molecules in the air or the food. It is the largest gene family in mammals, which are indeed able to smell very well. Humans have 853 genes that encode an olfactory receptor, but 466 of those are pseudogenes that have lost their function. The frequency of pseudogenisation is, as such, even higher than for the KRTAP genes. It is interesting that also Neanderthal and Denisova Man suffered from the loss of olfactory receptors. Analysis of the Neanderthal genome has shown that, in addition to the hundreds of genes that humans also lack, there are ten olfactory receptors that were only lost within the lineage of Neanderthals. Possibly Neanderthals experienced the smells in their environment in a different way than we do.

Finally, we point out the various chromosome mutations: inversions, fusions, duplications and translocations (table 4.1). Many of these chromosome mutations have medical relevance as they are associated with large phenotypic consequences (for example, *trisomy 21*, the triple presence of

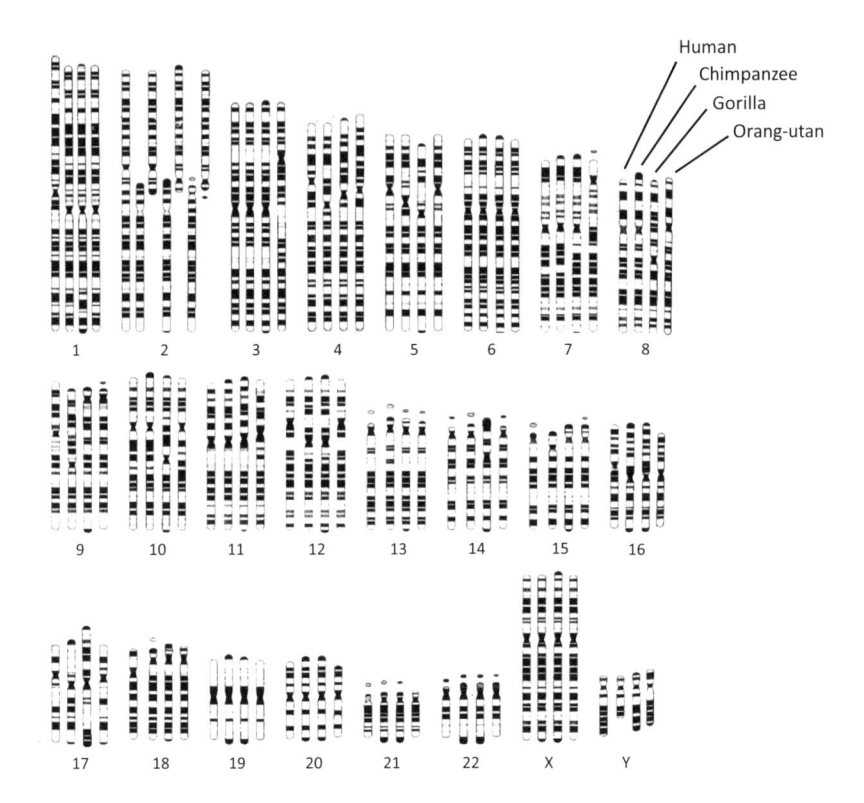

Fig. 4.8: Chromosome charts of human, chimpanzee, gorilla and orang-utan. The similarity in banding patterns is striking. Please note that human chromosome 2 is a fusion of two chromosomes of the great apes.

chromosome 21, and the *Robertsonian translocation*, a fusion between chromosome 21 and 14 or 15).

A chromosome mutation that might have been important for the evolution of the hominins is the fusion between two ape chromosomes, leading to our chromosome 2 (fig. 4.8). All apes have 23 autosomes, and we have 22. Recent research on the genome of Denisova Man shows that also this species had the chromosome fusion. This is evident from the position of a repeated TTAGGG motive, characteristic for telomeres. For the Neanderthal genome this has not yet been sorted out properly as the available genome sequences are not as good as the Denisova genome, but its very likely that Neanderthal had 22 autosomes, too, given the cross-breedings between humans, Neanderthals and Denisovans (to be discussed in chapter 5). Usually when two related species differ in the number of chromosomes the production of fertile offspring by the hybrid becomes a problem, as the

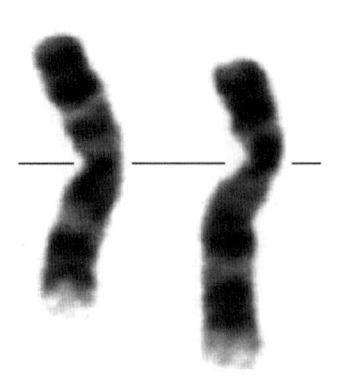

Fig. 4.9: Karyotype of human chromosome 9 with a common pericentric inversion: inv(9)(p12q13) (left). The horizontal line represents the position of the centromere.

synaptic pairing of chromosomes during meiosis is prevented. This is, for example, the case for hybrids of donkey and horse, and hybrids of goat and sheep. Perhaps likewise the chromosome fusion in human evolution was a mechanism that contributed to reproductive isolation between the first hominins and the great apes.

A relatively common chromosome mutation in humans is a pericentric inversion in chromosome 9 (fig. 4.9). An estimated 1-3% of the human population has this inversion (including the first author of this book!). Due to this inversion the two chromosomes 9 are unable to properly form a synaptic pair during meiosis. In principle this may lead to loss of chromosomal material in the gametes, but in this case this apparently does not happen, at least there are no clear phenotypic consequences (the author is also healthy). The fact that such a drastic chromosomal alteration remains without phenotypic consequences, while in other cases the substitution of one base in a gene can lead to a dramatic disease profile (fig. 4.4), indicates how difficult it is to predict the phenotypic consequences of a mutation.

Equilibrium between allele and genotype frequencies

Mutation leads to new alleles, of which the frequency changes as a result of selection or drift, hence the definition of evolution according to the Modern Synthesis. But selection operates on the phenotype rather than on an allele. That is why we should obtain more knowledge about the relationship between allele frequency, genotype frequency and the phenotype. *Allele frequency* is defined as the number of alleles of a specific type in a population, relative to the total number of all alleles at a particular locus, the latter equalling two times the number of individuals if every member of the population is diploid. Allele frequencies vary between 0 and 1. The term 'gene frequency' is often used as an equivalent of allele frequency, but in our experience this is a confusing term for students trying to grasp the principles of population genetics and we will not use it.

The easiest assumption is that the allele and genotype frequencies are in equilibrium, meaning that every allele in the population randomly combines with another allele to form a genotype; consequently, the genotype frequencies are completely determined by the allele frequencies. In that case the renowned principle of Hardy and Weinberg applies. Independently of each other, around 1908, the English mathematician G.H. Hardy and the German physicist W. Weinberg formulated this 'law', which can be described as follows:

> If, in the case of a locus with two alleles, *A* and *a*, the allele frequencies equal *p* and *q*, the frequencies of the three genotypes *AA*, *aa*, and *Aa* will equal p^2, q^2 and $2pq$, respectively.

The rule for a locus with three alleles states:

> If, in the case of a locus with three alleles, *S*, *M* and *F*, the allele frequencies equal *p*, *q* and *r*, the frequencies of the six genotypes *SS*, *MM*, *FF*, *SM*, *SF* and *MF* equal p^2, q^2, r^2, $2pq$, $2pr$ and $2qr$, respectively.

For loci with many alleles, the number of possible heterozygotes will rapidly rise. In order to still retain overview, the heterozygotes are often considered jointly. This is referred to as the *heterozygosity* of the population; the expected value can be estimated from the allele frequencies using the formula:

$$H_e = 1 - \sum_{i=1}^{n} p_i^2$$

in which H_e is the expected heterozygosity for a locus, p_i the frequency of allele *i* and *n* the number of different alleles present in the population for this locus. The summation measures the expected total frequency of homozygotes (*homozygosity*) and subtracted from 1, which gives the expected frequency of all heterozygotes together. Often heterozygosity is calculated for many loci and the average is then regarded as a measure of the global genetic variation of a population.

Using the Hardy-Weinberg principle, the expected genotype frequencies can be calculated if the allele frequencies are available. If the observed

Table 4.2: Fictive examples to illustrate the Hardy-Weinberg rule for a locus with two alleles, T and t, in three different populations, each consisting of one hundred individuals

	Numbers per genotype			Total	Allele frequencies	
	TT	Tt	tt		p = {T}	q = {t}
Population 1	24	72	4	100	0.6	0.4
Population 2	36	48	16	100	0.6	0.4
Population 3	20	80	0	100	0.6	0.4

For each of the three populations the frequency of allele T equals p = 0.6 and of allele t q = 0.4, but only population 2 is in Hardy-Weinberg equilibrium. Population 1 has too many heterozygotes, population 3 lacks any homozygous-recessives.

genotype frequencies are also available, one can test whether the population is in Hardy-Weinberg equilibrium. Table 4.2 gives an example of this test. For a locus with many alleles, for example, for microsatellites, it is common to focus directly on heterozygosity. Is the observed total number of heterozygotes in accordance with the heterozygosity expected from the allele frequencies? The test may give us an indication whether there is a something special with the genetic composition of a population.

As indicated, the principle of Hardy and Weinberg applies if the alleles of a locus randomly combine into genotypes. In biological terms it can be stated that this will happen if the following five conditions are met:
– The population is large.
– There is no selection.
– There are no mutations.
– There is no migration, neither from nor towards the population.
– There is panmixia, meaning that the chance for an individual to breed with another individual is equal for every individual (or in any case does not depend on the genotype).

The HW equilibrium formulates a null hypothesis for almost all population-genetic studies: the rule expresses what we 'normally' would expect. Despite the very stringent conditions listed above it appears that the principle holds widely in field populations including human populations. Even in cases where we know that the conditions are not met, it appears to be difficult to statistically demonstrate deviations from HW equilibrium. The most common cause for deviation is population subdivision, which leads to *heterozygote deficiency* (too few heterozygotes). This will be discussed in the next section.

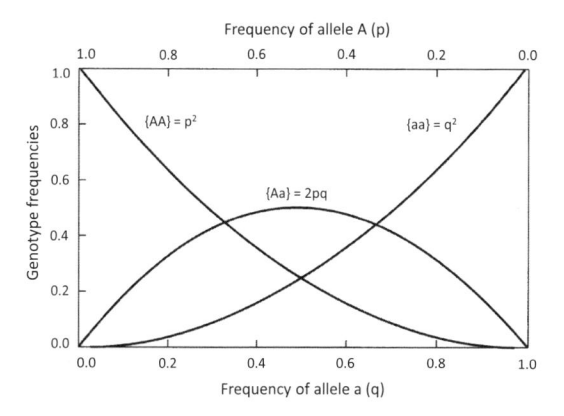

Fig. 4.10: This is how homozygote and heterozygote frequencies depend upon allele frequencies (q running from left to right, p from right to left) in a population that is in Hardy-Weinberg equilibrium.

According to the Hardy-Weinberg equilibrium, genotype frequencies are quadratic functions of allele frequencies. This is illustrated in fig. 4.10. The frequency of heterozygotes is always maximal if the two allele frequencies are equal ($p = q = 0.5$). If one of the allele frequencies is small (on the left and right side of the graph), the frequency of homozygotes becomes very small, much smaller than that of the heterozygotes (for example, when $q = 0.1$ q^2 equals 0.01 but $2pq$ equals 0.18). This principle underlies the common observation that disease alleles in a human population almost only occur in the heterozygotes who do not express the disease. Only few people are actually ill (the homozygotes), whereas a much larger part of the population is a 'carrier' (the heterozygotes). Conversely: if selection acts against an adverse allele, it may take a long time before it is removed from the population, because selection only affects the homozygous recessives. Selection should be strong and persist long enough to 'pull' the adverse allele out of the heterozygotes.

If there is selection the Hardy-Weinberg equilibrium of course doesn't apply. Usually this is noticed on the basis of too few homozygous recessives, but selection has to be very strong to cause a detectable deviation from HW equilibrium; it is actually only noticed when the homozygous recessive is lethal. Population 3 in table 4.2 is an example of this.

The degree of selection is quantitatively expressed as the *selection coefficient* s. This is the relative decrease of a certain genotype under the influence of selection in one generation. The opposite is the '*fitness*' of that genotype. The term 'fitness' leads back to Darwin ('*survival or the fittest*'), who, in later editions of his famous book *On the Origin of Species*, introduced

Table 4.3: Changes of genotype frequencies under selection assuming different dominance relations, for a locus with two alleles, A and a, where A dominates over a; genotype frequencies are indicated by brackets

Type of dominance and selection	Change of {AA}	Change of {Aa}	Change of {aa}
Complete dominance; selection against the homozygous recessive	1	1	$1-s$
Complete dominance; selection against the homozygous dominant and the heterozygote	$1-s$	$1-s$	1
Codominance; selection against the homozygous recessive and partly against the heterozygote	1	$1-\frac{1}{2}s$	$1-s$
Overdominance; selection against both homozygotes	$1-s$	1	$1-t$

s, t = selection coefficients

this phrase, which he copied from Herbert Spencer, who invented it in 1864. The fitness of an individual obviously depends on the combination of genotype and environment. The notion of fitness includes the ecology of the species. However, in population-genetic theory fitness has a quite simple and descriptive definition: the frequency of a genotype in the next generation, relative to its frequency in the current generation. Fitness is indicated with the letter W. This is to honour Sewall Wright (1889-1988), an American geneticist who substantially contributed to the Modern Synthesis.

The relationship between fitness and selection is: $W = 1 - s$. If in a population a genotype (for example, AA) produces 100 fertile offspring and another genotype (for example, aa) 90, selection against aa is said to equal $s = 0.1$, while the fitness of aa equals $W = 0.9$. In population genetics fitness is always regarded relative to the genotype in the same generation that has the highest number of offspring or the best survival; this is given the value 1.

The effects of selection depend on the dominance relations (table 4.3). In the event of complete dominance only the homozygous-recessive genotype is selected against. The recessive allele will only slowly disappear from the population, as previously discussed. If selection is against the dominant allele, this allele, depending on the degree of selection, will quickly disappear from the population (which could be counteracted by repeated mutation). An example of this is *achondroplasia*, a form of dwarfism that is inherited dominantly.

The third case in table 4.3 concerns *co-dominance*. This occurs if both alleles are expressed in the heterozygote. The heterozygote has a phenotype that is distinct from the homozygous dominant and can be recognised as such in the population. This is also referred to as *intermediary heredity*. In botany we know many examples where flower colours are concerned, whereby one homozygote has, for instance, red flowers, another white, while the heterozygote is pink. Co-dominance is crucially important at the molecular level, because in molecular genotyping the investigator wants the heterozygotes to be visualised separately from the homozygotes, like in fig. 4.6. In the case of co-dominance, selection against a recessive allele influences the allele frequencies more quickly than in the case of regular dominance.

A special case is *overdominance*, also called *heterosis* or *heterozygote advantage* (table 4.3, bottom line). In that case the heterozygote has a higher fitness than both homozygotes; the consequence is an overrepresentation of heterozygous genotypes in the population. Population 1 of table 4.2 illustrates this situation. This type of heredity leads to retention of genetic variation: as both alleles provide an advantage, they are both maintained and the population remains polymorphic. Heterozygote advantage of course implies co-dominance.

An example of heterozygote advantage is the human MHC II (*major histocompatibility complex II*), a collection of genes that codes for proteins involved in the immune system. These genes are co-dominant (both alleles are expressed). In addition, it is advantageous to be heterozygous, as this provides a better protection against infectious diseases. Because of this mechanism different alleles are maintained. The result is that human populations are often very polymorphic for these genes, and that in many populations there is an overrepresentation of MHC II heterozygotes.

Fig. 4.11 graphically illustrates how heterozygote advantage leads to stable polymorphism. If selection is against both homozygotes, allele frequencies will decrease if they are high (because the allele is then mainly present in the homozygotes), whereas allele frequencies will increase if they are low (because the allele is then largely present in the heterozygotes, which are favoured). The result is that allele frequencies reach equilibrium, somewhere in the middle, where the increase flips over to a decrease. The position of the equilibrium frequency (q_{eq}) can be predicted from the selection coefficients using some population genetic theory, with the formula:

$$q_{eq} = \frac{s}{s+t}$$

where *s* is the selection coefficient against homozygous genotype *AA* and *t* the selection coefficient against homozygous genotype *aa*, while *q* is the frequency of *a*. So if, for example, *s* = 0.2 and *t* = 0.3 the equilibrium is reached at q_{eq} = 0.4 (fig. 4.11).

Heterozygote advantage is an example of *frequency-dependent selection*: the fitness of a genotype depends on its frequency in the population. Often the advantage is higher when the frequency is low. This is the also the case, for example, when rareness protects against predators as these do not develop a search image for prey types that are seldom seen. Evolutionary biologists are very interested in frequency-dependent selection, because it can explain why genetic variation of a population is maintained. Also temporal variation (different genotypes alternately have an advantage) and spatial variation (in one habitat one genotype has an advantage, while another habitat favours another genotype) are mechanisms to maintain genetic variation. The different forms of selection that are distinguished in population genetics are summarised in table 4.4.

How does the frequency of an allele vary if it is subjected to selection and mutation at the same time? Population genetic theory shows that any deleterious recessive allele arising by mutation at rate *u* and being subjected to selection at rate *s*, will reach an equilibrium frequency q_{eq}, which is given by:

$$q_{eq} = \sqrt{\frac{u}{s}}$$

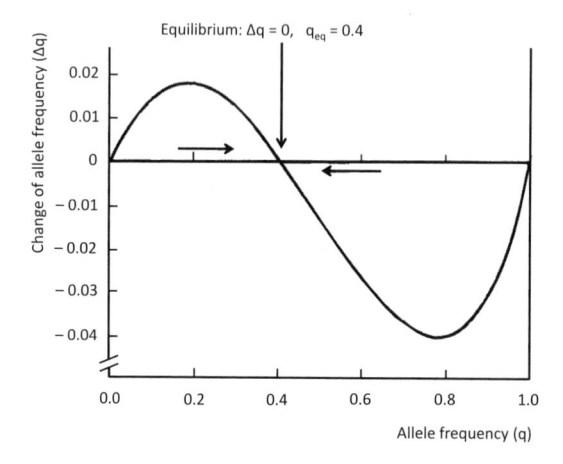

Fig. 4.11: In the case of heterozygote advantage, the recessive allele will increase at low frequencies (Δq > 0 at small q) and decrease at high frequencies (Δq < 0 at large q). Eventually an equilibrium frequency will be reached (here q_{eq} = 0.4 and p_{eq} = 0.6); both alleles will remain in the population, at frequencies depending on the degree of selection against the two homozygotes.

Table 4.4: **Summary of different modes of selection, and their consequences**

Term	Description	Consequence
Directional selection	Selection consistently favouring phenotypes with the highest (or the lowest) values for some trait	The average phenotypic value will increase (or decrease); genetic variation will decrease
Disruptive selection	Selection favouring the highest and the lowest phenotypic values at the same time	The average phenotypic value will stay in place, but the variation will increase, sometimes causing a split in the frequency distribution of phenotypes
Stabilizing selection	Selection favouring the average, median of modal phenotype	The average phenotypic value will stay in place, but the variation will decrease, as the extremes disappear
Purifying selection	Selection favouring the most common DNA sequence, disfavouring mutations	The most common DNA sequence will continue dominating the population
Balancing selection	General term to indicate all situations where selection results in the maintenance of polymorphism, which can be due to disruptive selection, heterozygote advantage, or temporally and spatially variable selection	Genetic variation (polymorphism) in the population will be maintained
Frequency-dependent selection	General term to indicate all situations where selection depends on genotype frequencies, favouring rare phenotypes and disfavouring common ones, for example, in heterozygote advantage	Genetic variation (polymorphism) in the population will be maintained

The equilibrium is called *mutation-selection equilibrium*. This type of equilibrium is rather important in human populations. It explains the persistence of deleterious alleles, even if they are strongly selected against. Under very strong selection ($s \approx 1$), an allele arising by mutation will continue to be present at a frequency equal to the square root of the mutation rate.

So far we have regarded the HW equilibrium for one locus. Things become more complicated if we look at two or more loci at the same time. In that case there is HW equilibrium if both loci are in equilibrium, independently from each other. Two loci will inherit independently from each other if they are on different chromosomes or are far apart on the same chromosome. But the closer two loci are to each other, the greater is their linkage. If they are very close to each other, recombination between them will not suffice to guarantee independent inheritance. The result is described technically as '*linkage disequilibrium*'.

Linkage disequilibrium can be expressed in different ways. The most common approach is to define it in terms of the frequencies of recombinant gametes produced by a double heterozygote. If we have a two-locus genotype *AaBb*, where *A* and *B* are on the same chromosome and *a* en *b* are on the other, the (haploid) gametes will be *AB* and *ab* if there is no recombination, but also include *Ab* and *aB* if there is recombination. The linkage disequilibrium parameter *d* is therefore defined as:

d = {AB}x{ab} − {Ab}x{aB}

where the brackets indicate the frequencies of the four different gametes ({*AB*} + {*ab*} + {*Ab*} + {*aB*} = 1). The recombination rate itself is expressed in a parameter *r*, defined as the relative number of recombinant gametes, so in this example

r = {Ab} + {aB}

There is complete linkage equilibrium if *d* = 0. This will happen if the number of recombinants is maximal, i.e. *r* = 0.5.

Linkage disequilibrium may also be expressed as the correlation between alleles of different loci. Fig. 4.12 gives an example for three human populations where this measure was used. Linkage between loci was examined here by means of a SNP analysis. The data show that linkage disequilibrium decreases with genomic distance between the loci, but the rate of decrease depends on the population. In the human genome linkage decreases to a background value in about 200 kbp of genomic distance, on the average.

Since recombination and linkage are closely connected, the genetic distance between two loci can be measured by the rate of recombination. This was the great insight of the American geneticist Thomas Morgan who worked on fruit flies in the beginning of the twentieth century. Morgan's work gave rise to a unit in genetics, called after him *centimorgan* (cM); 1 cM corresponds to a recombination frequency of 1%, or *r* = 0.01. The centimorgan is still used in breeding experiments, although now of course the actual (physical) distance is determined by means of genome sequencing. As long as linkage gradually decreases with genomic distance (like in fig. 4.12) there is good relationship between the distance in cM and the distance in base pairs, but in DNA stretches with low recombination frequency, the cM provides an underestimate and in stretches with high recombination ('recombination hotspots') the cM gives an overestimate of the genomic distance.

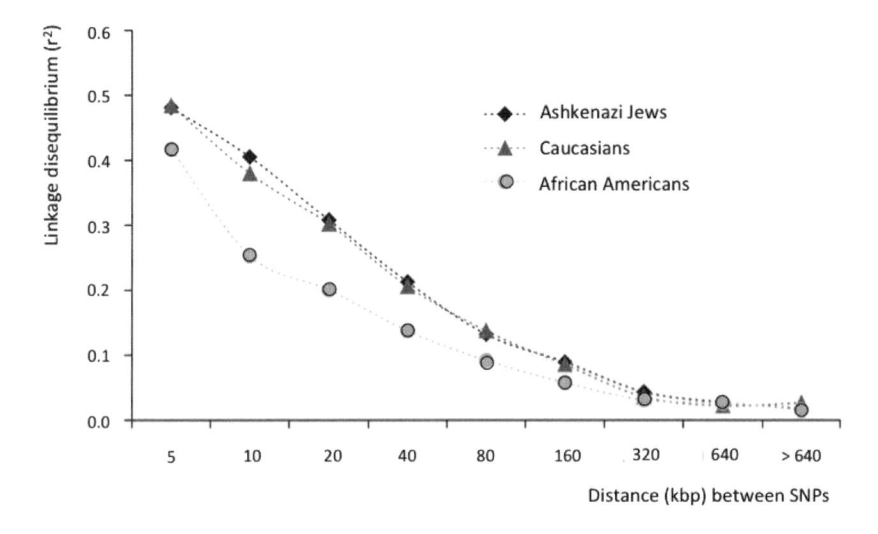

Fig. 4.12: Relationship between linkage disequilibrium between loci (measured as r²) and the distance between these loci in the genome (measured in kilobase pairs, kb), for three populations. If the loci are far from each other (> 200 kb), linkage disequilibrium decreases to a background value. The analysis was conducted for ninety persons per population, who were genotyped by means of SNPs.

For loci that are far apart, the chance of recombination occurring between them is so high that those loci inherit independently from each other. Still, this is strongly dependent on the DNA region. The above-mentioned MHC II, for example, is located in an area of the genome where recombination frequency is extremely low, causing genetic linkage over several millions of nucleotides. This also applies to the *Hox* clusters. A stretch of genome sequence that is not subject to recombination and thus inherits as a single unit, is called a *haplotype*. For example, the entire mitochondrial genome is a haplotype because it does not recombine. This also applies to a large part of the Y chromosome.

There is also linkage that is not related to the distance between loci; this is called '*long-range linkage*'. This may happen when two loci have a functional connection, for example, if a certain combination of alleles of two separate loci provides a specific advantage. In population genetics this is called a '*co-evolving gene complex*'. Another well-known phenomenon is that migration between genetically different populations leads to enlargement of linkage disequilibrium. It takes a while before a new allele, introduced in a population by migrants, is in equilibrium with the existing alleles in its genomic environment. Conversely, the analysis of genetic linkage provides important information about the evolutionary history of a locus. In chapter 5 the convenient use of this will be discussed.

Linkage may cause certain loci to 'hitch-hike' on selection acting upon neighbouring loci. An example of this is a mutant form of the G6PD locus. This locus, positioned on the X chromosome, encodes *glucose-6-phosphate dehydrogenase*, a well-known enzyme in the glycolysis. A mutant form, G6PD-A is disadvantageous due to functional deficiency of the enzyme, but as a side effect causes a substantial degree of resistance against malaria, especially in women. Since malaria, as a disease, causes a significant level of morbidity and mortality in Africa, in that continent there is an advantage for the mutant allele. The frequency of G6PD-A in some populations may be as high as 20%. But the allele is never fixed, because it also causes a disadvantage, namely G6PD deficiency. This is an example of frequency-dependent selection (*cf.* table 4.4). But regarding linkage: due to the limited recombination frequency in the chromosomal area where G6PD is situated, also the alleles of neighbouring loci, linked to G6PD-A, are selected. The effect stretches over 1.5 million base pairs and covers nine flanking loci. This is sometimes called a '*selective sweep*': a non-equilibrium situation in the genome due to a recent strong selection on a specific locus. The example also shows a possible mechanism for the selection of 'by-products', as previously discussed in this chapter.

A special case of linkage is *sex linkage*. This happens to all loci that are located on a sex chromosome. In humans, like in all mammals (but not in many other animals) the male sex is *heterogametic*, that is, he has two different sex chromosomes, X and Y, while females are *homogametic* (XX). Because the Y chromosome is small and hardly carries any genes, almost all sex-linked genes are on the X chromosome.

Because genes located on the X chromosome have a counterpart on the other chromosome only in females, their phenotypic expression may differ between the sexes. This is most apparent for recessive traits, such as red-green colour blindness. Males have a much greater chance of having red-green colour blindness because they express the mutant allele whenever they have it. Females would have to be homozygous recessive in order to have red-green colour blindness. The male condition is called *hemizygotic*.

The reason why the Y chromosome has so few genes is that it does not recombine with X. The larger part is called *NRY* (*non-recombining region of Y*). The suppression of recombination is due to the sex-determining principle being located on the Y chromosome. The gene that initiates the male program and is activated after about 6 weeks of pregnancy is called *SRY* (sex-determining region of Y) or *TDF* (*testis-determining factor*). In any system with a sex-determining master switch on a chromosome, it becomes profitable to express all the genes in the sex-determining program as one co-adapted complex. This can be achieved by avoiding recombination with the other chromosome. This will lead to accumulation of mutations and finally inactivation of the other genes on the chromosome carrying the switch. This is the most common explanation for the remarkable phenomenon, seen in all mammalian species, why the Y chromosome is small and practically devoid of coding sequence.

Neutral evolution

As we have seen above the theory of population genetics generally predicts the establishment of equilibria (mutational equilibrium, heterosis equilibrium, mutation-selection equilibrium, stable polymorphisms, *etc.*). However, in practice such equilibria can hardly be observed because they are completely overruled by random fluctuations of allele frequencies, especially in small populations. Three types of chance effects play a decisive role:
– 'Drift' of allele frequencies in small populations
– Bottleneck effects due to catastrophes
– Colonisations and migrations

In all three situations there is change of genetic composition, so there is evolution, but there is no selection. This is called *neutral evolution*. The theory of neutral evolution was described in a classic book by Motoo Kimura (1924-1994), *The Neutral Theory of Molecular Evolution* (1983).

In the previous section we saw that one of the conditions for the presence of Hardy-Weinberg equilibrium is a large population. The reason is obvious: only in the case of a large population can we assume that all possible combinations of alleles will actually be realised. If the population is small, a locus can become fixed for a certain allele due to a succession of chance effects: polymorphism will be lost and the population will become monomorphic.

From a genetic perspective, the size of a population is not simply a count of the number of individuals. Each population has individuals who do not have children and thus won't pass on allele combinations. This is why population geneticists define the '*effective population*', which will always be smaller than the actual population. The difference is quite significant for populations in which reproduction is in harems. In that case, there are a large number of men who do not contribute genetically to the offspring; the next generation derives from a limited number of dominant men, the harem leaders. Using neutral theory it can be demonstrated that the effective population size in this case equals:

$$N_e = \frac{4 N_f N_m}{N_f + N_m}$$

in which N_e = effective population size, N_f = the number of females participating in reproduction, and N_m = the number of males participating in reproduction. This shows, for example, that in a population of 200 individuals, with a sex ratio of 1, where all women have children, while five of them live together with one man, the effective population size does not equal 200, but only 67.

A harem structure also causes the effective population size of males to be smaller than that of females. The reason is that the offspring are more equally divided over women than over men: although most women will have children, only a limited number of men have children, and they have many, while a significant number of men don't have children at all. Since harem formation was seemingly important during early human evolution, especially in *Australopithecus* (see chapter 1), this may give rise to complications in the reconstruction of demographic events in the past, if we use markers which inherit exclusively via the female line (for example, mitochondrial DNA) or which inherit exclusively via the male line (markers on the Y chromosome).

In small populations genetic drift will play an important, sometimes dominating, role. The effect of drift can be well illustrated by use of computer simulations. Suppose a computer program simulates a population consisting of individuals that cross and reproduce, but every generation is started with, for example, twenty individuals (fig. 4.13). If such a population is allowed to breed for a number of generations, it will show that alleles tend to go extinct due to chance effects: the population becomes fixed for one allele, it has become monomorphic. In fact, this effect will always occur, but only in small populations is the likelihood so large that it becomes a real phenomenon. In the simulations of fig. 4.13 only one out of ten populations was still polymorphic after one hundred generations.

Genetic drift gives rise to the concept of '*fixation time*'. This is the expected (average) time that it takes for a neutral allele arising by mutation to become fixed in the population. The great majority of new alleles will die out soon after they arise. However, every now and then, a neutral allele goes to fixation by chance effects (*cf.* fig. 4.13). The theory developed by Kimura shows that the expected fixation time, T_x, depends only on effective population size via a very simple formula:

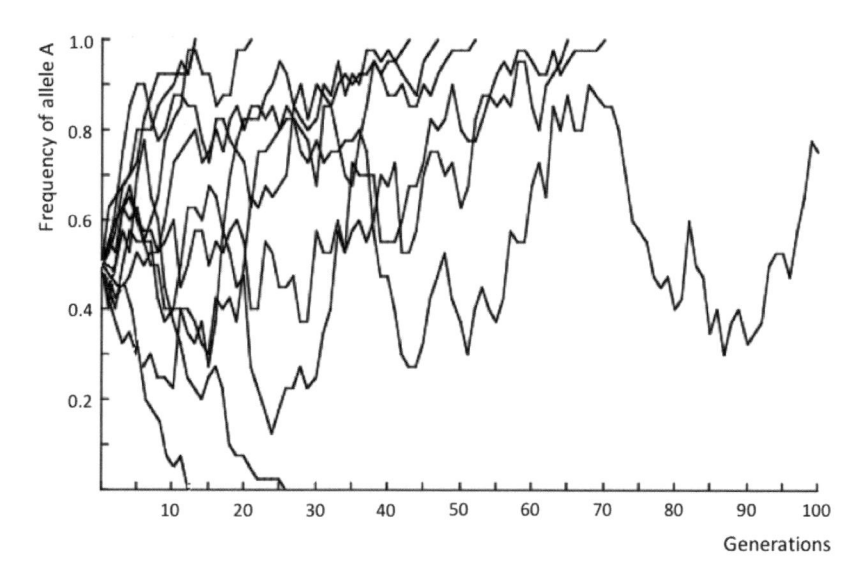

Fig. 4.13: Results of ten computer simulations in which a population of twenty individuals was allowed to breed over one hundred generations. For each of the ten simulations the allele frequency of one of two alleles (A and a) of a locus is plotted; in all cases the initial frequency was 0.5. In seven out of ten cases, the population got fixed for allele A, and in two cases for allele a. In only one out of ten cases, the population was still found to be polymorphic after one hundred generations.

$$T_x = 4 N_e$$

where T_x is measured by the number of generations. This shows that without selection, fixation of an allele introduced by mutation will hardly ever happen in a large population, but becomes a real phenomenon when the population is small. The formula is also used conversely, to define the effective size of a designated population as the size corresponding to an observed fixation time.

The effect of drift can be so strong that selection is negated. The Japanese theorist Tomoko Ohta has done extensive research on this, using various population genetic models. Building on the work of Kimura, he demonstrated that an allele in a population can become fixed, even when it is selected against. This is called the '*near neutral theory of evolution*', an extension of the original neutral theory of Kimura, in which not only purely neutral phenomena are considered, but also the behaviour of alleles with small selective disadvantages.

A crucial quantity in the 'near-neutral' theory is the product of the selection coefficient and the effective population size, multiplied by 4: $4N_e s$. If $4N_e s$ is significantly large, so with large populations or very strong selection, the effect of drift is negligible and a harmful allele will always be removed from the population. But if $4N_e s$ is small, smaller than 1 in practice, the drift effect will be dominant. In that case there is a chance that an adverse allele fixes, despite it being selected against.

A second phenomenon of neutral evolution happens if the population goes through a *bottleneck*. This means that the number of individuals at a certain point in time strongly decreases, due to a catastrophic event, upon which the population recovers again. In the case of a dramatic decrease, rare alleles are likely to get lost and the new population will show less genetic diversity (fig. 4.14).

Bottleneck effects may also have played an important role in human evolution. In 1998, the American anthropologist Henry Harpending conducted a population genetic analysis, which suggested that *H. sapiens* faced a catastrophic reduction in its population around 70-75 ka BP. This bottleneck was associated with the eruption of the Toba volcano on Sumatra, a super-eruption that shot an enormous amount of dust into the atmosphere and perhaps caused a worldwide volcanic winter. At the location of the Toba volcano there is now Lake Toba. Later analyses by a Swedish group, however, didn't confirm this idea. The evidence for demographic disasters in the early evolution of *H. sapiens* is unclear at the moment.

The third phenomenon of neutral evolution we find in *colonisations*. If a group of migrants abandons an area to settle elsewhere, the founder population starts with a significantly smaller genetic variation than was present in the large population that they came from. Due to chance effects alleles that just happened to be present in the colonisers, can reach a frequency that deviates strongly from the mother population. The so called '*founder effect*' is enhanced when no crossbreeding takes place with the local or original population, either because these are absent, for instance, on a deserted island, or for cultural reasons. Due to these types of effects, even now we can detect a strongly deviating genetic composition in villages that were isolated for a long time (like Volendam in The Netherlands), on islands like the Azores (*cf.* the case of Machado-Joseph disease, discussed earlier in this chapter) and with relatively isolated demes such as the Basks in Spain and the Amish in the United States.

In human evolution, genetic founder effects most likely accompanied the various 'out of Africa' migrations that started about 180 ka BP. It is assumed that these migrations initially consisted of small bands of humans that went to explore new land or were isolated from their source population by drift on water. The genetic consequences of these colonisations can still be recognised in present-day human populations as a 'serial founder

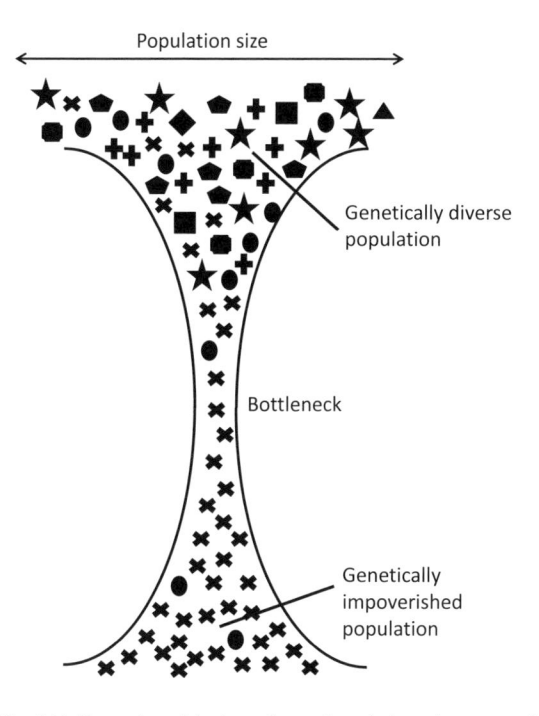

Fig. 4.14: Illustration of the loss of genetic variation when a population experiences a bottleneck.

effect', even though the initial colonisations were later supplemented by new migrations and crossbreeding.

Another issue of population size is *inbreeding*. This is defined as the probability that two alleles identical by descent are combined in one genotype. There will always be a certain degree of inbreeding in any population, but it will be most pronounced in small populations. Inbreeding leads to enhanced homozygosity and therefore increases the likelihood that recessive diseases are expressed. Negative fitness effects due to inbreeding are known as '*inbreeding depression*', a real phenomenon in small and isolated human populations. The effects of inbreeding depression are also well known for cultural traditions that practice arranged marriages with relatives. Inbreeding is promoted by a population being subdivided over several small subpopulations isolated from each other. This mechanism is sometimes assumed in explanations for the demise of the Neanderthal: their population became fragmented and each of the subpopulations was too small to remain genetically viable.

A population with inbreeding obviously does not obey the Hardy-Weinberg equilibrium. For a locus with two alleles with frequencies p and q the expected frequencies of the three genotypes are not p^2, q^2 and $2pq$, but $p^2 + pqF$, $q^2 + pqF$ and $2pq - 2pqF$, where F is the *'inbreeding coefficient'*. Thus it can be seen that the number of homozygotes is increased (each by pqF) and the number of heterozygotes is decreased (by $2pqF$). F is a number between 0 (no inbreeding) and 0.5 (maximal inbreeding by self-fertilisation, only possible in hermaphrodites).

The neutral theory is a very powerful part of evolutionary biology. It serves, just as the Hardy-Weinberg equilibrium, as a null hypothesis in the explanation of population genetic composition: if you want to demonstrate the effects of natural selection, you should prove that they go beyond neutral evolution. Alleles originating by mutation in a small population can become fixed by genetic drift. As this happens in the absence of selection, such populations can evolve by neutral processes, while mutation is, in fact, the driving force behind the change, just like Masatoshi Nei claimed in his theory on 'mutation-driven evolution' (see above).

Geographical distance causes genetic differences

We saw above that populations can be subdivided in several small populations. The genetic consequences are seen as a deviation from Hardy-Weinberg equilibrium for the whole population: there are too few heterozygotes because alleles specific to the subpopulation do not mix with the rest of the population. In fact, this is the most common cause for deviations from the Hardy-Weinberg rule.

Conversely when a small immigrant population mixes with a larger population, the frequency of inborn diseases in the immigrants is seen to decrease due to loss of homozygosity. This phenomenon is called the *Wahlund effect*, after the Swedish population geneticist S.G.W. Wahlund (1901-1976). It is also called 'isolate breaking'.

These considerations lead to the logical idea that we need a measure for something like 'genetic distance' between two populations. The most common parameter is the *F-statistic*, F_{ST}, which is due to the American statistician Ronald Fisher, who contributed a lot to the Modern Synthesis of evolution in the beginning of the twentieth century. F_{ST} for two populations, 1 and 2, is defined as:

$$F_{ST(1,2)} = \frac{(p_1 - p_2)^2}{2P(1-P)}$$

where p_1 is the frequency of an allele in population 1, p_2 is the frequency of the same allele in population 2, and P the allele frequency in the world population, or, in case this is not available, the average allele frequency of the populations considered. This calculation applies per locus, but can easily be extended to a large number of loci at the same time, as a result of which the estimate of genetic distance of course becomes more reliable.

F_{ST} can be calculated for a large number of pairs of populations: between the Dutch and the French, between Dutch and Germans, between French and Germans, *etc*. If, in addition, we record the geographical distance between those populations, we obtain a so-called IBD diagram. IBD is an abbreviation of *'isolation by distance'*, the most important mechanism to explain genetic differences between people. Fig. 4.15 shows an example of this. This is based on genotyping 377 microsatellite loci in a large number of population groups across the world, conducted by a research team directed by Luca Cavalli-Sforza at Stanford University. The data obviously show that genetic distance increases with geographic distance. This is due to independent drift processes in different populations, causing two populations to attain different allele frequencies. As the chance of genetic exchange will be smaller in populations that are geographically dispersed, the effect of drift is not compensated by the homogenising effect of migration in populations that are far apart. IBD is also due to serial founder effects, as discussed in the previous section.

In the graph geographical distance is measured in two different ways: via the shortest distance between two points on the globe (upper figure) and via five obligatory waypoints. These fixed points reflect the assumed migration routes of *H. sapiens*, which will be discussed in chapter 5. For example: Europe was colonised from Africa via the Middle East, not by crossing the Mediterranean; America was colonised from Asia, via the Bering Strait and not via the Atlantic Ocean. By taking account of these migration routes, the correlation in the IBD diagram was improved.

The exceptions in the graph can be explained rather easily. For example, the Mayas are less different from Europeans than you might expect on the basis of their large geographic distance (to be measured via the Bering Strait and North America!). This is probably a result of mixture with Spanish DNA after 1492. It is very difficult to obtain DNA samples of the original inhabitants of North and South America that are free from European influences from the days of Columbus and later. Conversely, the Mbuti pygmies in northeast Congo are more different from their neighbouring peoples than was expected, which could be attributed to (culturally defined) isolation.

Geographical distance is the most determining factor for the explanation of genetic differences between humans. By and large, you look most like your neighbours and least like people on the other side of the globe.

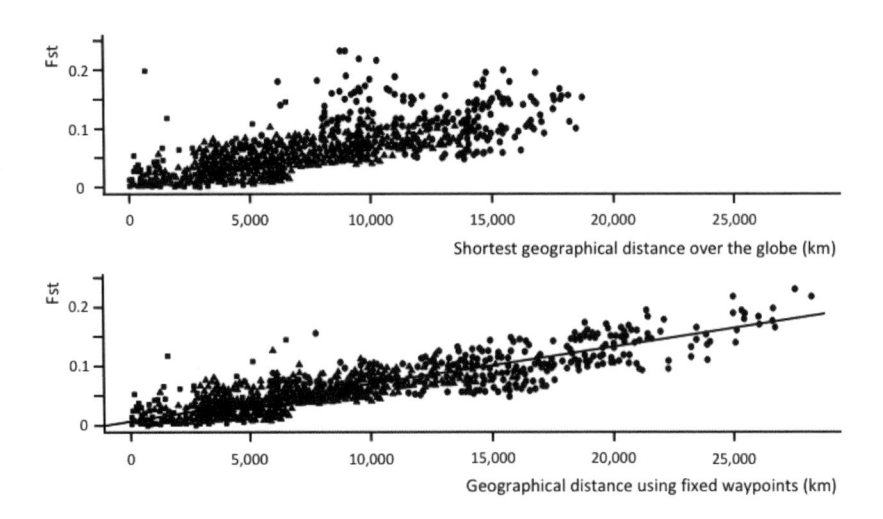

Fig. 4.15: Genetic distance, expressed as F_{ST}, as a function of geographical distance for a large number of pairwise comparisons of populations across the world. F_{ST} was calculated on the basis of 377 polymorphic microsatellite loci. In the upper graph the geographical distance was measured as the shortest arc length over the earth's surface; in the lower one, the distance was measured with arc lengths forced through a number of fixed waypoints, so as to do justice to the migrations of *H. sapiens* over land. Squares: within-region comparisons, triangles: Africa-Eurasia comparisons, circles: comparisons including America and Oceania.

On top of genetics

So far, in this chapter we have used the classical framework of population genetics to discuss evolutionary processes like mutation, selection and drift. It has now become clear, however, that supplementary mechanisms exist that are not strictly genetic, but may be essential in terms of evolution.

To understand the discussion we will need to go back to the early days of evolutionary thinking, to the French zoologist Jean-Baptiste Lamarck (1744-1829), who published an influential treatise on evolution entitled *Philosophie zoologique*, in 1809. Lamarck argued as follows: if an animal species exercises a certain organ during his life, or uses it intensively, this will be enhanced or increased in size. Subsequently the offspring will inherit that improved trait, and show it when they are exposed to the environment that induced the trait in their ancestors. This became known as the principle of *'inheritance of acquired characters'*. In table 4.5 a quote is shown from Lamarck's book indicating his reasoning.

In the time of Darwin, Lamarck's argument was the only scientifically based theory of evolution, but Darwin argued against it and proposed the principle of natural selection to replace the Lamarckian dogma, although

Table 4.5: Quotation from Jean-Baptiste de Lamarck, Philosophie zoologique (1809), p. 261

| Or, tout changement acquis dans un organe par une habitude d'emploi suffisante pour l'avoir opéré, se conserve ensuite par la génération, s'il est commun aux individus qui, dans la fécondation, concourent ensemble à la reproduction le leur espèce. Enfin, ce changement se propage, et passe ainsi dans tous les individus qui se succèdent et qui sont soumis aux mêmes circonstances, sans qu'ils aient été obligés the l'acquérir par la voie qui l'a réellement créé. | Well, every change that is acquired in an organ due to a habit of sufficiently intensive employment is preserved in subsequent generations, if this habit is common to all individuals that contribute to the propagation of their species by means of reproduction. Eventually this change will spread and thus be transferred to all subsequent individuals exposed to the same conditions, without them being obliged to acquire it in the way in which it was actually created. |

at that time a proper alternative theory of heredity was still lacking. Real evidence against the Lamarckian principle came from the famous experiments by the German zoologist August Weismann (1834-1914), published in 1888. Weismann conducted multi-generation breeding experiments with mice. He clipped the tails of his mice in every generation. Each time the tailless mice were allowed to give rise to a new generation, but even after 20 generations no mouse was born without a tail. Weismann also pointed out that in the Middle East, where the tradition of male circumcision has been in place for centuries, no male babies are born without a foreskin. To explain such results Weismann argued that in a relatively early stage of development, a separation is established between the *germline* (the parts of the body that are involved in the production of offspring) and the *soma* (the rest of the body). This separation was later called the '*Weismann barrier*'. His principle proved to be entirely correct. Weismann in fact invented the right concept of heredity even before Mendel's laws were rediscovered.

In the Modern Synthesis, Lamarckism has been renounced definitively, as no biological mechanism was found to exist to support it. After the discovery of the structure of DNA by James Watson and Francis Crick in 1953 and the genetic code in the 1960s it was shown that information can only be transferred in the order: DNA > RNA > protein > phenotype and not the other way around. This is often cited as the '*central dogma of molecular biology*', a theorem proposed by Francis Crick in 1970. We will see below, however, that even this dogma is showing cracks now.

Going back to Darwin, we have to realise that at the time several scientists found it difficult to accept that random undirected mutations acted upon by natural selection were sufficient to explain the evolution of adaptive characters. Among them was the American psychologist James Mark Baldwin

(1861-1934). He was a true adherent of Darwin, but sought a way to include a stronger influence of the environment on the generation of evolutionary relevant variation. Baldwin argued that many species are phenotypically plastic, that is, they can develop different phenotypes from one genotype. This will allow natural selection to act upon phenotypes that would not have been present without phenotypic plasticity. So *phenotypic plasticity* and environmental variation increase the scope of variation for natural selection to act upon. This became known as the *'Baldwin effect'*.

Arguing along similar lines, the British biologist Conrad Waddington (1905-1975), introduced the term *'epigenetics'* in 1942. The Greek prefix *'epi'* means 'upon', so epigenetics does not replace genetics but is added to it. Waddington imagined the way in which a certain genotype develops into a phenotype as an *'epigenetic landscape'*: a sloping terrain with several trenches and channels through which a ball finds its way (fig. 4.16, left). The landscape allows only a limited number of developmental pathways, which is expressed in the term *'canalisation'*. A highly canalised development is one in which the epigenetic landscape allows only one outcome, that is reached along a single deep trench, buffered against perturbations. In his view the shape of the landscape would be determined by a network of underlying genes, that interact with each other and the environment (visualised as strings or poles under the landscape in the right panel of fig. 4.16). Changes in the landscape, allowing another developmental pathway and an alternative phenotype, would be due to mutations in the underlying genes, or strong influences from the environment affecting their expression.

Waddington also introduced the term *'genetic assimilation'*. He argued that if a phenotypically plastic species is subjected to a specific set of environmental conditions for many generations, the same phenotype will be induced in every generation, although others are possible. Under such circumstances mutations that fix the development towards that optimal phenotype will be advantageous. This is assuming that the maintenance of plasticity towards other phenotypes (that are never induced) is costly and can be better removed from the developmental repertoire. So, natural selection will promote genetic assimilation of the pathway towards the optimal phenotype. This pathway would then become more and more fine-tuned by stabilising selection, which removes the 'developmental noise' and deepens the trenches in fig. 4.16. Waddington called this *'developmental homeostasis'*.

Canalisation will contribute positively to fitness because it provides robustness against environmental perturbations. Developmental instability is seen in bilateral animals if the left and right morphologies of an organ

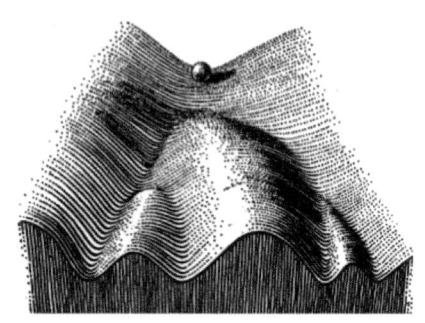

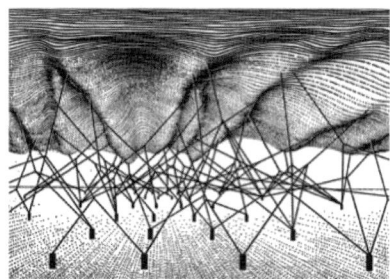

Fig. 4.16: The 'epigenetic landscape', a metaphor proposed by Conrad Waddington for the canalisation of development towards a limited number of outcomes (left). On the right a picture that illustrates how the genes of an organism (the blocks on the floor), by interactions with each other and the environment (the strings), would fix and modulate the epigenetic landscape.

are not precise mirror images of each other. This is called *'fluctuating asymmetry'*, and it is often measured as an indicator of environmentally induced stress. Conversely, facial symmetry in humans is interpreted as an indicator of health and this is also the reason why it is seen as an aspect of beauty.

In addition to stabilising the phenotype against a variable environment, canalisation is also assumed to provide robustness against genetic perturbations, *i.e.* mutations in developmental genes. This explains why new body plans evolve only rarely because mutations will have to move the rolling ball in fig. 4.16 to a totally different trench. Theoretical models of gene networks have shown that genetic canalisation may provide significant fitness advantages if mutations are frequent and have large effects.

Waddington's genetic assimilation is related to the Baldwin effect and the two concepts are often confused, but they are not the same. In the Baldwin effect phenotypic plasticity is not costly (it is maintained and may even increase), while genetic assimilation assumes that plasticity is costly and is eliminated by selection. The two concepts are sometimes seen as Lamarckian, but this is also incorrect: both Baldwin and Waddington were true Darwinians and they rejected Lamarckism.

The American ecologist Mary-Jane West-Eberhard working at the Smithsonian Tropical Research Institute in Costa Rica has also contributed significantly to our thinking about developmental plasticity and evolution. In the tradition of Waddington, she argued in her book from 2003, and in an influential article published in 2005, that phenotypic changes anticipate genetic changes. Genes are to be considered 'followers' rather than 'initiators' of change. Reproductive isolation is not where speciation begins, she says. Evolution starts from plasticity, environmentally induced alternative phenotypes, followed by genetic assimilation.

In modern biology, the term 'epigenetics' is used in a much more narrow sense than originally introduced by Waddington. Modern epigenetics is the study of mechanisms that regulate the gene expression profile of a cell in such a way that this cell devotes itself to a certain function. Epigenetics acts through markers, or 'imprints' upon the DNA, that do not change the DNA itself but affect its activity. Epigenetics explains why gene expression differs between cell types, how cell-dependent expression profiles are transferred to other cells of the same lineage, and in which way external conditions influence that profile. A great deal of information on molecular epigenetic mechanisms has become available in the last ten years. Here, we confine ourselves to the three most familiar mechanisms, which are summarised in table 4.6 and fig. 4.17.

The complexity of epigenetic regulation is enormous. The best-known mechanism is *methylation* of cytosine residues in the DNA. However, fig. 4.17 shows that not only cytosines may be methylated; also the 'tails' of *histone* proteins, protruding outside the nucleosomes, can be modified in many different ways. These modifications, which also affect each other, change the local chemical environment of the DNA, as a result of which the proteins of the chromatin complex start reacting with each other or are, on the other hand, kept at distance from each other. The most important function of the epigenetic markers thus seems to be deactivation or activation of certain parts of the DNA as a result of changes in the structure of *chromatin*, the entirety of proteins and DNA in the chromosomes. Chromatin condensation is accompanied by inactivation as the DNA becomes inaccessible for DNA-binding proteins.

Epigenetic mechanisms as studied in modern developmental biology and stem cell research would not be mentioned in a book on human evolution, if it hadn't been for the fact that sometimes epigenetic markers are inherited. Normally all epigenetic markers are erased in the germ line. During the production of gametes, first all methyl groups are removed and subsequently specific parts of the DNA are methylated again. After fertilisation the combined genome is once again entirely demethylated and then methylated again (fig. 4.18). The methylation status of the parental genomes thus does not permeate to the embryo. This principle is the basis for the Weismann barrier.

However, at the beginning of the twenty-first century, exceptions were found to the central dogma. In experiments with mice, scientists demonstrated that the diet of the mother was able to influence the phenotype of the

Table 4.6: Overview of the three most familiar epigenetic mechanisms

Mechanism	Explanation
DNA methylation	An enzyme, DNA methyltransferase, can transfer a methyl group (CH_3-) from a donor molecule to the C5 atom of cytosine. The methylation does not interfere with base pairing, but the methyl group sticks in the main groove of the double helix, changing the local chemical environment. Normally, a certain fraction of the DNA is methylated all the time, but this is strongly dependent on the tissue and on external factors. Hypomethylation or hypermethylation of DNA has a large effect on the expression of nearby genes.
Histone modification	The DNA in the cell is wound around proteins, so called histones, of which there are five different types. Depending on whether the gene is expressed or not, one histone can be replaced by another one, for instance, histone H3 is often replaced by histone H3.3 in active parts of the genome. Furthermore, the N-terminus of a histone, protruding from the DNA-protein complex like a 'tail', can be modified in various ways, by attaching molecules to it (acetate, phosphate, methyl, ubiquitin). As a result, the local chemical environment is changed, with consequences for the spatial structure of chromatin.
Small interference RNAs (siRNAs)	These regulatory RNAs are transcribed from micro-RNA genes and loop upon themselves to form a double-stranded molecule. This dsRNA is cut into smaller parts by a specific enzyme in the cytoplasm. Subsequently, these short sequences can be included in a protein complex, which is directed to a specific location in the DNA, aligning with the siRNA sequence. In that location the complex can then attract and activate histone modification enzymes and methyltransferases.

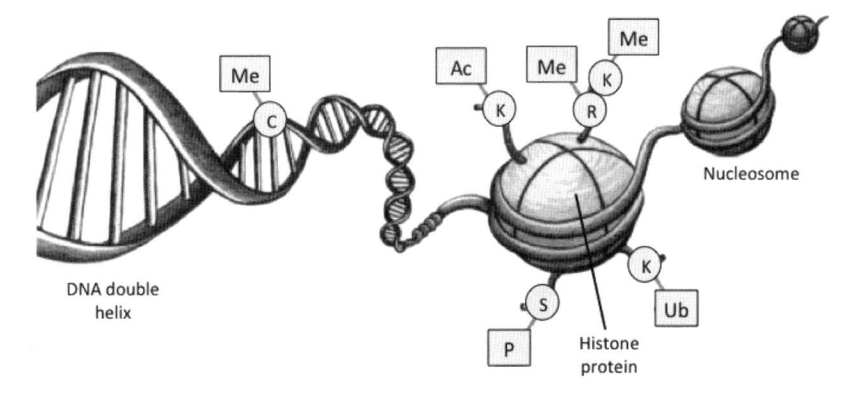

Fig. 4.17: Diagram of the different epigenetic markers that may appear in chromatin: methylation (Me) of cytosine (C), phosphorylation (P), ubiquitination (Ub), acetylation (Ac) and methylation of amino acids (S, K, R) in the 'tails' of histone proteins protruding outside the DNA helix.

offspring. By feeding the mother a diet that lacked any compounds that could act as methyl donor, the mother's gene *Agouti* was not methylated; it turned out that *Agouti* was also unmethylated in the offspring. Hypomethylation of *Agouti* results in a mouse with a deviating, light-yellow skin and a strong tendency to develop obesity. That the phenotype of offspring could be induced by nutrition of the mother is a clearly Lamarckian way of evolution and a seemingly serious violation of the central dogma.

A similar example in humans is due to the effects of the *Hongerwinter* (Hunger Winter) of 1944-1945 in the Netherlands. Children that grew in their mother's womb while the mother suffered a serious shortage of food, turned out to carry the burden of this for their entire life. Genes involved in growth and energy metabolism, are tuned differently, due to a different pattern of the epigenetic markers in the vicinity of those genes. The result is that later in life the 'Hunger Winter children' suffered from health problems, including an increased cholesterol level in the blood. The phenomenon is related to the 'thrifty gene' hypothesis that we will discuss in chapter 7.

The Israeli geneticist Eva Jablonka, conducting active research on these mechanisms, has meanwhile collected approximately one hundred proven cases of transgenerational transfer of acquired characters. In fungi and plants it seems to be more common than in animals. In addition it is remarkable that in animals the examples of transgenerational epigenetics are mostly due to adaptation to environmental factors, for instance, nutrition and toxic substances. But also for the behaviour of animals (for example, nursing the young) heritable epigenetic patterns have been described. For that reason, in 2006, Jablonka postulated her model of 'Evolution in four dimensions':

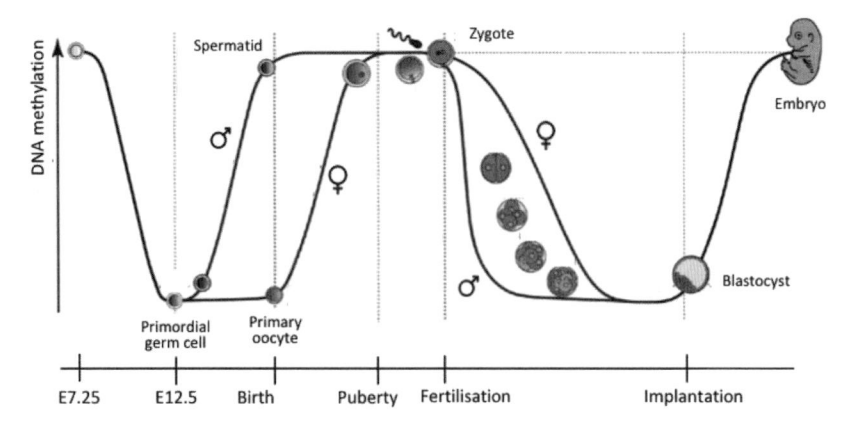

Fig. 4.18: Global DNA methylation of gametes, zygotes and early embryonic stages as a function of the life cycle of men and women.

in addition to the classical Darwinian evolution there would be three other evolutionary principles operating for human beings: epigenetics, transfer of behaviour (social learning and similar), and transfer of symbolic communication (language, art, religion and similar).

Epigenetic inheritance is also associated with the evolution of *homosexuality*. It has been known for a long time that male homosexuality has a genetic component. The question is why homosexual behaviour can continue if it is partly genetic. Homosexual men, after all, have fewer biological children; as such, you would expect it to become extinct soon. Still it doesn't. Between 3% and 10% of males are homosexual, and it's been like that for centuries.

In 2012, an American-Swedish team postulated that homosexuality could be a consequence of the fact that methylation markers in the germ line are not always completely removed. Genes that channel the development of the embryo in a male or female direction would still carry epigenetic markers, which are the consequence of the feminising influence of the mother. A strongly feminising influence during the development of a female embryo stimulates the fertility of the later adult and is, as such, beneficial. Yet, a side effect is that this influence is more difficult to erase if the female gets male descendants herself. In the event of an XY zygote, a male embryo will develop that is less sensitive to the testosterone pulse that takes place in the later embryo, as a result of which the brain does not develop in the usual male direction. According to this story, homosexuality itself has no evolutionary advantage or disadvantage, but it is a by-product of selection for high fertility in females.

Despite these examples it remains to be seen whether inheritance of epigenetic markers is an important evolutionary mechanism. The current collection of cases seems too arbitrary, and a convincing mechanistic principle, other than a series of exceptions, is lacking.

5. The past in the present

The current manifestation of humankind is the result of a long history, of which the last episode started some 7 million years ago, as we saw in chapter 1. But since we cannot travel back in time, we are not aware of many of the processes that took place in the past. Still, we can reconstruct the crucial events by looking at present humans. To that end we have three approaches at our disposal: comparing body plans between humans and related animals (this is what we did in chapter 3), analysis of human development from embryo to adult (see chapter 2), and comparing DNA analysis of human populations. In this chapter we will use the third approach, building on the framework of population genetics presented in chapter 4. It turns out that this will enable us to answer questions like: Where did *Homo sapiens* emerge? How did they migrate across the globe? Which older human species were they in contact with?

Phylogenetic reconstruction

Mapping out and describing the evolutionary history of groups of organisms is what we call phylogenetic reconstruction. At the centre of this is the phylogenetic tree, a graphical display of the relationship between species or other groups of organisms. Fig. 5.1 gives an example of a tree that will often recur in this chapter. Each split provides a point of divergence, called a *node*. In the case of species a branching point indicates that speciation has occurred, whereby one species splits into two new ones. The way in which the species are connected with one another, independent of how the tree is drawn, is called the '*topology*' of the tree.

The common ancestor of all organisms in a tree is called the *root* of the tree. Since we do not know this ancestor, let alone bring it back to life, the ancestral features are estimated using an *outgroup*: one or more species that we assume to display the characteristics of the ancestor. The rest of the tree, for which we want to reconstruct the evolution, is called an *ingroup*. It is still possible to draw a tree without an outgroup, but in that case the tree is *unrooted*: it only shows the relatedness among species and does not specify where evolution begins.

The branching times are indicated by the positions of nodes relative to the horizontal axis (usually a tree is drawn from left to right). The time period from a divergence to the present is what we call the *divergence time*. The

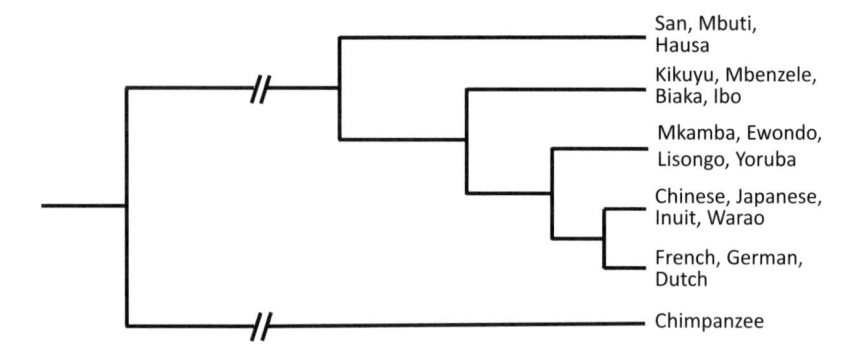

```
                               ┌─────── San, Mbuti,
                          ┌────┤        Hausa
                      ┌───┤    └─────── Kikuyu, Mbenzele,
                  ┌───┤ //                Biaka, Ibo
                  │   └────────────── Mkamba, Ewondo,
                  │                     Lisongo, Yoruba
              ────┤                ┌── Chinese, Japanese,
                  │                │   Inuit, Warao
                  │                └── French, German,
                  │                    Dutch
                  └──────//────────── Chimpanzee
```

Fig. 5.1: Phylogenetic tree of five groups of people, with the chimpanzee as an 'outgroup'. The phylogeny shows that Africans are not monophyletic; they consist of multiple evolutionary lineages. Europeans and Asians appear to branch from the third African group. The tree is a greatly simplified representation of a phylogeny derived from mitochondrial DNA sequences.

horizontal axis may have an absolute time scale (measured in Ma BP or ka BP), or it may represent relative distances without an explicit scale, such as in fig. 5.1. The vertical axis has no meaning. The pattern of branching is important, but each branch of the tree can be rotated around the horizontal axis through its ancestor; such rotations do not affect the topology of the tree. In fig. 5.1 the three groups of Africans were put on top, but they might as well have been positioned at the bottom (putting the Europeans on top).

The organisms at the tips of the phylogenetic tree are called *operational taxonomic units* (OTUs). Usually an OTU is a species, but they might also be individuals, families or higher taxonomic groups. All OTUs with a common ancestor are collected in a *clade*, a branch of an evolutionary tree. Closely related species that are directly adjacent in a tree are called *sister species*. The associated branches are called sister clades.

An important feature of a phylogenetic tree is that it has *monophyletic groups*. A monophyletic group is a collection of OTUs that meets two conditions: (1) all OTUs in the group derive from the same ancestor, and (2) all descendants of that ancestor are in the group. If a group meets criterion (1) but not (2), we call it a *paraphyletic group*, which is in fact an incomplete clade. In fig. 5.1 the Africans form a paraphyletic group, because in terms of evolution the Europeans and Asians cluster within the Africans. Finally, we mention the term '*polyphyletic group*'; this is a collection of OTUs that derive from different ancestors, meaning OTUs with ancestors inside as well as outside the tree.

Ideally, the biological classification of species and higher categories is such that each taxon represents a monophyletic group. In that case systematics coincides with evolutionary descent, but that is not always the case. For example: in the past, humans were the only species placed in the family of Hominidae, whereas chimpanzee, gorilla and orang-utan were considered together in a different family, Pongidae (fig. 5.2, left). In this way, however, Pongidae is paraphyletic as humans derive from the chimpanzee lineage and thus should also fall under Pongidae. The currently chosen solution is to consider chimpanzee, gorilla and human together as Hominidae and leave only orang-utan in the Pongidae family. In this way, both families are monophyletic (fig. 5.2, right).

A tree is developed from *characters* that may have different *character states* in different species. For example, 'number of limbs' may be a character and 'six', 'four', and 'two' may be character states. The character states for each species are conveniently listed in a table such as table 5.1. These tables can become quite big if we are dealing with many species and many characters. If a character state differs from its ancestral state, it is called an acquired or *derived* condition; the scientific name for this is *apomorphy*. If such a newly acquired character state is shared by two or more species, we call it a *synapomorphy*. The term 'synapomorphy' is a combination of three Greek words: '*syn*' meaning together, '*apo*' meaning from and '*morph*' meaning form. Synapomorphies are crucial for every phylogenetic reconstruction, since they define the commonalities between species that were acquired by changes within the tree.

The biggest problem in phylogenetic reconstruction is that not all corresponding character states are synapomorphic. Identical character states may have evolved several times, independently from each other, so their similarity should not be regarded as indicating common ancestry. This is referred to as a *homoplasy*. There are two important causes for this. The same character state may emerge twice in different species, if they are exposed to a similar selective force. We call this *convergent evolution*. The fins of whales

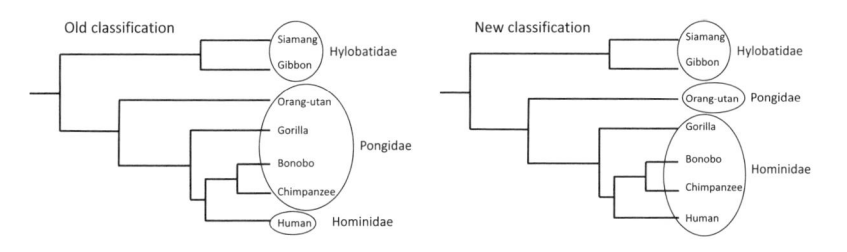

Fig. 5.2: The former family classification of Hominoidea (left), in which humans were the only representatives in the Hominidae family, is less correct in evolutionary terms, as it made the family Pongidae paraphyletic. In the new classification (right) all three families are monophyletic.

are not homologous with the fins of fish, because whales descend from a mammalian ancestor, which is not the ancestor of fish. A second cause for homoplasic character states is that a character can revert or 'fall back' into its original ancestral state. Such a character state is called an evolutionary reversal. If it concerns a morphological character it is known as an *atavism*. For instance, the appearance of hind limbs in whales, or supernumerary nipples in humans are atavisms.

In addition to being apomorphic, character states can also be *plesiomorphic*. This means that they have not changed compared to the ancestor. If two OTUs have a common character state, that they share with the ancestor, we call their similarity symplesiomorphic. A symplesiomorphy is not very informative in phylogenetic reconstruction; at most such similarities tell us that the two OTUs are in the correct group, but they do not provide information about the relative position of such OTUs within the tree.

The science that deals with the reconstruction of relations between organisms on the basis of phenotypic characters is called *cladistics*. Meanwhile, this has become a classical part of evolutionary biology, but as a matter of fact cladistics isn't that old yet. The methodical framework was developed by the German entomologist Willi Hennig (1913-1976). He was mainly interested in how to classify species in the best way possible. Later, the cladistic method was used not only for biological systematics, but also for the reconstruction

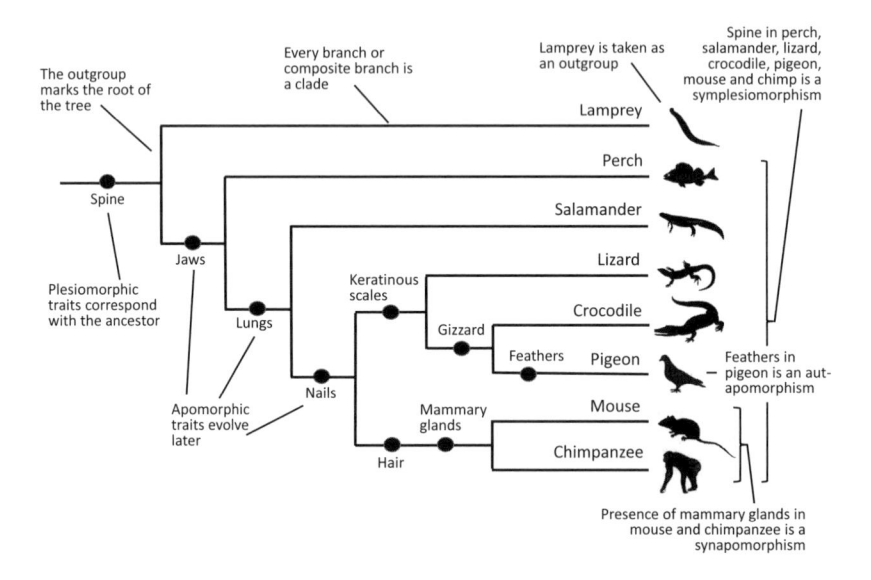

Fig. 5.3: Clarification of the different terms used in cladistics, illustrated by a phylogenetic tree of eight vertebrates. See table 5.1 for a list of character states.

of ancestral relationships, phylogeny. That is why phylogeny and cladistics have become more closely related. In the 1970s, cladistics and phylogeny were given a boost with the introduction of powerful computers. At that time, cladistics was called 'numerical taxonomy'. The modern phylogenetic reconstruction methods are very computationally intensive and make use of advanced computer programs that go beyond the scope of this book.

As illustrated above, a multitude of technical terms is used in phylogenetic reconstruction. All of these have a very specific meaning, which, in our experience tends to utterly confuse the student of evolutionary biology when exposed to them for the first time. Still, they are all very important terms, because each of them describes a specific aspect of a phylogeny. The terms are explained again in fig. 5.3.

The way in which a phylogeny is reconstructed can be illustrated by means of the evolution of eight vertebrate animals: sea lamprey, perch, pigeon, chimpanzee, salamander, lizard, mouse and crocodile (fig. 5.3, table 5.1). Let's initially assume that all acquired characteristics emerged only once from a common ancestor and did not undergo reversion (no homoplasies). As illustrated in table 5.1, the lamprey has no derived feature whatsoever in relation to the other seven taxa. Therefore, lamprey is chosen as an outgroup for the reconstruction of our phylogenetic tree. This is important, because, in this way, we can conclude from the similarity or non-similarity with sea lamprey, whether a character state is apomorphic or plesiomorphic. It is also evident that chimpanzee and mouse have many derived character states in common, whereby mammary glands and fur evolved only in chimpanzee and mouse. These are acquired character states that have evolved in the

Table 5.1: Character table for eight different vertebrates, illustrating the principle of phylogenetic reconstruction (compare with figure 5.3)

Taxon	Derived traits							
	Jaws	Lungs	Claws or nails	Gizzard	Feathers	Fur	Mammary gland	Keratinous scales
Lamprey	−	−	−	−	−	−	−	−
Perch	+	−	−	−	−	−	−	−
Salamander	+	+	−	−	−	−	−	−
Lizard	+	+	+	−	−	−	−	+
Crocodile	+	+	+	+	−	−	−	+
Pigeon	+	+	+	+	+	−	−	−
Mouse	+	+	+	−	−	+	+	−
Chimpanzee	+	+	+	−	−	+	+	−

ancestor of mouse and chimpanzee after they separated from the other vertebrates. It can be concluded that mammary glands and fur are synapomorphic character states that link chimpanzee and mouse. We can derive the other characters in the same manner. Keratinous scales on the skin are synapomorphic for lizard, crocodile and pigeon. A gizzard is only found in pigeons and crocodiles and is indicative of the relationship between these two vertebrates. Another remarkable fact is that the pigeon has a unique feature that cannot be found in the other vertebrates, namely feathers. This character doesn't provide any information about the relationship between the pigeon and the other vertebrates in this example. Such a character is called an *autapomorphy* (auto meaning 'for/by itself'). An autapomorphy is a character state that is found in one OTU only.

By integrating all this information it is possible to reconstruct a phylogenetic tree that graphically illustrates the evolutionary relationship among the eight vertebrates (fig. 5.3). This tree can be reconstructed manually and relatively easily from the data in table 5.1 but it is not always done so easily. For example, had we included snakes in the analysis, we would have come across a problem. Just like lizards, snakes are reptiles, but they have no limbs, neither do they have claws or nails. Therefore they would be clustered with salamanders, which are not reptiles but amphibians. In such cases we need extra information about multiple derived characters, but as we add more characters, we will be bothered by convergences and reversals. This is a typical dilemma of biologists working with phylogenetic reconstruction. How can we distinguish synapomorphic character states from characters that show homoplasy?

A way to avoid this to a certain extent is by making use of the principle of *maximum parsimony*. Parsimony states that the simplest explanation is the most probable one. In a phylogenetic context this means that the tree assuming the smallest number of evolutionary changes, is the most probable one. This guarantees that the number of homoplasies is always minimised. The principle is based on the logic of *'Ockam's razor'*, the idea of the fourteenth-century monk William of Ockham, who stated that the best explanation is always one with the least assumptions. This principle appears to work very well when explaining the evolution of life.

Technically, the principle of maximum parsimony works with the *length of the tree*, the sum of all evolutionary changes in the tree (summed over all characters). By shifting the OTUs and assuming different branching patterns, tree length can be calculated for all possible trees. The tree with the shortest length will then be chosen as the most probable representation of evolutionary history. In the example discussed here, this can more or

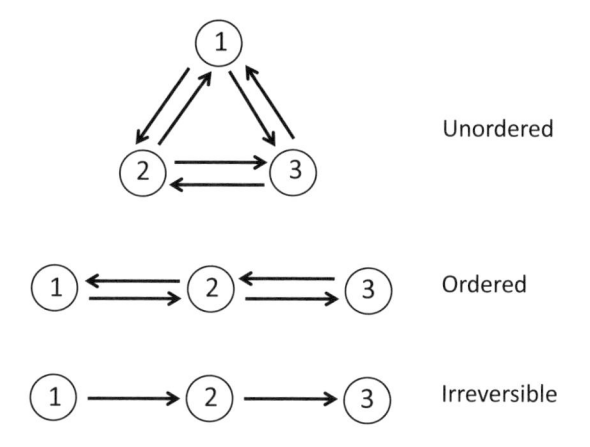

Fig. 5.4: Depending on the transitions that are allowed between character states, characters can be classified as unordered, ordered or irreversible.

less be done manually, but for more extensive and more complex situations computer programs will be required.

The tree length also depends on the transitions allowed between the different character states. The simplest assumption is that any state can change into any other. In that case the character is said to be *unordered* (fig. 5.4). But usually the character states must follow a certain sequence. The length of a limb can, for example, have character states small, medium or large, but you cannot jump from small to large at once, you need to go via medium. In that case the character is said to be *ordered*. Finally, there are also characters that can be regarded as *irreversible* (fig. 5.4). Usually this is assumed for complex morphological structures (for instance, the placenta of mammals), which, once they emerged, cannot easily evolve back to an ancestral state. The principle of evolutionary irreversibility is known as *Dollo's rule* after the Belgian palaeontologist Louis Dollo, who argued in 1893 that once a species has evolved a particular complex structure, it will not return to the original state along the same trajectory.

Phylogenetic trees are also used to visualise the evolution of individual genes, or any stretch of DNA, including whole genomes. Nowadays by far most trees are based on DNA sequences. This provides a number of great advantages over morphological and anatomical features. Firstly, the scoring of morphological characters is often difficult; it is the work of specialists, conducted by people who are scarce, because a lot of experience is required. A second important point is that in DNA analysis there is no argument about heredity. Besides, it is relatively easy to find DNA sequences that are homologous. Using the current automatic DNA sequencers, molecular

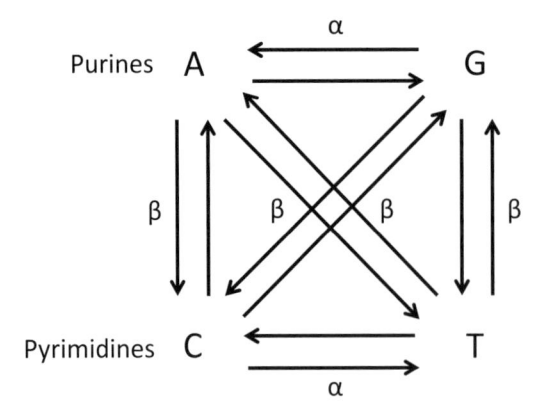

Fig. 5.5: Substitutions in the DNA may be transitions (changes within the purines or pyrimidines) or transversions (between purines and pyrimidines). Often different frequencies (α and β) are used for transition rates and transversion rates.

biologists can generate an incredible amount of data within quite a short period of time. Finally, based on experimental evolution with viruses and bacteria in the laboratory, it has been demonstrated numerous times that the changes (mutations) that take place in DNA are mostly synapomorphic. The biggest problem of DNA analysis though is that DNA positions, once mutated, can relatively easily mutate back, leaving hidden reversals in the sequence that are difficult to identify.

From a cladistic point of view, a DNA sequence consists of a long series of characters (nucleotide positions) each of which can occur in four states, A, C, G or T (fig. 5.5). The transitions between them are not entirely equivalent because A and G are purines and C and T are pyrimidines. In other words, a position in the DNA cannot be regarded as an entirely unordered character. The chance that an A changes into a G (a transition) is considered larger than the chance that an A changes into a C (a transversion). In fact, there are two transition probabilities, indicated by α and β, whereby α > β. This assumption is called *Kimura's two-parameter model*. Obviously other assumptions are possible that go much further, leading to complex substitution models.

Before using DNA sequences for phylogenetic purposes, one should make sure that the compared positions are indeed homologous, which means that they go back to the same ancestor. You can do this by making an *alignment*. Table 5.2 shows a sequence of 30 bp from the mitochondrial genome of human, chimpanzee, gorilla, orang-utan and gibbon as an example. It is obvious that the selected 30 bp sequence, although short, is homologous between the species, otherwise the similarity couldn't be that significant. But there are also differences and these can be used to reconstruct the evolutionary

Table 5.2: Alignment of a 30 bp sequence in the mitochondrial genome of four apes, compared with human (the differences between ape and human are shown in bold)

Species	Sequence
Human	AAGCTTCACCGGCGCAGTCATTCTCATAAT
Chimpanzee	AAGCTTCACCGGCGCA**AT**TA**T**CCTCATAAT
Gorilla	AAGCTTCACCGGCGCAGT**TGT**TCT**T**ATAAT
Orang	AAGCTTCACCGGCGCA**ACC**ACCCTCAT**G**AT
Gibbon	AAGCTT**T**ACAGGTGCA**ACC**GTCCTCATAAT

relationships. This alignment shows that human and chimpanzee are indeed more closely related than human and gorilla, human and orang-utan *et al*. Making an alignment is a very basic technique in molecular biology, for which a program is used that was developed in the 1980s by Stephen Altschul of the American NCBI, named *BLAST (Basic Local Alignment Search Tool)*. This program, which was later refined in all kinds of different ways, has become so common that 'to blast' (carry out a BLAST search query) has become a verb.

Due to the size of DNA data for phylogenetic reconstruction, dedicated computer programs were developed to make the calculations, as a result of which phylogenetic trees become increasingly more reliable, while also increasingly larger groups of organisms can be studied. Most of the programs also include the option to test the reliability of the calculated clades. This is done by use of a numerical method, pseudo-sampling with replacement. This so-called *bootstrap* method is due to the American computer scientist Joseph Felsenstein, who published it in 1985. 'Bootstrapping' refers to the story of Baron Munchausen, who bragged that he had drawn himself plus his horse out of a swamp by pulling on his own bootstraps (although in the original story the baron is said to have pulled himself by his hair); in computer sciences 'booting' has obtained the meaning 'to start from nothing'.

The bootstrap algorithm generates pseudo-sequences by randomly drawing homologous DNA positions across species, until a pseudo-dataset is generated that contains as many base pairs as the original. These pseudo-sequences are treated exactly the same as the original dataset. The result is a new phylogenetic tree. If the data are very reliable, the branching of this new tree will hardly, or not at all, deviate from the branching of the original tree. Bootstrapping is usually repeated 500 to 1,000 times. The frequency by which a specific clade is observed is displayed as a percentage for that clade. If, for example, exactly the same clade appears in all bootstrap

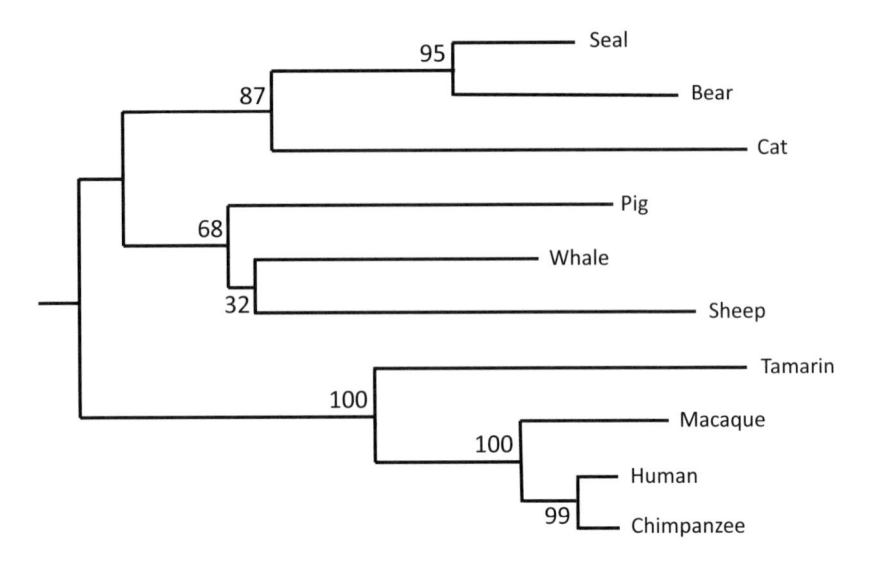

Fig. 5.6: Phylogeny of a number of mammals, illustrating the principle of statistical reliability of clades. The numbers at the base of a clade indicate bootstrap values, a measure for confidence of the clade. These numbers show that the sister group relationship between sheep and whale, and between these two and pig, are insufficiently supported. The branching within the primates and the Carnivora, however, do have a high statistical reliability.

replications, the number 100 will be attached to it. This indicates that this clade is highly reliable. In general, in published phylogenetic trees, only clades are displayed with bootstrap values exceeding 50% and only the clades with values exceeding 80% are considered reliable.

Fig. 5.6 shows a phylogeny of the class Mammalia to explain the use of bootstrap values. The tree shows that some clades (particularly the position of whales relative to ungulates) are not properly supported, whereas there is a lot of support for the relationships within the primates. It often happens that not all parts of a tree show the same statistical reliability, especially when the reconstruction involves highly diverse groups of species, whereby the rate of evolution is not the same in different parts of the tree. Also it appears that computer programs tend to place two groups into a single clade if they are both subject to many changes. This is called 'long branch attraction'. Special algorithms are required here to avoid these artefacts.

Nowadays phylogenetic reconstruction has developed into a specialised area of bioinformatics where a multitude of numerical approaches are tested and continuously improved. Rather than the simple principle of maximum parsimony bioinformaticians use other optimality criteria to guide the selection of the best tree such as *neighbour-joining* and *maximum likelihood* models. A discussion of these techniques falls beyond the scope of this book.

The molecular clock

The principle of phylogenetic reconstruction can be taken one step further by assuming that the differences between species accumulate at constant rate over time. Looking at orthologous DNA sequences across species the number of differences would be proportional to the time since these species diverged. This is called the *molecular clock hypothesis*.

The hypothesis is controversial because we can ask ourselves whether DNA indeed changes in a uniform, strictly constant manner over time. Still, the concept turns out to be useful if the hypothesis is limited to certain areas of the genome for which we know the mutation frequency.

Walter M. Fitch, professor of molecular evolution at the University of California, Irvine, tested the molecular clock hypothesis in a classic study published in 1976. He wondered whether a linear relationship existed between the evolution of vertebrates and the degree of change in seven selected proteins (globines, cytochrome c and fibrinopeptides). He created a phylogenetic tree for eighteen vertebrates and correlated the branching points in the phylogeny with geological time settings. The use of fossil calibration points made it possible, for each pair of vertebrates, to determine the time elapsed since the split from a common ancestor. Fitch decided to compare this divergence time with the differences he found in the seven proteins and strangely enough this turned out to provide a practically linear relationship (fig. 5.7). In other words: the average number of differences between two species is a measure for the elapsed time since their evolutionary divergence. The average rate of molecular change was 0.47×10^{-9} substitutions per year, per nucleotide position, which is an estimate of the rate of evolution of the vertebrates in these loci.

Another striking result in Fitch's study was that the most recently evolved vertebrates, the primates, showed a delay in the molecular clock (the so-called *primate slowdown*). Later it was suggested that this could be due to the more advanced DNA repair system of primates. However, other scientists suggested this was a matter of resolution: when using enough proteins, and multiple organisms over different time intervals, the differences between primates and other mammals would disappear.

Fitch's proposal to calibrate the molecular clock on phylogenetic divergences in the fossil record is not the only method. Two other approaches have been proposed. One is to estimate mutation rates from a comparison of well-dated ancient DNA samples. Counting the number of nucleotide differences in orthologous sequences spanning a time period of several thousands of years has provided estimates for the mutation rate in mitochondrial DNA that that

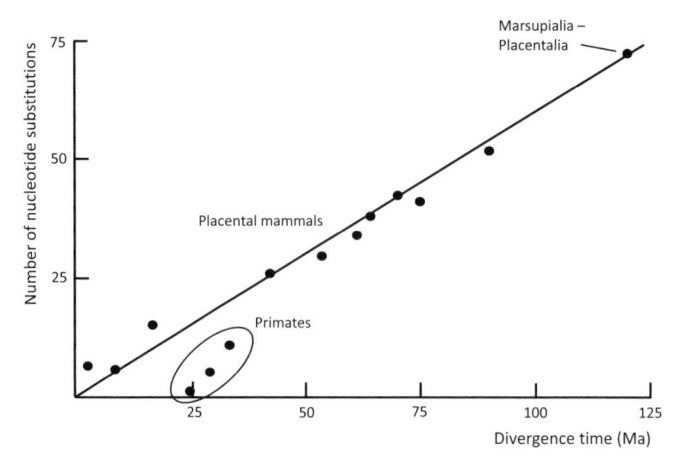

Fig. 5.7: Representation of the molecular clock hypothesis for mammals as a linear relation between the time since divergence (measured against fossils) and the number of non-synonymous nucleotide substitutions in four proteins. Each dot is a pairwise comparison. For example, the dot on the upper right is the comparison between marsupials and placental mammals, which diverged 120 million years ago and have a different amino acid in 74 locations. In this graph, the primates deviate significantly from the regression line: genetically they differ less from each other than expected based on their geological age.

are well in line with the phylogeny-based estimates. Another approach is to compare the genomes of children with their parents; any DNA changes that appear in the child but are not present in one of the parents must be due to new mutations. This method yields estimates that are about half the rates assumed in the phylogenetic and ancient DNA approaches. Why different methods can produce such diverse estimates is unclear at the moment. The molecular clock hypothesis is thus subject to much discussion and uncertainty, yet it is used in many publications to date evolutionary events.

Another phylogenetic reconstruction approach related to the molecular clock is so-called *coalescence* analysis. This analysis, which originates from the neutral theory of evolution, estimates the time that, using a certain mutation model, is necessary to coincide two or more orthologous DNA sequences. Looking at the example of table 5.2 (30 bp of mitochondrial genome sequence, aligned for four ape species and humans) you could define an ancestral sequence in which the different observed sequences 'coincide', meaning, become equal, using overall as few substitutions as possible. The ancestral sequence is called the *coalescent*, and the calculated required time is called the *coalescence time*. The words are derived from the Latin word '*coalescere*', which means 'to merge'.

The coalescent process can be viewed as genetic drift in reverse. If a mutation took place in the past and is now observed as a difference between species, this mutation was fixed after its emergence in one of the two species. In chapter 4 we saw that the chance of fixation strongly depends on population size: in small populations this happens a lot quicker than in large populations. As long as the mutation frequency is constant, a certain number of substitutions between two species, for a small population size, indicates a recent divergence (a short coalescence time). In the case of a large population the same difference indicates a large coalescence time. The coalescence theory therefore produces a surprisingly simple equation for the relationship between coalescence time and population size:

$$T_C = 4 N_e$$

where T_C is the coalescence time (measured in generations) for all the variants of a gene in a certain generation, and N_e is the effective population size (see chapter 4 for the latter term). It should not come as a surprise that the same equation holds for fixation time (see chapter 4). In fact, coalescence

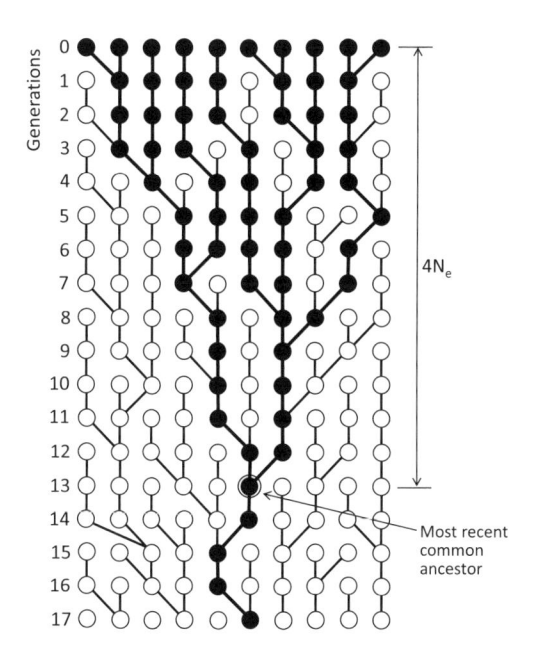

Fig. 5.8: The concept of coalescence illustrated by a gene genealogy. All variants of a gene present in generation 0 have a coalescent in one of the past generations. Here the most recent common ancestor is in generation 13. Theoretical analysis shows that under certain simplifying conditions the expectation (average) for the generation that has the coalescent is $4N_e$, where N_e is effective population size. It is assumed here that the transmission of genes is a purely stochastic (neutral) process, however, every gene leaves maximally two copies of itself in the next generation (but some leave one or zero, to make sure that population size is constant).

and fixation are two sides of the same medal: coalescence is divergence run in reverse gear. Both are drift processes governed by the rules of neutral evolution. Coalescence can be pictured as a process of tracing down the genealogy of a set of homologous DNA sequences (fig. 5.8). But please note that, despite the fact that the various versions of a gene will always coalesce in one single ancestral sequence, the individual carrying that ancestral sequence was not the only one in his or her population. There were others in the ancestral population whose version of the gene just by chance somewhere along the line did not leave a descendant. Coalescence theory makes extensive use of mathematical population genetics and probability calculation (this is too complicated to be discussed in detail here).

Out of Africa or multiregional evolution?

The modern methods of phylogenetic reconstruction and coalescence were deployed to answer fundamental questions on the origin of *H. sapiens*. Where did human beings emerge and when? For a long time, scientists assumed that hominins evolved via an anagenetic process: from *Australopithecus* to *H. habilis*, to *H. erectus*, to *H. neanderthalensis*, to *H. sapiens*. The German-American evolutionary biologist Ernst Mayr justified this process in the 1950s and 1960s. In the tradition of the Modern Synthesis and keeping to a strict interpretation of Darwinism, Mayr considered species to be gradually changing entities and saw their fossil remains as snapshots in a continuous evolutionary process. At each moment in time there could be only one species; *H. sapiens* developed in a straight line from its ancestor.

Due to the prevailing paradigm of phyletic gradualism, palaeontologists felt compelled to merge different types of fossils together into an organically evolving species concept, a situation that obviously didn't comply with fossil reality. Thus, Neanderthal had to be regarded as an ancestor of modern humans (although there were obvious differences) and all fossils that resembled Java Man, even if the resemblance was remote, were lumped into *H. erectus*.

Only in the 1970s another image of human evolution developed. Influenced by Niles Eldridge, Stephen J. Gould and Ian Tattersall, all American palaeontologists, a cladogenetic alternative started to develop (*cf.* the introduction of 'punctuated equilibria' in chapter 4). New species did not develop in small steps from other species, but emerged at discrete moments in time and branched from existing species. Human evolution became a process of branching, splits and extinctions, a process of cladogenesis, a view matching the fossil record much better.

Still, the concept of human evolution as a linear process remained popular for a long time, until in the 1980s. The American palaeontologists Allan Templeton, Milford Wolpoff and Tim White were important advocates of the anagenetic model, whether or not supplemented by some degree of exchange between the geographical regions. According to this model, a *'sapiens* stage' developed in at least three different regions on earth: in Africa, in Europe and in Asia, from the older humans present there. This was called *multiregional evolution. H. sapiens* was considered an advanced version of *H. erectus*, and the advance took place in different regions of the distribution range of *H. erectus*.

However, uneasiness with the concept of multiregional evolution began to rise since the 1970s. Research by the British palaeontologist Chris Stringer largely contributed to this. Applying quantitative analysis of skull features, he was able to show that modern *H. sapiens* in Asia shows more similarity with ancient *H. sapiens* from the Middle East and Africa than with *H. erectus* in Asia. Also, he demonstrated that Neanderthal should be considered a sideline and wasn't part of the ongoing lineage towards modern humans.

The fossil of the *Dali Man*, a primitive hominin excavated near the touristic town of Xi'an in China (fig. 5.9), played an important role in this discussion. The age of the Dali fossil is estimated at 209,000 years. The supporters of multiregional evolution regard it as a transitional form between *H. erectus* and *H. sapiens*, confirming the regional continuity of the Asian fossil hominins. However, by studying the skull in detail (a model is displayed in the Shaanxi Historical Museum in Xi'an) it is easy to identify several typical *erectus* features, such as a sagittal keel and quite a strong torus

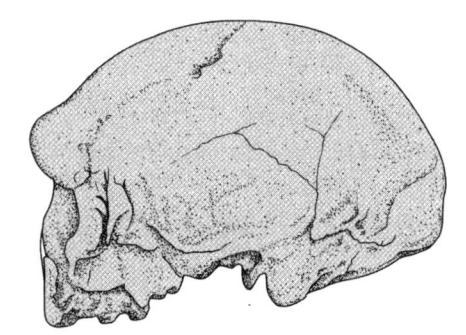

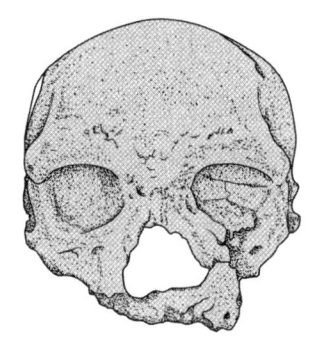

Fig. 5.9: Lateral and frontal views of the Dali skull, excavated near the town of Weinan, 60 km from Xi'an in China. The fossil is dated about 200 ka BP. According to the adherents of multiregional evolution it is to be regarded as a transitional form between *H. erectus* and *H. sapiens*.

supraorbitalis. According to Chris Stringer, a supporter of the *Out of Africa* theory, Dali Man can be considered a late *H. erectus* or a *H. heidelbergensis*.

The final blow for the concept of multiregional evolution came in 1987, with the publication of a sensational article by an American research group headed by Allen Wilson (1934-1991), with Rebecca Cann as the first author. These scientists based their ideas on DNA data, which, at the time, was a revolutionary approach. In American hospitals, DNA samples had been isolated from placentas of women in childbed with different ethnic backgrounds: Africans, Asians, Europeans, Australians, and women from New Guinea. The analysis was focused on restriction patterns (DNA fragments obtained after cutting with endonuclease enzymes) of the mitochondrial genome. A greatly simplified form of the phylogenetic tree, made from these patterns, is shown here as fig. 5.1. Four new insights, which back then were revolutionary, can be derived from this:

1. The Africans are the first groups to branch from the ancestor.
2. The genetic variation among Africans is much larger than among Asians and Europeans. African lineages differ much more from one another than, *e.g.*, Germans and Chinese.
3. The European and Asian sequences are part of an originally African lineage.
4. The coalescence time of the mtDNAs is between 166 and 249 ka.

These genetic data immediately and obviously pointed out that humans originated in Africa, and that all *H. sapiens* derive from an African ancestor. After migration from Africa, *H. sapiens* replaced all previous species, such as *H. erectus* in Asia and *H. neanderthalensis* in Europe, and did not mix with them. This theory was soon known worldwide as *Out of Africa*. At the time (1985) there was a famous movie by that name, featuring Meryl Streep and Robert Redford. Undoubtedly, this played a role in the popular name given to the evolutionary theory.

The analysis of Cann was relatively simple as it was based on mitochondrial DNA. We know that mitochondrial DNA only inherits through the female lineage. The sperm cell does have mitochondria but little cytoplasm, which, during fertilisation does not contribute to the cytoplasm of the egg. Due to maternal inheritance, mitochondrial DNA is subject to the coalescent process that we discussed in the previous section (fig. 5.8). According to coalescence, all current mtDNAs can be traced back to one ancestral sequence, or one woman who lived in the first group of *H. sapiens* that all of us derive from. This woman was – rather appropriately – called '*mitochondrial Eve*'. However, despite her name not all present humans descend from that woman. Our nuclear DNA derives from the gene pool of all women and men in the first group of *H. sapiens*. Only the mtDNA of 'Eve' happened to leave descendants up to the present day. Because 'mitochondrial Eve' was certainly African, the *Out of Africa* theory is also called the *Black Eve* hypothesis. In a further wordplay of biblical spirit, the *Out of Africa* theory is sometimes called the '*Garden of Eden hypothesis*'. In addition, a specific version of '*Out of Africa*' goes under the name '*Weak Garden of Eden hypothesis*'. This was to indicate that not only gene genealogies but also changes in population size affect our estimates of human origins.

Cann's article was heavily criticised in the 1980s, which, in hindsight is logical because it was the first implementation of a DNA analysis in a field that had been dominated by bones. The original article also contained a number of imperfections in the statistical analysis, which were widely covered. The American palaeontologist Milford Wolpoff, in particular, a supporter of the multiregional evolution theory, took a strong stand against the *Out of Africa* ideas of Stringer and Cann.

Fig. 5.10 shows the difference between *Out of Africa* and multiregional evolution. An important consequence of multiregional theory is that the differences between Africans, Europeans and Asians go back to the migration of *H. erectus*. A genetic comparison of these population groups should result in a coalescence time of more than 1 million years (the time when *H. erectus* started to migrate from Africa). The genetic data produced by Cann are simply in conflict with this, because they imply a coalescence time of around 200 ka.

Also analyses of male markers support the *Out of Africa* theory. As we have seen in the previous chapter, the largest part of the Y chromosome does not recombine with the X chromosome and thus inherits as a single haplotype (NRY). In the 1990s, the American medical geneticist Michael Hammer started characterising the different NRY haplotypes. At the time he distinguished ten main types, which were subdivided into various subtypes. Nowadays hundreds of different haplotypes are catalogued, allowing each Y chromosome that is found somewhere in the world to be allocated to a specific group. In The Netherlands, the team of Peter de Knijff (Leiden University Medical Centre) made important contributions to modern NRY haplotype systematics, its correlation with geography and languages, as well as the use of NRY haplotypes as a diagnostic tool in forensic science.

The haplotype tree developed by Hammer (fig. 5.11) clearly shows that the original haplotype (1A) is African and that haplotypes of men in other parts of the world are all derived from an African ancestor. Hammer's analysis therefore supports *Out of Africa*.

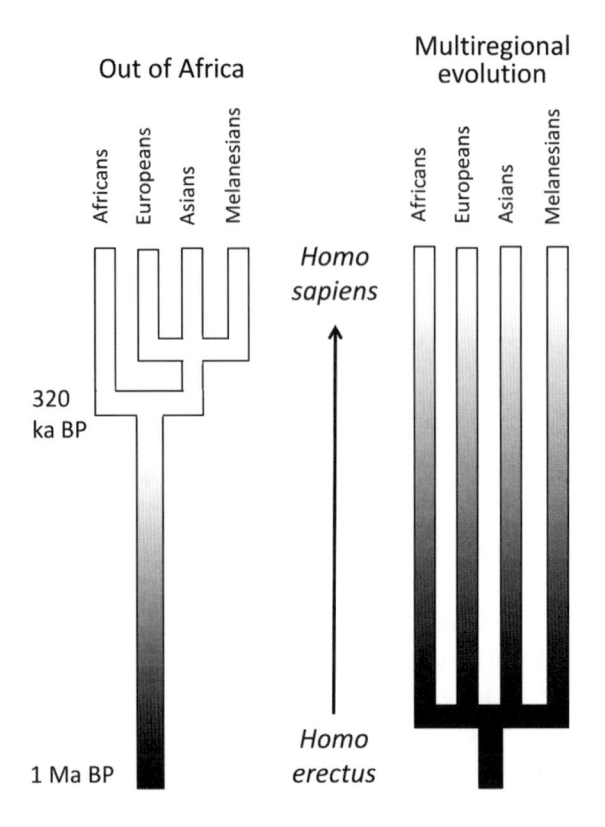

Fig. 5.10: Representation of the two conflicting theories that dominated the ideas about the origin of *Homo sapiens* for a long time: 'multiregional evolution' versus 'Out of Africa'.

In Hammer's original analysis the coalescence time for the human NRY was estimated as 147 ka. In other cases an even more recent date, 59,000 years, was estimated. Is it possible that the coalescent along the male lineage is younger than that of the female lineage? The difference in estimated coalescence times between mtDNA and NRY is attributed to a certain degree of polygyny in the first human populations. This made the effective population size of men smaller than that of women. As we saw in the previous section, a smaller effective population implies a shorter coalescence time. Also the degree of exchange between groups could have differed between men and women. It is assumed that coalescence estimations via the female line represent the real origin of *H. sapiens*.

That the 'cradle of mankind' is to be found in Africa is quite widely accepted now, but the timing of the origin has shifted. Following the publication by Cann several mitochondrial DNA analyses have been conducted and they all confirmed Cann's coalescence time of approximately 200 ka.

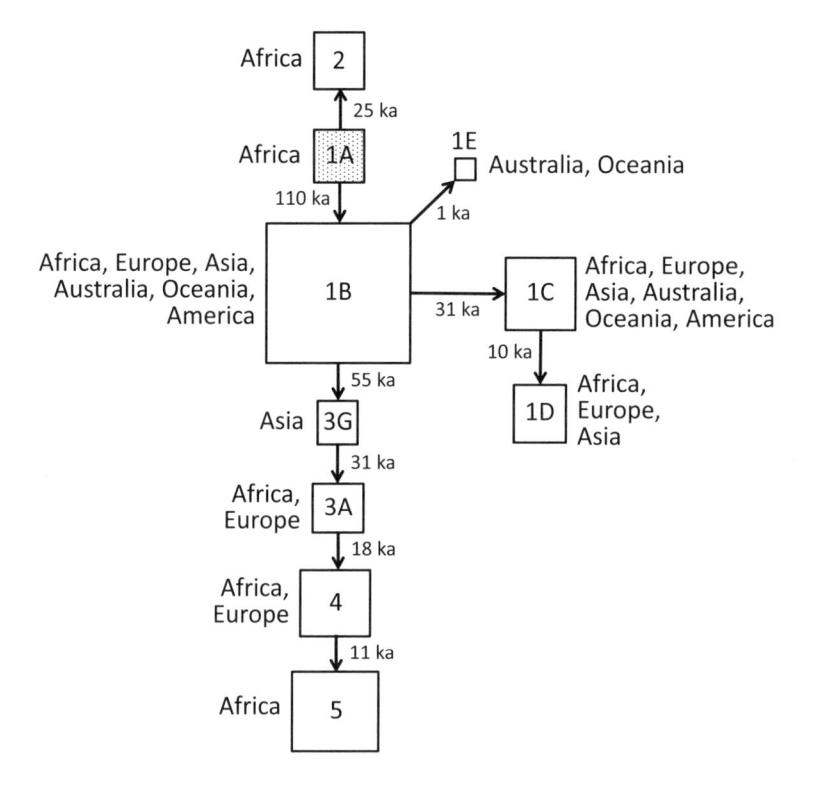

Fig. 5.11: Relationships between ten different haplotypes on the basis of nine polymorphic positions in the non-recombining part of the Y chromosome (NRY). The numbers show the original numbering of Michael Hammer, which is not being used any longer. The size of the squares measures the frequency; the arrows show the evolutionary relationships, while the years indicate the estimated times of the mutations. 1A is the ancestral haplotype. The distribution of haplotypes across the world supports the 'Out of Africa' model.

However, whether that time should be regarded as the dawn of *H. sapiens* has become less certain. A coalescence time indicates the start of an expansion, a demographic event, but it does not exclude that there might have been a period of slow population growth before that. In addition, we shouldn't forget that the earlier estimates of mutation rates were all derived from the calibration of genetic differences on divergences in the fossil record (*cf.* fig. 5.7). The next-generation sequencing methods that became available by about 2010 allowed DNA to be analysed on a much larger scale than was possible before.

A much lower mutation rate was estimated from large-scale genome sequencing of parents and offspring; this study, published in 2012 by scientists at the Sanger Institute in Cambridge, UK, pushed back the origin of *H. sapiens* to 250-300 ka BP. Another recent study by a Swedish group at Uppsala University, who compared ancient DNA isolated from South African remains, resulted in a time window for modern human divergence between 350 and 260 ka BP. In the light of recent fossil discoveries outside Africa, to be discussed in the next section, the lower estimate of this range is likely to be more close to the truth than the upper. Based on what we know at this point, we have in this book (chapter 1, fig. 1.14) tentatively set the emergence of *H. sapiens* at 320 ka BP.

The genetic studies have evidently put an end to the discussion about multiregional evolution versus *Out of Africa* in favour of *Out of Africa*. However, at the same time the new methods of DNA analysis have pushed back the date for human origins far deeper in history than originally proposed in the Cann paper.

Migrations in all directions

The scenario for the migrations of *H. sapiens* across the world is largely based on DNA analysis of people living today in different geographical regions. This approach assumes that the history of human migrations can be derived using reconstruction methods such as coalescence analyses. The trick is to find DNA samples of people who still have an original signature, meaning ancestry without influences of other ethnic groups. Both in America and Australia it is difficult to find such samples because of the significant influence of the (later) European DNA. In addition, as a result of the enormous increase of international trade and touristic travelling, the DNA of the entire world population is gradually becoming one large gene pool. One example of this is the population of Brazil, in whom a new phenotype seems to have emerged, one that contains influences of all kinds of ethnic backgrounds that can no longer be traced. The reconstruction approach is reaching its limits in such mixed populations.

Since about 2010 the classical method has been supplemented by *ancient DNA* analysis. Rather than trying to reconstruct events in the past from present variability, it has become feasible to analyse DNA samples from human remains and fossils on a large scale. As put into words by David Reich, one of the leading scientists in the field and head of the ancient DNA lab at Harvard Medical School, 'ancient DNA discoveries opened the floodgates, producing a torrent of findings that have disrupted many of the comfortable understandings that we had before'. This especially concerns the recent history of the human species, since about 20 ka BP, which is traditionally described on the basis of artefacts such as pots, tools, and decorations, but is now supplemented with an extremely rich source of new data on the DNA of the people who actually made and used these materials. A recurrent theme in ancient DNA research is the fact that hardly any people that live in a place today were living there in the past. Almost all peoples in the world are descendants of those who experienced mass migrations, accompanied by significant replacements and genetic mixing. No people in the world can boast of the 'purity' of their heritable material.

In this section we will focus on the prehistoric human migrations before 10 ka, covering the period from the emergence of *H. sapiens* up to the Neolithic transition. This period covers about 97% of our evolutionary existence as a species, so it may be considered responsible for many of our present-day genetic features.

As we saw in the previous section, the migration pattern following from genetic analysis is basically an *Out of Africa* scenario, but modified in many details. We call it the '*Adjusted Out of Africa*' model (fig. 5.12). Starting from an origin of *H. sapiens* in East Africa around 320 ka BP, at first a distribution within Africa took place. The fossils found in 2017 at Jebel-Irhoud, Morocco, suggest that the geographical area of early human evolution was not limited to the Great Rift Valley, but included large parts of Africa and maybe even the nearby parts of Arabia. The first out of Africa migration must have been very early, about 180 ka BP. This is based on the discovery in 2018 of a fossil partial maxilla undoubtedly belonging to *H. sapiens* and found in the Misliya cave near Mount Carmel in Israel. Whether these migrants went any further than the Middle East is uncertain; maybe the humans from the African continent just expanded their territory to the Middle East every now and then.

The second wave of migration to the Middle East took place about 120,000 years ago. Firstly, a group branched off who are the ancestors of the current (original) Australians and a part of the Melanesians. This acknowledges the evidence for the presence of humans in northern Australia 65,000 years ago. Indeed, recent DNA analyses conducted on a hair sample dating before the time of European arrival in Australia showed that the Australian DNA is more in line with African DNA than with Asian DNA (fig. 5.13). In addition, it is very well possible that fossils found in 2015 in the Fuyan cave near Daoxian, in south-east China, also belong to the early Asian migrants. The latter finds include 47 teeth and molars with ages between 80,000 and 120,000 years. The early colonisation of Australia was followed by a second one, starting around 40 ka BP, which replaced the previous migrants and gave rise to the present native Australian peoples.

On the basis of geographical proximity and the IBD model, one would expect Australians to be related to Chinese. The dark skin colour of the

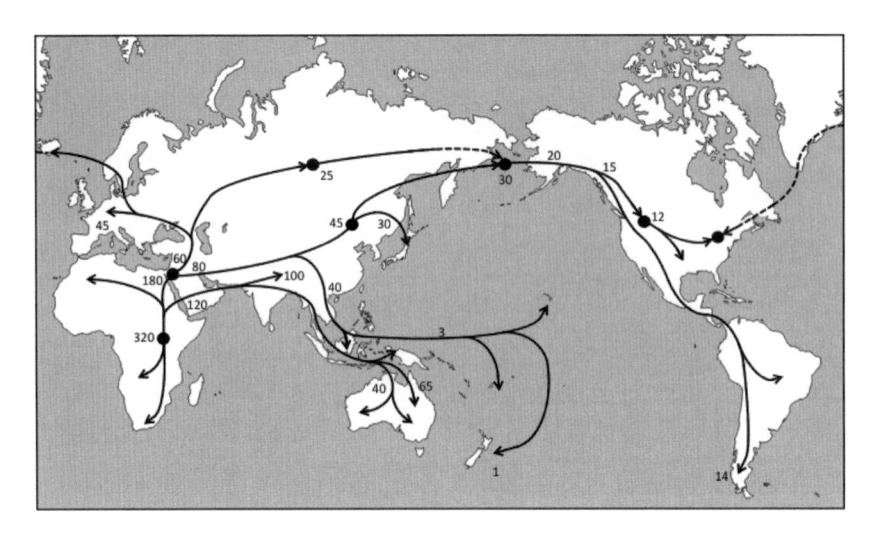

Fig. 5.12: According to the adjusted 'Out of Africa' model, *Homo sapiens* originated about 320 ka BP in Africa and had colonised Africa and the Middle East by 180 ka BP. Australia was first colonised by about 65 ka BP and this could have been due to an early migration that also contributed to human remains in China, aged 100 ka. A second migration took place around 80,000 years ago from the Middle East; one group migrating to Central and Eastern Asia, penetrating the Malay Archipelago and Melanesia, another one taking a northernly route, colonising the European continent and northern Asia by about 40 ka BP. Beringia was colonised by Central Asian peoples but received a significant North Asian input. North and South America were then colonised commencing about 20 ka BP from Beringia. The latter migrations proceeded very quickly and gave rise to a number of Native American populations that remained isolated from each other for a long time. In Europe, Siberia and Asia *H. sapiens* interbred with ancient hominins that lived in these places already (Neanderthals, Denisovans). The numbers indicate thousands of years before present.

Australians, corresponding with the skin of Africans, would then be a *convergence*; they would be two separate adaptations to a sunny climate. This interpretation is still frequently found in evolutionary textbooks. But the genetic comparison of Europeans, Africans (Yoruba), Chinese (Han) and Australians clearly puts the Australians with the Africans, not with the Han Chinese (fig. 5.13). The similarity in skin colour between Australians and Africans must be attributed to common ancestry. Most likely, the first humans that colonised Europe and Central Asia were also still dark-skinned. Skin lightening is a more recent phenomenon (starting only 11-19 ka BP), which occurred independently in Europeans and Chinese. Molecular analysis has shown that different alleles of the pigmentation genes *SLC24A5*, *MC1R*, *KITLG* and *OCA2* underlie the convergent skin lightening in northern Asia and Europe, as adaptations to reduced UV radiation at higher latitudes.

More detailed analysis of native Australian and Papuan DNA samples has revealed that both groups descend from the same African migration

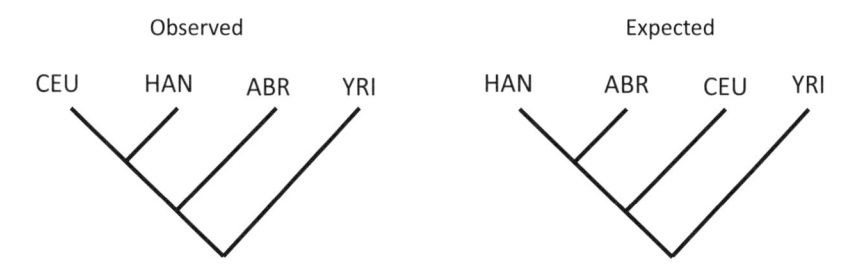

Fig. 5.13: Genetic analysis shows that the original inhabitants of Australia are directly related to Africans (left). Based on geographic proximity, a sister relationship with the Chinese was expected (right). CEU = Central Europeans, HAN = Han Chinese, ABR = Aboriginals (Australia), YRI = Yoruba (Africa).

but split between 25 and 40 ka BP. Subsequently, the Australian branch split in a lineage that colonised north-east Australia and another one that colonised south-western Australia. Remarkably, these two groups of native Australians hardly mixed with each other but remained separated up to modern times, which in some way matches with the importance attached to ecosystems and the landscape that can be found in Aboriginal cultures.

Meanwhile in the Middle East other groups of *H. sapiens* migrated to Europe and Asia. The exact route towards Europe is unknown, but usually it is expected that first a northward movement took place, to the current Armenia and further towards Russia, and that western Europe was colonised from the east. We know for a fact that *H. sapiens* had definitely established in western Europe around 40,000 years ago; we call them the Cro-Magnon Man (see chapter 6).

Around the same time we see the first *H. sapiens* appearing in China. From there, the various parts of the Asian continent were colonised, including the Eastern Islands, Indochina and the Malay Archipelago. The colonisation of the Japanese islands happened relatively late, not sooner than 30,000 years ago. Another group of Asians headed north and colonised Beringia, the area at the level of the current Bering Strait, which was a fertile steppe with a moderate climate 20,000 years ago. Because the sea level was lower back then, Beringia was dry land.

DNA analysis has demonstrated unambiguously that the original inhabitants of North and South America are descendants of a north Asian (Mongolian) group. The best-known archaeological proof for *H. sapiens* in America is based on stone tools that were found near the town of Clovis. The *Clovis culture*, with characteristic pointed hand axes, is approximately 11,500 years old. Pre-Clovis artefacts of 13-15 ka old were reported in from

Buttermilk Creek, in Texas. Suggestions for pre-Clovis presence of humans is also coming from Monte Verde, a coastal site in Chile, which dates back to 14 ka BP. Whether a coastal migration pathway, from Beringia to the south, all along the Pacific, should be considered a separate arrival route of humans in America is disputable because it is not in line with reconstructions based on modern and ancient DNA.

The oldest North American fossil from which DNA was retrieved is a boy from a Clovis excavation, 10,500 years old. Another one is the so-called *Kennewick Man*, a skeleton of 9,000 years old. Both genomes have the greatest similarity to modern Native American DNA. The sequencing of genomes from ancient Native Americans created a lot of debate in the United States, because native tribes demanded that the remains be left alone and thought it was sacrilege to grind the bones of their ancestors. A U.S. act stipulates that remains should be returned to Native American tribes whenever there is evidence of a cultural or biological connection to present-day peoples. Independent of that connection, the critical issue is that collecting samples from old graves without consulting the peoples who might be related to the people buried should be considered unethical. In the case of Kennewick Man, after much debate, the skeleton was released for research and a publication from 2015 proved that he was indeed related to a Native American population group.

In summary, the '*Out of Beringia*' scenario runs more or less as follows: Around 36 ka BP a 'Mongolian' group separated from the ancestral Asian population of *H. sapiens* and moved to Beringia. They lived there for several thousand years, up to 25 ka BP, while maintaining a high level of gene flow from their Asian ancestor, plus a significant introgression from Siberian populations. The latter is evident from a northern European signature found in all Native American DNA, which matches the DNA of a fossil from Mal'ta, a cave near Lake Baikal in eastern Siberia, giving the Beringians a dual ancestry. Due to changes in climate the Beringians became isolated and they split into two lineages around 21 ka BP, one remaining in Beringia (and finally going extinct), the other one colonising America. Around 15 ka these migrants again split into two populations, one staying in the north of the continent, the other one migrating to the south. Subsequently, the northern populations quite extensively mixed with indigenous populations from Greenland and eastern Siberia, that themselves descended from a north Eurasian ancestry, separate from the East Asian ancestors of the Native Americans. This scenario is depicted in fig. 5.14.

The genomic data match with the analysis of Native American languages by the American linguist Joseph Greenberg (1915-2001), who had argued

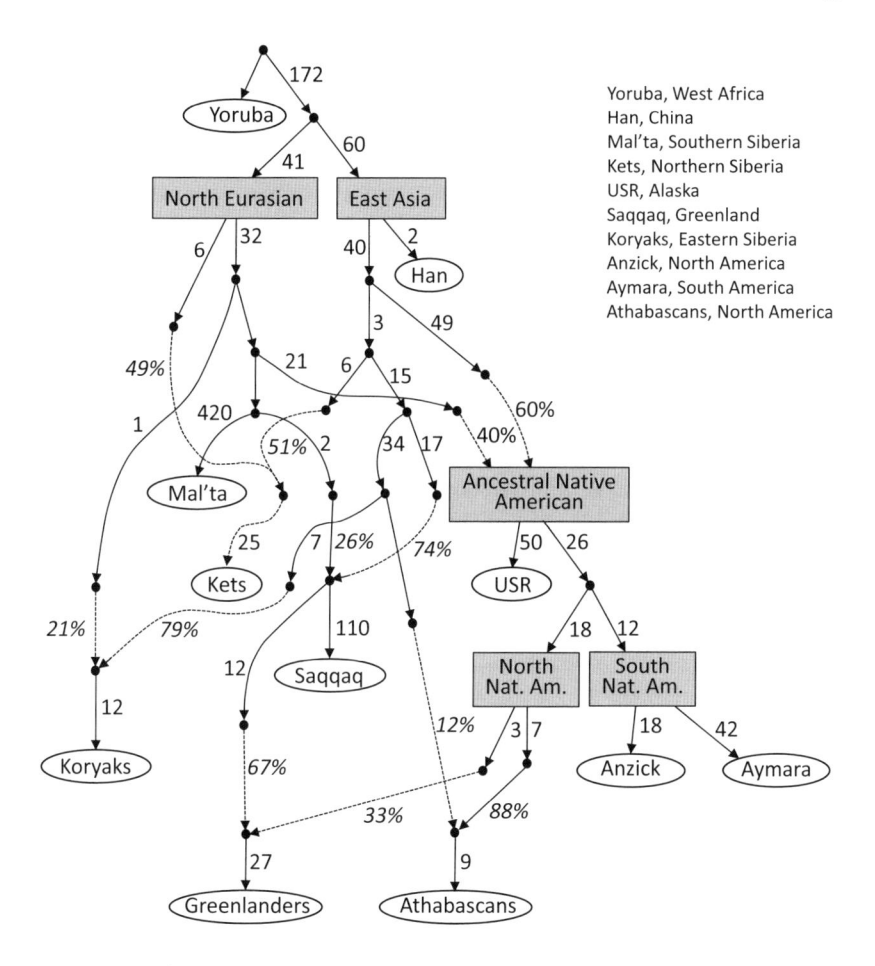

Fig. 5.14: Model for genealogy and genetic mixing underlying the different Native American lineages, in relation to Yoruba (Africa), Han (China) and Eurasian origins. The results are consistent with a model in which America was colonised from Beringia by a single-founder population ('*Ancestral Native Americans*'), which quickly split in North American and South American founder populations. The numbers along the solid lines indicate F_{ST} values (x 1000). Genetic mixing is indicated by dashed lines and percentages indicate relative contributions. The data are based on both ancient and modern DNA; the geographic areas of the various samples are listed in the upper-right panel.

that there are three main language groups in America that are separated by very deep splits. The genomic scenario is in accordance with this since after the very rapid colonisation of the Americas the various tribes lived in the same place for a long time; every language finds its root in the same founding population that peopled America.

Hybridisations between ancient humans

In May 2010 evolutionary biologists suffered the severest shock since the publication of the *Out of Africa* article by Rebecca Cann: the human genome appeared to contain traces of Neanderthal DNA. For years, scientists, including the authors of this book, had insisted that human and Neanderthal were two separate species that were reproductively isolated from each other. This statement was crystal clear, and was not only supported by the fact that the morphologies consistently differ at numerous points (see fig. 1.12), but also from DNA data. DNA sequences from Neanderthal fossils had been known for a number of years. This concerned parts of the mitochondrial genome. MtDNA is better preserved than nuclear DNA because the mitochondrial double membrane protects it from degradation. In addition, a large number of mitochondria appear in every cell. Phylogenetic analysis of those sequences clearly demonstrated that they fell apart into two monophyletic groups: human and Neanderthal. A human sequence was never found to cluster in the midst of Neanderthal sequences.

But due to technological innovations DNA sequencing from fossils (ancient DNA or *aDNA*) has spectacularly advanced since 2005. One was no longer dependent on mtDNA; nuclear DNA could also be studied. The isolation of aDNA takes place in an extremely clean work environment, to prevent contamination with human DNA. This is particularly important because ancient hominin DNA so strongly resembles modern human DNA. All scientists working in the lab must also undergo a DNA analysis to make sure that the sequences found are not theirs. In addition, it appears that 95% to 99% of the DNA isolated from fossil material is not from the fossil itself, but from bacteria that grew on the body after death or during the fossilisation process. This DNA can be removed by use of carefully selected enzymes that cut only the bacterial DNA. Furthermore, special methods have been elaborated to correct the degradation of DNA, the deamination of cytosines, in particular. All in all, the sequencing of aDNA is extremely specialised work with numerous checkpoints that only a few laboratories in the world can reliably perform. The researchers working under the Swedish scientist Svante Pääbo at the Max Planck Institute for Evolutionary Anthropology in Leipzig, Germany, are considered the founders of *palaeogenetics*, but nowadays several other laboratories, in Germany (Johannes Krause), the United States (David Reich), Denmark (Eske Willerslev), and Sweden (Pontus Skoglund), are part of the ancient DNA community and the high throughput sequencing of ancient DNA has led to many discoveries in the past years.

The success of aDNA isolation strongly depends on the way the fossil is preserved. Drying is very harmful to DNA. The best material is frozen; that is why the most complete ancient DNA samples were obtained from fossils found in Siberia. It is very difficult to isolate DNA from tropical fossils. For example, so far nobody has managed to retrieve DNA from *H. floresiensis*, although this is a relatively recent fossil (60 ka old, see chapter 1). The record for sequencing ancient DNA was 430,000 years in 2016. It is not expected that we will get much further than 1 Ma, because older fossils contain very little intact DNA. That means, for the moment only *H. sapiens, H. neanderthalensis* and the Denisovans are accessible to ancient DNA research, although DNA results from Heidelberg Man, *Homo naledi* or even late *H. erectus* are not to be excluded in the near future.

Back in 2010, Pääbo estimated the proportion of Neanderthal DNA in human DNA based on a statistical comparison of SNPs (see chapter 4 for a clarification on SNPs). Due to the historical importance, we reproduce the approach here, in fig. 5.15. In the genome, you look for all polymorphic SNP positions at which Neanderthal has an allele different from the chimpanzee. Those positions have changed in the lineage towards human and Neanderthal. Then you check what allele Europeans and Africans have at those positions. In all cases in which these two differ from each other, there should be no preference for one or the other allele. In practice there appears to be a slight bias: positions for which the Europeans correspond with Neanderthals occur more often than positions in which Europeans correspond with Africans, their ancestors. The ratio, known as the D-statistic,

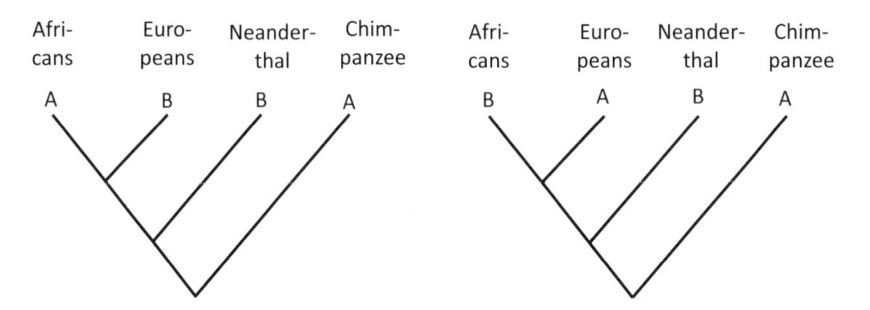

fig. 5.15: Diagram of a SNP analysis applied by the group of Svante Pääbo, which for the first time indicated that Neanderthal DNA entered the human genome via hybridisation. Examine all polymorphic positions where Neanderthal has an allele different from that of chimpanzee (B instead of A). Which allele do Europeans and Africans have in that position? You would expect BA and AB to occur equally often, but ABBA occurs more often than BABA. The relative overrepresentation of ABBA over BABA is a measure for the frequency of (horizontally obtained) Neanderthal DNA in the European genome. The statistical test is called D-test or (in advanced form) a 'four-population test'.

was estimated to 1.01 to 1.04. Undoubtedly this is caused by crossings between Neanderthals and humans.

So, from this method it appears that between 1 and 4% of the human genome derives from hybridisation with a Neanderthal. Later the percentages were revised slightly. The final percentage seems to be somewhat higher for Asians than for Europeans (1.38%, versus 1.15% for Europeans). Africans have virtually no Neanderthal DNA (< 0.1%). The hybridisation is dated at 60,000 years ago and thus took place when a group of people left Africa and met Neanderthals in the Middle East. Since there is also a Neanderthal signature in native Australians, the group that migrated to colonize Australia descends from the interbreeding Middle East group also.

Initially, it was believed that the Neanderthal-human hybridisation was a unique event, limited to the Middle East. Recent analyses of fossil *H. sapiens* from Siberia and Romania, however, indicate that in some human fossils higher percentages of Neanderthal DNA are present. In a fossil from Romania up to 9.4% Neanderthal DNA was found, which, in addition, was present in large segments. The distribution across the genome indicated a hybridisation that must have taken place within six generations before the person died. Since the fossil is about 40,000 years old, this must have been an event different from the one in the Middle East. It seems as though crossbreeding between humans and Neanderthals frequently occurred, but was followed by selection against the hybrids, causing the percentage of Neanderthal DNA to decline with the generations. The relative lack of Neanderthal DNA in X chromosomal regions encoding proteins involved in male fertility suggests that there was selection against male hybrids that carried Neanderthal genes. This is supported by a near-linear decline of the percentage of Neanderthal DNA in ancient genomes of different ages.

Another fascinating fact is that the Neanderthal signature is only apparent from nuclear DNA, not from mitochondrial DNA. The mtDNA of humans is always from the *H. sapiens* type. That was also the reason why we believed for so long that Neanderthals and humans were reproductively isolated: all that was available was mtDNA. It is unclear at the moment why mtDNA should tell another story than nDNA. In addition, the mtDNA of the oldest Neanderthals, excavated at the Sima de los Huesos location in Spain, is of a different type. One possibility suggested in the recent literature is that there was, at a certain moment an introgression into Neanderthals from *H. sapiens*, by which the old mitochondrial haplotypes were replaced. That should have happened between 413 and 218 ka BP, so long before the presently assumed first *Out of Africa* migration (180 ka BP). How this fits into the

African origin of *H. sapiens* and the '*Adjusted Out of Africa*' scenario is not yet clear at the moment.

The year 2010 brought a second discovery. In a cave in Siberia, in the Altai Mountains, a fossil phalange was found, thought to be a Neanderthal's. A proper DNA sample was isolated from that phalange, whereupon the mtDNA was sequenced. It appeared to differ substantially from both Neanderthals and *H. sapiens*. Therefore it had to be a new species, which was given the provisional name '*Denisovan*', after the cave. Initially, based on the mtDNA, Denisova Man was positioned as a sister species of both humans and Neanderthal, as a very early branching from Heidelberg Man. But when the nuclear DNA was sequenced, it turned out that Denisova Man is more related to Neanderthal than to humans; it had to be considered a sister species of Neanderthal only. Again, the reason why mtDNA tells a different story than nuclear DNA is unknown. We already met '*Homo denisovae*' in chapter 1 and based on the nuclear DNA analysis gave it a place in the hominin tree next to Neanderthal.

Further, aDNA analysis has confirmed this reconstruction but also pushed back the splits deeper in time. A sample from a very old Neanderthal fossil from Sima de los Huesos, a cave in Spain, dated the split between Neanderthals and Denisovans to 470-380 ka BP. The divergence between these two species and humans was also pushed back in time, to 750-550 ka BP. This implies that *H. heidelbergensis*, after branching from the Neanderthal-Denisovan lineage, lived on in Africa for a considerable time, at least until the origin of *H. sapiens* (320 ka BP). This interpretation was also discussed in chapter 1 (fig. 1.14).

To complicate things further, a new analysis of the Neanderthal and Denisova DNA from Altai published in 2016 showed that that *H. sapiens* DNA is present in Altai Neanderthals, but not in European Neanderthals and not in Denisovans. This introgression possibly dates from 100,000 years ago and reportedly also took place in the Middle East. In addition, there is Denisovan DNA in *H. sapiens* living in Melanesia and Papua New Guinea, even more than Neanderthal DNA (around 4-6%, although some estimates are lower). This could be due to a hybridisation dated around 45,000 years ago when *H. sapiens* was on its way to South-East Asia. It also indicates that the distribution range of Denisova Man was not limited to the cave where it was first found, but included large parts of Asia. The Melanesian peoples thus have a double heritage of hybridisation: from Neanderthals and from Denisovans. Finally, hybridisations also took place between Denisovans and Neanderthal. The first sequenced Denisovan genome appeared to have a small percentage of Neanderthal DNA, but in 2018 Pääbo's group published a genome sequence isolated from a bone fragment found in

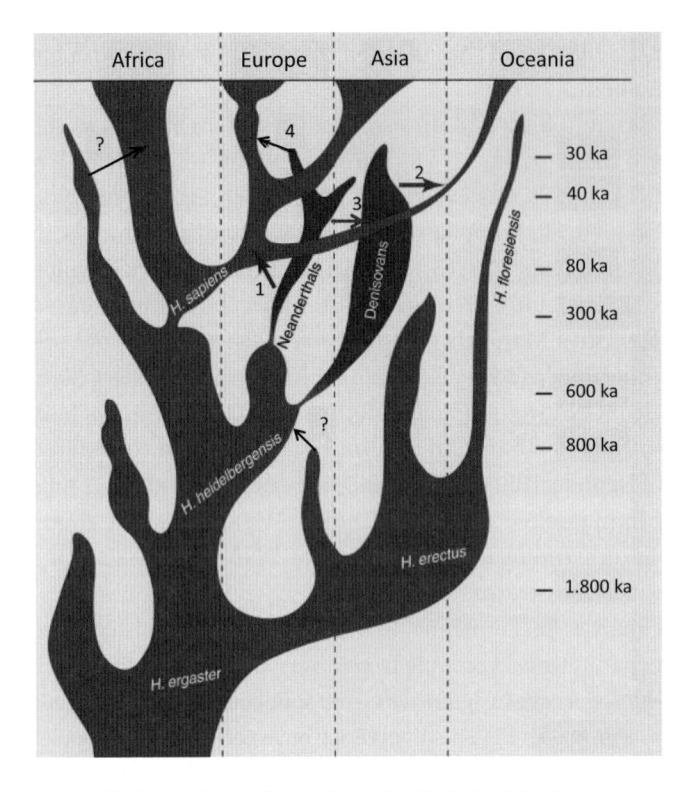

Fig. 5.16: Phylogenetic tree of recent hominins displaying hybridisations between Neanderthals and *H. sapiens* in the Middle East (1), between Denisovans and *H. sapiens* in South-East Asia (2), between Neanderthals and Denisovans in Asia (3), and between Neanderthals and *H. sapiens* in Europe (4). In addition, there are indications for two more hybridisations (indicated with question marks).

Denisova Cave, 50,000 years old, which turned out to be a direct offspring from a Neanderthal mother and a Denisovan father.

The complex pattern of hybridisations between old hominins is depicted in fig. 5.16. It shows that recent human species (at least *H. sapiens* and Neanderthal), despite differing so clearly morphologically, are mutually compatible in terms of reproduction. This is a fascinating fact. According to the definition of a biological species, humans and Neanderthals should therefore not be separated: they should be classified as a single species because they crossbred and produced fertile offspring. Due to the clearly distinguishable morphological characteristics, however, most scientists prefer to continue to distinguish the two. Apparently, the morphology provides insufficient leads to characterise the species. If we extend this to the entire hominin phylogeny, many more species may in fact need to be considered together, as suggested by the scientists from Dmanisi, who considered a

Table 5.3:   Genes with alleles of proven Neanderthal or Denisovan origin in the genome of *Homo* sapiens

Gene	Allele/haplotype	Possible functional significance
Microcephalin	Haplogroup D	Brain development
Dystrophin	Haplotype B006	Structure of muscular tissue
STAT2	Haplotype N	Communication in the immune system
HLA Class I	Alleles B*06, B*51, C*7:02 and C*16:02	Immunity
BNC, POU2F3	Apomorphic haplotypes	Skin colour, skin structure
EPAS1	Allele with 5 SNP motive	Viscosity of blood in case of low oxygen tension

morphologically very diverse collection of skulls to be one species, *H. erectus* (see chapter 1). Some leading palaeontologists, such as Ian Tattersall of the American Museum of Natural History, find this idea abhorrent.

In 2010 it was still questionable whether the different hybridisations had contributed to functional genes in the human genome. Now we are certain that this is the case. Neanderthal genes in Europeans and Asians prove to be distributed across the entire genome, except the X chromosome. The diverse genes represent functions in the immune system, the skin and the brain (table 5.3). Interestingly, many medically relevant phenotypes are correlated with Neanderthal alleles, such as neurological, psychiatric, dermatological and immunological phenotypes. For example, a significant association was found with the tendency to develop depression. However, the effects explained by these associations, although statistically significant, are quite small.

There are also cases in which the Neanderthal version of a gene was specifically advantageous to humans. A beautiful demonstration applies to *EPAS1*, a gene involved in the regulation of blood viscosity. The current Tibetans have a rather deviating variant of this gene, which gives them an advantage when living at great heights. It is known that the body produces more erythrocytes in the case of low oxygen pressure, leading to an undesired increase of the viscosity of the blood. The Tibetan allele prevents the increase of viscosity.

In 2014, a Spanish-Chinese research group found out that the Tibetan *EPAS1* allele is very similar to the allele of Denisova Man. Probably an Asian population group obtained this allele via introgression from Denisova Man, after it was strongly positively selected when these people started living high in the mountains. So far this is the strongest evidence for the claim that introgression of Neanderthal DNA has enhanced the evolutionary potential of *H. sapiens*.

With the study of ancient DNA, we're just at the beginning. Within the coming few years, we are bound to discover a lot more.

6. The cultural human

Does the reach of evolution extend to cultural expressions? Can cultural traditions, such as music, dance, visual art, clothing, language, literature and religion, be seen as human behaviours that can be explained as adaptations to the environment? In this chapter we will find that answering these questions will introduce new evolutionary concepts, such as cultural evolution, group selection and kin selection. But first we will go back to the origin of human culture in prehistory.

Prehistoric tools and cave drawings

All kinds of animals change their habitat and create structures, such as nests, bowers, dams or holes. Such structures can sometimes be considered artefacts, as they didn't come about through abiotic processes. But commonly the term 'artefact' is reserved for objects created by humans. Such objects fundamentally differ from the structures made by animals. Typical for a human artefact is that an object from the environment is changed in accordance with a predetermined plan. This method uniquely applies to human beings. Even chimpanzees are not able to turn flint into stone tools (fig. 6.1). They are basically able to perform the actions to get there and they use stones to crack nuts, but they do not readily change objects in their environment with the aim to use them as a tool. It seems as though they lack an image of the object to be made.

When did humans start creating artefacts? Old stone tools were first found at Olduvai Gorge in the north of Tanzania. These tools, described in the 1930s by Mary and Louis Leakey, are simple stones that were processed on one side, hence they are called *unifacial*. Apart from Tanzania, they were also discovered during excavations in Ethiopia and Kenya, for instance, at Gona, Hadar, Omo and Lokalalei. The oldest objects are 2.6 million years old. This collection of artefacts is called the *Oldowan culture* (fig. 6.2).

Little is known about the exact use of such artefacts. Possibly the sharp side of the stone was used to chop plants, bones or other objects. But not only the core stone itself, also the flakes that were cut from the cores using a hammer stone might have been used, especially for cutting through meat or hide. This would imply that the hominins who used these tools brought a hammer stone and a core with them to make flakes on the spot where they were required.

Fig. 6.1: Chimpanzees use stones to crack nuts, but will never spontaneously process stones to create tools.

The Oldowan culture is traditionally attributed to *Homo habilis*. Leakey classified the fossils that he found with the stone tools into the genus *Homo*, as he believed that making tools marked the moment when hominins ceased to be ape-men. Over the last few years, however, this interpretation is disputed, as we already saw in chapter 1. Currently other stone tools are known which are older than *H. habilis*, 3.3 million years, originating from West Turkana, in Kenya. These so-called Lomekwi stones are larger and more primitive than the Oldowan culture. They appear to have been used mainly to break things and are therefore called '*knappers*'. Dating from approximately the same era are traces of scraping and chopping that were found on bones of bovine animals in the Dikika excavation in Ethiopia. These marks seem to be the result of attempts to use stone tools for separating meat from bone to get access to the marrow.

Still, as yet, there is only fragmentary proof of stone tools older than the Oldowan culture (2.5 million years). If indeed there were pre-Oldowan tools, artefacts would not be limited to *Homo*, but could also have been used by *Australopithecus*. The hand of *Australopithecus* certainly allows this. Analysis of hand bones shows that the long, opposable thumb typical

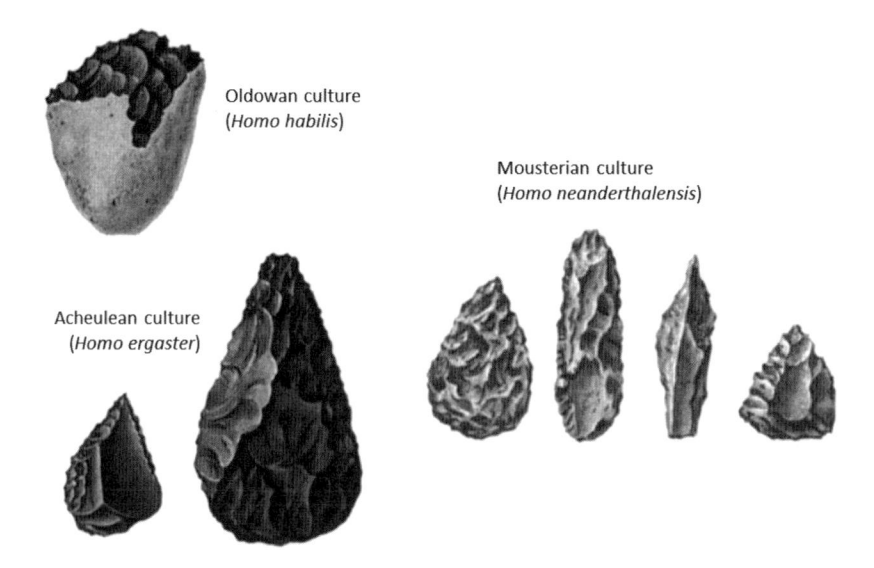

Fig. 6.2: The three most important cultures of stone tools from the Lower and Middle Palaeolithic, indicating the hominin species that made them.

for *Homo*, which allows the renowned forceps-like grasp of the fingers, already existed with *A. afarensis*. The long thumb should not be regarded as an adaptation to tool use because it predated systematic toolmaking. The shape of the human hand is rather an exaptation, a structure that already existed in *Australopithecus* and was co-opted in *H. habilis* for tool use.

From approximately 1.5 million years ago, we notice the appearance of an entirely new type of stone artefacts in Africa. These tools have the shape of a typical hand axe, with a round and a pointed side, and are processed on two sides, hence they are called *bifacial*. They are larger than the Oldowan stones. They were described in the nineteenth century for the first time after they were found during excavations near the town of Saint Acheul, close to Amiens, in France. Later they were also found in Africa and the Middle East. The Acheulean culture is named after the original site (fig. 6.2).

The Acheulean hand axes were clearly 'invented' in Africa and Europe, by *Homo ergaster*, but were also used by *H. erectus* in Asia and perhaps also by *H. heidelbergensis* and the first Neanderthals. Oddly enough, the culture has not always travelled along with the species itself. The American archaeologist M.L. Movius drew a line over eastern India, east of which, according to him, no genuine Acheulean hand axes are found, but only the more primitive artefacts, similar to the Oldowan culture. The *Movius line* is explained by assuming that the migration of *H. erectus* to Asia preceded the invention of the new stone tools. The Acheulean culture was invented later in Africa and passed on, by cultural transmission, eastward, but ended in India. Another explanation is that the technique of *H. erectus* did not last in Asia and that people proceeded to use bamboo, material that leaves no fossils. Recent data from Attirampakkam, an archaeological site in south-east India, suggests that the Acheulean culture in India went through a transition at around 385 ka BP, to smaller and finer stones that looked more like the *H. sapiens* tools found in Africa and Europe at the time. However, since there is no evidence for the presence of *H. sapiens* on the Indian continent that old this transition must be due to local innovation.

A remarkable aspect of the Acheulean culture is that the hand axes, at least the African and European ones, seem almost industrially manufactured: they show a high degree of mutual resemblance. For more than a million years the shape hardly changed. Perhaps the tools were made by certain members of a tribe, craftsmen, who had a workshop and passed on the technique from parent to child. The Acheulean hand axe probably served several goals, and is therefore called the 'Swiss Army knife of prehistory'.

With the arrival of the Neanderthal and *H. sapiens*, the spectrum of artefacts changed significantly. The classical Neanderthal culture is the *Mousterian* (fig. 6.2), named after the town of Moustier in the Dordogne Valley, France. It is a culture that can be viewed as a refinement of the heavy Acheulean hand axes. Stones were further processed, cut on multiple sides and polished. A well-known example is the *Levallois technique*, a method in which flakes are chopped off from a prepared core; those flakes were then further processed into refined scrapers, knives, arrowheads and the like. The core itself is a flat structure with sharp edges all around, as a result of the processing. The Mousterian culture can be found throughout the Neanderthal distribution range, from Iran to Spain and, to the north, up to Siberia. Tools of this culture has also been found in the Netherlands (fig. 6.2).

The oldest culture that can undoubtedly be attributed to *H. sapiens* is the Aurignacian (40,000-25,000 years ago). This consists of so-called *microliths*, thin arrow-shaped stone objects that could be used for meticulous cutting, but also to pierce hides and clothing. The variety of tools and other artefacts strongly increased; each location and period had its own style. In addition, in the case of *H. sapiens*, we notice the usage of materials other than rock, such as bone and ivory. Fig. 6.3 gives an impression of the diversity of *H. sapiens* artefacts, collected in the excavation of Geissenklösterle, in southern Germany.

Between the Mousterian of the Neanderthals and the Aurignacien of *Homo sapiens*, over a short period of time, 45,000-40,000 years ago, we find a culture that is difficult to classify, the *Châtelperronian*, discovered during

THE CULTURAL HUMAN 187

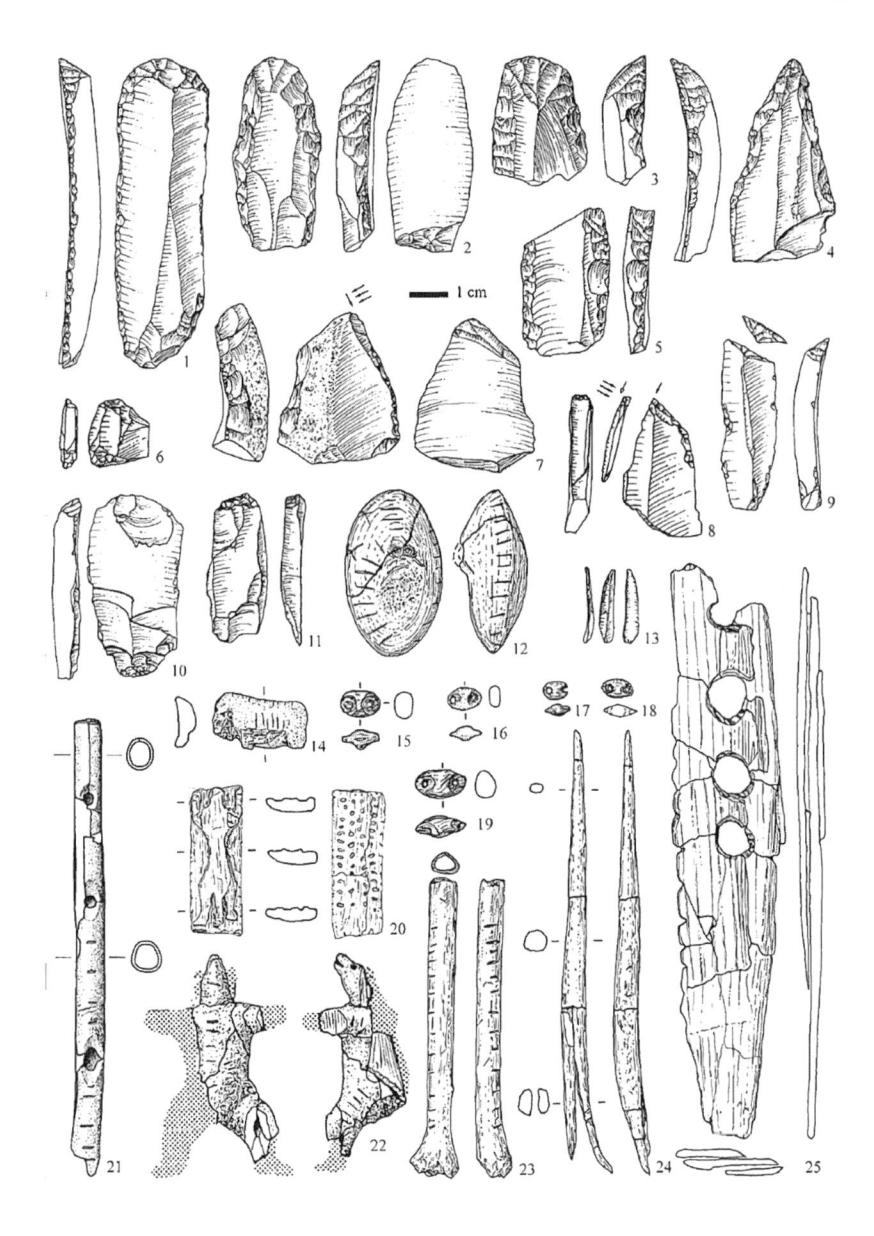

Fig. 6.3: Illustration of the diversity of stone tools, utensils and decorations (scraping stones, cutting tools, flutes, sculptures, beads) that were discovered in the Aurignacian (Upper Palaeolithic) of Geissenklösterle, a cave near Blaubüren in southern Germany.

excavations in France and northern Italy. This culture is characterised by a processing method that differs from the Mousterian as well as the Aurignacien. The collection of artefacts, named after the first finding place,

Châtelperron, consists of typical, somewhat curved and serrated stones with a sharp and a blunt side. There has always been a lot of discussion about their makers. As the culture was initially found in association with Neanderthal fossils, scientists believed that this was a Neanderthal culture, influenced by *H. sapiens*. The last Neanderthals would have learned new techniques from modern humans and drastically innovated their Mousterian culture. Châtelperronian also includes ivory items with a hole in them, probably used for decoration. Such decorations, indicative of symbolic behaviour, would shed a new light on the cognitive abilities of the Neanderthal. The association with Neanderthals, however, became uncertain after scientists showed, in 2010, that the stratigraphy of the 'Grotte de Rennes' in Arcy-sur-Cure and of the St. Césaire cave in La Roche à Pierrot, where the culture was found, had been subject to disruption during the first excavations in the nineteenth century. As a result, the Neanderthal fossils ended up in the wrong layer. There is now a tendency to attribute the Châtelperronian to the first *H. sapiens*.

The English archaeologist Grahame Clark classified the successive cultures of stone tools in four stages, or modes (table 6.1). The classification of the 'stone age' is of course based on the stone tools, but we should realise that the succession of cultures did not take place at the same pace everywhere. There were peoples who still made use of stones at the time of Columbus, while at the same time in other places in the world, the use of bronze and iron had been common for a while. Table 6.2 shows a simplified classification of the 'stone age', largely based on the cultures in Africa and Europe.

In addition to the artefacts themselves, and the layers they were found in, also the spatial distribution of materials used by people is important,

Table 6.1: Clark's classification for the different cultures of stone tools during the 'stone age'

Class	Examples of culture	Typical elements	Attributed to
Mode 1	Lomekwi, Oldowan	Simple hammer stones and chopping stones, processed on one side	*Australopithecus?* *H. habilis*
Mode 2	Acheulean	Hand axes, processed on multiple sides, ending in a point, 'industrial' manufacture	*H. ergaster, H. erectus, H. heidelbergensis*
Mode 3	Mousterian	Refined structures, knives and scrapers, sharpened stones, processed on each side	*H. neanderthalensis, H. sapiens*
Mode 4	Châtelperronian, Aurignacian	Microliths, arrowheads, polished stones, great variety in shapes	*H. sapiens*

because it provides insight in the lifestyle of the ancient hominins. As an example, fig. 6.4 shows the spatial arrangement of mammoth bones and fireplaces, as they were found in an excavation in Moldavia. These remains are attributed to *H. sapiens*. We notice that there are multiple fireplaces, suggesting that several families lived together there. The large bones are concentrated more or less in a circle on the outside, indicating that these people organised the space around them in a systematic way. In Neanderthal sites such organisation is not seen; Neanderthals tended to scatter the bones of mammoths and the like around them in a more disorganised fashion.

Apart from the gradual increase of diversity of stone implements and tools, in the case of *H. sapiens* in Europe we also notice a sudden increase of artistic expression. The hominins living in those days are known as the Cro-Magnon man, a variant of *H. sapiens* with modern anatomic features in every respect, hardly distinguishable from ourselves. The cultural expressions of Cro-Magnon Man appear so suddenly, approximately 40,000 years ago, that their appearance is referred to as a 'cultural explosion'. Various objects being used as decoration, such as sculptures that were carved in bone or sandstone, accompanied this revolutionary change, but the most familiar are cave paintings, which can still be seen in the caves of Chauvet, Lascaux and Altamira. The cave drawings of Lascaux were applied with such attention to artistic expression that they are referred to as the 'Sistine Chapel of prehistory'. Many cave drawings refer to hunting scenes (buffalo, European bison and deer are often portrayed), yet the exact meaning is not clear. A striking fact is that relatively few pictures exist of humans themselves. The many claviform signs in the caves of Altamira are mysterious (fig. 6.5). They look like abstract symbols; we can only guess what they mean.

The cultural explosion is also characterised by sculptures produced from bone or sandstone. The Venus of Willendorf, a female figure with excessively large breasts and buttocks, is widely known. Willendorf is a small village on

Table 6.2: Simplified classification of the 'stone age', based on the succession of stone tool cultures in Africa and Europe

Age	Years before present		
Neolithic	10,000	to	1,000
Palaeolithic	2.7 million	to	10,000
Late	40,000	to	10,000
Middle	250,000	to	40,000
Early	2.7 million	to	250,000

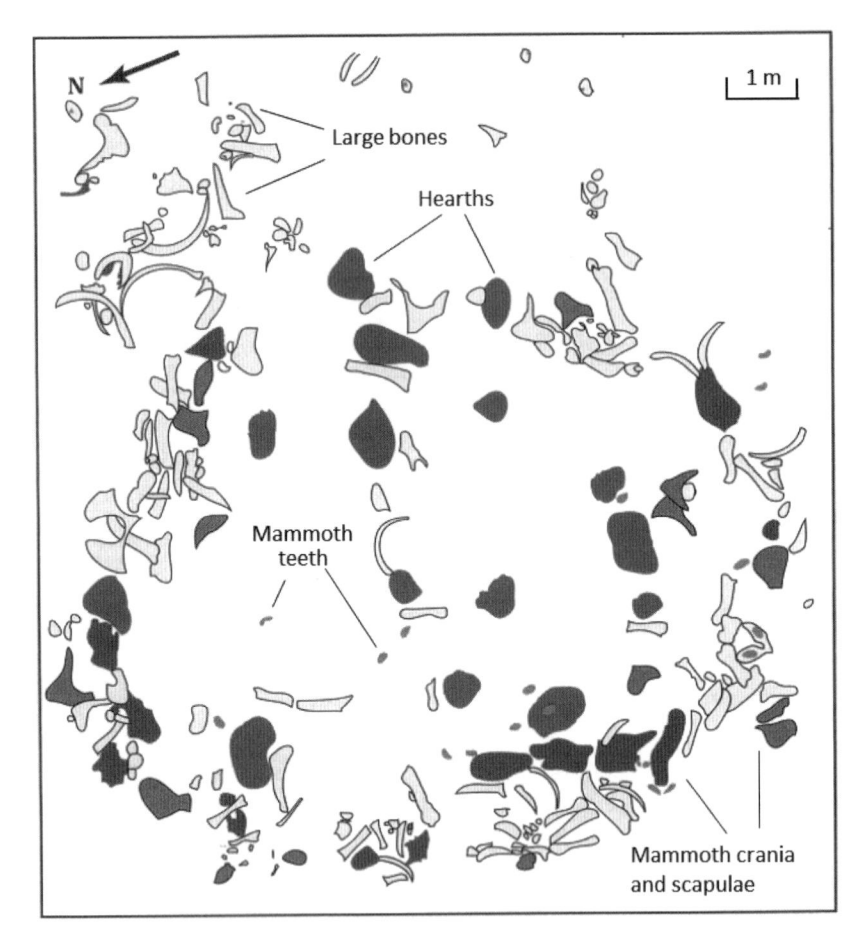

Fig. 6.4: An archaeological excavation in Moldavia, from the Upper Palaeolithic, shows how families of H. sapiens lived together and used the space.

the Danube, to the west of Vienna, where this figurine was found in 1908, in a clearance to make way for a railroad. It dates back to 24 ka BP. Even more explicit is the Venus of Hohle Fels (Baden-Württemberg, Germany), also with large breasts, as well as an outspoken vulva. The Hohle Fels figurine is the oldest representation of a human (35 ka BP). These figurines might have been used as fertility symbols during ritual gatherings, but perhaps they were also meant as a sex object. The fact that different figurines found in different place in Europe show some resemblance to each other could mean that the human groups that made them were in contact with each other and could have exchanged such figurines. This also explains why the materials from which the artefacts were made are not always found in the same place.

THE CULTURAL HUMAN

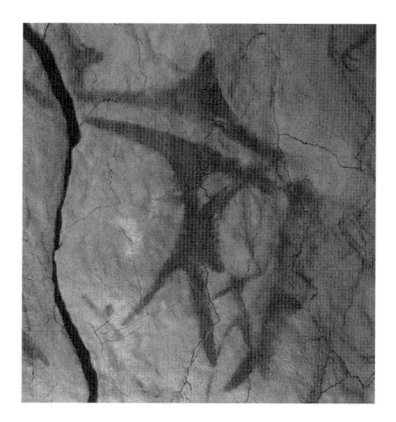

Fig. 6.5: Example of 'claviform' signs of the Magdalenian culture, discovered in the caves of Altamira (northern Spain), 16,500-14,000 years old. The meaning of these abstract figures is uncertain.

Where and when did human beings start depicting the world around them? The traditional view is that this happened in Europe alone and only starting 40,000 years ago. This hypothesis has increasingly been put under pressure over the last few years. In the Blombos cave in South Africa a collection of snail shells has been found that were all pierced in a characteristic way, apparently to be used in a string of beads. This material is 75,000 years old and is regarded as evidence of symbolic thinking before the cultural explosion in Europe. In the Liang Timpuseng cave in Sulawesi, Indonesia, a picture of a pig was found, 40,000 years old, indicating that old cave drawings were also made outside Europe. Finally, scientists of Naturalis in Leiden, the Netherlands, have reported that 480,000-year-old shells, collected on Java by Eugène Dubois, show scratches that must have been applied deliberately. *H. erectus* must have been responsible for that, because *H. sapiens* didn't exist at the time. These observations indicate that the emphasis on the European cultural explosion of 40,000 years ago, as a unique new phase in human evolution, might be misplaced. On the other hand, the evidence for self-expression and symbolic thinking older than 40,000 years is still rather fragmentary at this time.

The Neolithic transition

The Canadian anthropologist Ronald Wright wrote in his book *A Short History of Progress* (2004): 'The human career divides in two: everything before the Neolithic Revolution and everything after it'. It is a fascinating idea that so many of our present achievements started only in the last 3% of our existence. Geologists consider the moment of 10,000 years ago, the transition from the *Pleistocene* to the *Holocene*, as a tipping point in the climate. The Pleistocene was characterised by repeated periods of extreme cold, the ice ages, while the Holocene was granted a more moderate climate. But why do evolutionary biologists find this transition of such extreme importance?

The transition from the *Palaeolithic* ('Old Stone Age') to the *Neolithic* ('New Stone Age') basically involves three drastic changes:
- Humans started to settle in one location, instead of a leading a nomadic life; they became *sedentary*.
- People started farming instead of hunting and gathering. This also implied that certain animal species were domesticated and some were used as pets.
- The population underwent an exponential growth and society got a more complex character; a social hierarchy started to develop.

These three changes occurred somewhat simultaneously between 10,000 and 6,000 years ago, and in different places in the world: in the Middle East (the area from East Turkey to Iraq, known as the Fertile Crescent), in Central America, in Amazonia and the Andes, in the east of North America, in the Sahel and West Africa, in New Guinea and in the east of China. At each location humans made use of the plants and animals present there. In China rice was domesticated, wheat in the Middle East, in Africa it was millet and in Central America they used maize.

Not all plants and animals are equally easy to domesticate. The bitter flavour of almonds, for example, can be attributed to a relatively simple Mendelian system that can be removed by selection. But the bitter flavour of acorns is a complex property that is not easily selectable. Thus the almond tree has become a domesticated plant, whereas the oak tree hasn't. In fig. 6.6 a comparison is made between pairs of related species, one of which can be easily domesticated while the other cannot.

Regarding animals, it is known that zebras (*Equus quagga*) do not easily adapt to humans, while the closely related species *Equus ferus* (horse) and *Equus asinus* (donkey) can very well be domesticated. The natural instinct of an animal and the behavioural flexibility it shows are crucial. In Central America horses do not naturally occur, therefore llamas (*Lama guanicoe*) were used, which are suitable as pack animals, but offer fewer possibilities to ride on, like on a horse. For similar reasons, in China the pig and the chicken were domesticated. At a later stage, chickens, pigs, horses and donkeys were spread across the world.

Domestication of plants and animals can be viewed as a selection process that is not performed by nature, but by human beings. Charles Darwin was inspired by the large variation in external appearances of pigeons, that can be obtained by selection and breeding. Darwin introduced his term 'natural selection' as a counterpart of *'artificial selection'*. In similar ways, humans utilised rare mutants of certain plants and animals that spontaneously appeared in natural populations. In chapter 4 we saw that the current maize derives from a macro-mutation in teosinte, a grass species from Central America. All current maize plants derive from this mutant, which is estimated to have emerged around 9,200 years ago.

Also spontaneously or deliberately conducted crossings between species played a role. Our wheat (*Triticum aestivum*) is the result of two successive hybridisations, followed by genome duplications: *T. aestivum* is hexaploid (2n = 42), while the two species from which it originated, *T. searssi* and *T. speltoides*, are diploid (2n = 14). Due to further selection on large seeds, synchronous seed development and loss of spontaneous seed scattering, the different modern wheat varieties were created. In addition to common wheat, also einkorn wheat (*T. monococcum*) was

Fig. 6.6: Comparison of two species at a time, one of which (horse, reindeer, sheep, cow, goat, almond) proved to be relatively easy to domesticate, while for the other, related species, this turned out to be impossible.

domesticated at an early stage. This is a separate species that is diploid. Einkorn was also found in the stomach of Ötzi, the *'Tyrolean Iceman'*, a mummy 5,300 years old, discovered in 1992 in the melting ice of the Ötztal, in Tirol, Austria.

Regarding animals, the domestication of the dog (*Canis lupus familiaris*) is a model system. The dog genome shows many signatures of long-term human-mediated selection, for example, the ability to digest starch was improved by copy-number expansions of amylase genes. In contrast to what many people think, the European grey wolf (*Canis lupus*) is not our dog's ancestor. Our dog is derived from a wild dog that was common in the Middle East, but has now become extinct. The wolf descends from a dog species from the Pleistocene, of which many existed at the time. Crossbreeding among varieties also played an important role in animal domestication For example, the genome of European breeds of pig (*Sus scrofa domesticus*) shows evidence of selection acting upon Asian genes that were introduced by human-mediation introgression.

The question is: Why did human populations began to practice farming? The first agricultural communities must have faced numerous disadvantages: there was hard work to be done, abundant yields were by no means guaranteed and humans suffered from diseases that thrive on livestock. In a hunter-gatherer society there isn't as much hard work; nature always provides sufficient food and when that is depleted in one place, you simply move elsewhere. In addition, the diet of the hunter-gatherer is much more versatile than that of a farmer. While a farmer gets his carbohydrates from just a few plant species (potatoes, wheat, rice, maize, cassava) and his proteins from a few animal species (chicken, cattle and pig), the diet of hunter-gatherers consists of no less than fifty animal species and a hundred plant species, which, in addition are rich on minerals and low on fat.

Farmers were also confronted with diseases that jumped from livestock onto human beings. The entirety of organisms able to exchange between man and animal is called a *zoonosis*. Meanwhile, we have become immune for different zoonotic disease agents, but initially that wasn't the case. As late as in the eighteenth century, the British physician Edward Jenner discovered that girls who milked cows were rarely struck by smallpox. His hypothesis, which turned out to be correct, was that a disease agent in the cow, related to the human disease agent, protected the girls. Subsequently, he developed a smallpox vaccine (the word 'vaccine' is derived from *'vacca'*, the Latin word for 'cow'). Zoonoses still plague modern society (table 6.3).

Table 6.3: **Examples of zoonoses in modern society**

Disease	Pathogen	Original animal host
Rabies	Lyssavirus (Rhabdoviridae)	Dog, fox, bats, cat
Toxoplasmosis	Toxoplasma gondii (Apicomplexan parasite)	Cat, feline carnivores
Q fever	Coxiella burnettii (Gammaproteobacteria)	Sheep, goat
Influenza	Influenza virus (Orthomyxoviridae)	Chicken, wild birds
Acquired immuno-deficiency syndrome	Human immunodeficiency virus (HIV) (Lentiviridae)	African primates
Severe acute respiratory syndrome	SARS Coronavirus (Coronaviridae)	Civet
Middle East respiratory syndrome	MERS Coronavirus (Coronaviridae)	Dromedary camel

So, the exact advantages of farming while entering the Neolithic were not so clear at all. There are three theories about it:

– Population pressure: Due to increase of the human population and continuous local depletion of resources, the lifestyle of hunter-gatherers could no longer be maintained and more efficient ways of food production were needed to provide solutions.
– Climate change: As the climate in the temperate areas warmed after the last ice age, the steppes changed into extensive savannahs with many animal and plant species that were suitable to use in an agricultural lifestyle.
– Social complexity: The groups in which people lived became increasingly complex and stratified, with different layers in the population. This complexity forced human beings to settle in one place; agriculture was a result of this.

Many scholars at the moment consider the social complexity hypothesis most likely. In that case human population growth must be seen as a consequence. Once settled, agricultural communities could raise larger families, the net birth rate began to exceed the death rate and the population started to grow exponentially, while the hunter-gatherers from the Pleistocene were forced back into a marginal role. The world population, which was below 5 million at the start of the Holocene, started to increase exponentially 5,000 years ago, reaching 170 million at the start of the Roman calendar (fig. 6.7). A new phase of accelerated exponential growth began with the industrial revolution that is still ongoing. U.N. census data evaluated in 2014 predict that the world population is unlikely to stabilise this century and will reach an expected number of 10.9 billion in 2100, with an upper 95% confidence limit of 13.2 billion.

The term 'Neolithic' is linked with the start of agriculture, so does not indicate a fixed moment in time. In the Middle East, it followed directly on the Palaeolithic, but in other parts of the world, for example, in Europe, it took another 3,000 to 5,000 years before agriculture settled. The intervening period is called the *Mesolithic*. It is a time when humans still lived as hunter-gatherers but often came together in one place, sometimes building structures such as the temple complex at Gobekli Tepe in south-east Turkey, which is 9,500 years old.

Agriculture in the Middle East gradually expanded to Europe. This happened via Anatolia, currently known as Turkey. During excavations in northern Turkey, obvious clues were found that even 6,500-6,000 years before Christ cattle and goats were being milked around the Sea of Marmara and the Bosporus. This is demonstrated by traces of milk ingredients on potsherds and the like. From the Middle East, agriculture also expanded towards East Africa. As a consequence, in the DNA of different African peoples today a Euro-Asian signature can still be identified (6-7%).

The question is, How did this expansion work? Was it a matter of *cultural diffusion* ('learning from your neighbours'), during which the structure of a population was maintained, or was it a process of replacement (new

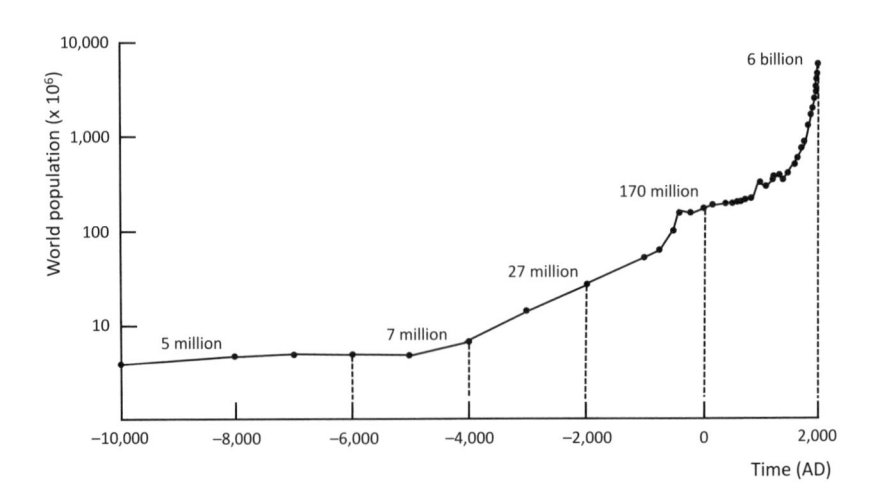

Fig. 6.7: Growth of the world population between 10,000 B.C. and the year 2000. Please note the logarithmic Y-axis! Until 5,000 years ago the world population was rather stable. After that time an exponential growth commenced, which showed a significant acceleration in the second half of the last millennium.

peoples appeared with new habits, the previous ones disappeared)? The model of *'demic diffusion'* was popular for a long time, which holds that a new culture is introduced by a limited number of pioneers coming from elsewhere that mingled with the current population without replacing it entirely. However, this model was mostly based on blood group variation in present-day Europeans. Modern genome-wide data and ancient DNA now show that, at least in Europe, replacement has been a much more important mechanism than cultural transmission.

That agriculture gradually spread from the Middle East to Europe in the period from 9,000 to 4,000 years ago, can still be noticed by gradients of genetic variation. Modern Europeans have a mixed genetic background: on the average 28% of their DNA is of Neolithic origin (related to peoples in the Middle East) and 78% is Palaeolithic (related to the Cro-Magnon Man). However, the gradient across the European continent does not run from east to west, as you would expect, but rather from south to north, which indicates that the pastoralists from Anatolia first replaced the peoples in southern Europe, or mixed with them, and then moved to the north. A study published in 2012 by the Swedish ancient DNA specialist Pontus Skoglund analysed the genome of a Neolithic farmer from the Funnel Beaker culture, in southern Sweden, 4,921 a BP; this genome was similar to southern European genomes and not at all to ancient Anatolian DNA or DNA from hunter-gatherers that lived in Scandinavia previously. This is very much in line with a migration and replacement model from the south, in which farming reached northern Europe by about 4,000 a BP.

Recently, it has become clear that the agriculture of the Middle East also experienced an eastward expansion. In the Zagros Mountains in western Iran fossil remains were found of agricultural peoples from 7,500 to 7,000 years old, who branched at an early stage (46,000-77,000 years ago) from the peoples in the Middle East. DNA samples isolated from these fossils reveal genetic similarities with modern Pakistani and Afghans and also with the Zoroaster, a tribe that lived in the current Iran before the existence of the Persian Empire. The spreading of agriculture towards the east thus is a result of diffusion, because the ancient DNA of the Zagros humans strongly differs from the Middle East peoples.

The Neolithic transition is also a lesson in human cultural history. We see that the cultural diversity of humans has a strong geographic and historical imprint. Which culture develops in a certain area depends on the landscape, plants and animals present, and the opportunity of exchange with other peoples. This theme was elaborated in the magnificent book by the American biologist Jared Diamond, *Guns, Germs and Steel* (1998).

Language: early or late?

One of the most characteristic differences between humans and animals is our language. As language is a cultural behavioural trait unable to fossilise, it is quite unclear when hominins started to speak or use language. There are two theories:
- Language evolved at an early stage, approximately with *Homo habilis*, 2 million years ago.
- Language evolved rather late, not much earlier than 40,000 years ago, and only with *Homo sapiens*.

The fact that there is such an extreme difference between these theories indicates how uncertain we are. The fossil data provide evidence that at least *Australopithecus* did not yet have the right anatomical features to be able to speak, but the data do not exclude that *H. erectus* and Neanderthals could have had the necessary morphological traits (table 6.4). If the anatomical features are decisive, language could be as old as *H. erectus*, and did perhaps even develop with *H. habilis*, although not enough fossil material is available for this species to be decisive.

Evolutionary biologists, who claim that language is very old, emphasise the continuity between the language of animals and that of humans. Many animals make sounds to communicate with each other; well-known are the warning signs of meerkats and the close harmony vocalisations of gibbons; also bird song is obviously an advanced form of communication. If language emerged primarily as a means of communication, there is no reason to assume that the predecessors of modern humans didn't know language. In that case language could have gradually evolved; sounds initially meant to convey simple warnings or utterances of threat or pleasure, could gradually have gained more meaning, allowing abstract concepts to be communicated.

In great contrast with the '*language early*' theory is the argument pointing out that our language is not primarily a means of communication, but a consequence of our inner world. Humans are capable of imagining the world around them in their brain. The advantage of this is that you are able to predict things like the movements of predators, the intention of peers and the consequences of your own actions, by simulating them in your inner world. As such, you would be a better hunter, being able to imagine the flight of your spear even before you throw it, and you would be better able to predict the action of group members and imagine their intentions. Because group life was very important in primitive humans, having an inner world could have contributed to fitness. This theory points out that we not only

THE CULTURAL HUMAN 199

Table 6.4: Overview of hominin morphological, genetic and behavioural traits, used as arguments in the debate on early or late evolution of language

Observation	Conclusion
Homo habilis (2 My BP) already had stone tools.	If language is required for the making of tools, language must be 2 million years old.
The position of the larynx of *Australopithecus* is comparable with that of chimpanzee.	The older hominins (until 3 Ma BP) didn't use any language.
The skull of *Australopithecus africanus* does not show any impression of Broca's area, while in *H. habilis* it does.	The older hominins (until 3 Ma BP) didn't know any language; language started with *H. habilis*.
The spinal canal of *H. erectus* from Africa does not show dilatation in the thoracic vertebrae, while that of the *H. erectus* from Georgia does.	If innervation of the respiratory muscles is crucial to the ability to speak, language originated in *H. erectus*, after the migration out of Africa.
The human hyoid bone resembles the ones from Neanderthal and *H. heidelbergensis* more than *Australopithecus* and chimpanzee.	If the shape of the hyoid bone is important for the flexibility of the tongue, it does not limit the use of language to *H. sapiens*.
The hypoglossal canal in the basicranium is wider for humans and Neanderthals than for chimpanzee.	As the hypoglossal nerve is enlarged at humans as well as Neanderthals, the Neanderthal also had extensive innervation of the tongue and was able to speak.
The human FOXP2 gene shows an accelerated evolution (two non-synonymous mutations); Neanderthals show the same apomorphic state as humans.	If the modified version of FOXP2 is crucial for the ability to speak, Neanderthals also possessed language.
Cave drawings, body decorations and sculptures are found only in association with *H. sapiens*.	The evolution of language is recent and came about only with *H. sapiens*, 40,000 to perhaps 100,000 years ago.

speak, but also think in language. We can even talk to ourselves without making a sound. The function of language as a means of communication with others can be considered as a derivative of our own thoughts.

An exponent of this 'psychological' theory is the Swedish philosopher Björn Peter Gärdenfors, who, in 2003, described his arguments in a remarkably readable book, *How Homo Became Sapiens*. Gärdenfors states that the evolution of language has seven stages. It begins with sensing the external world and adding meaning to the sensations; *sensations* become *perceptions*. A next step is the 'internalisation' of the external world, which turns perceptions into *imaginations*, which in fact are perceptions detached from the external world. In the inner world people can fancy about reality or imagine things that are not there. And, in a next step, people obtain an image of someone else's inner world. This level is essential for being able to show *empathy*: the ability to place yourself in someone else's state of mind, to understand why someone else is sad or excited and to respond to it. This cognitive ability is also called '*theory of mind*'. Self-consciousness is derived from this and, according to Gärdenfors, this step is essential for language.

The question is whether the *theory of mind* is an inborn feature of humans or that it is culturally defined (that is, it must be taught). Either view has a point. The American psychologist Cecilia Heyes emphasised in an article from 2014 the similarity between reading a text, looking at pictures and reading someone's mind. Just like reading from paper, according to her, reading someone's mind must be taught and this is done differently in different cultures. If children do not learn how to read, they won't be able to do it when they are older, and if they are not taught how to place themselves into someone else's perspective, neither will they show that capacity later in life. Reading someone's mind, according to Heyes, comes before reading of text and pictures. However, she still recognizes that there is a genetic basis for the capacity to read, either from paper or someone's mind. With this view Heyes shares the theory that the *theory of mind* is essential for language abilities.

The *theory of mind* also assumes the presence of self-consciousness. You cannot place yourself in someone else's mind if you are not aware of your own thoughts. Do animals also have self-awareness? The '*mirror test*' is considered the crucial experiment for this purpose. Can an animal recognise itself in the mirror? So far only great apes, dolphins, elephants and magpies have passed this test. In his books, the Dutch-American primatologist Frans De Waal discusses extensive material evidence that stresses the continuity between great apes and humans, including the mirror test and the ability to show empathy. De Waal emphatically stresses that the dawning of typically human characteristics, such as moral behaviour, took place before humans evolved. He argues that various characteristics that we consider as typically human are also observed in highly developed animals.

The psychological theory about the evolution of language corresponds to some extent with the development of children. A soon as children are born, they can evidently communicate, although they are not able to speak yet. At the age of 22 months children pass the mirror test and clear indications of self-consciousness are evident only after their second year. The emergence of an inner world is shown when children play fantasy games, for example, with dolls. The development of language comes afterwards. In accordance with this book's theme, the cognitive development of a child thus can be seen as supporting the late evolution of language.

Fig. 6.8: Illustrating the similarity between reading someone's mind (mind reading, theory or mind) and reading printed text or images (print reading).

Another line of evidence usually advanced in favour of *'language late'* is the observation that utterances of abstract and symbolic thinking, for instance, cave drawings, body decorations and sculptures are only found with *H. sapiens*. This would prove that only *H. sapiens* exhibits such signs of an inner world; even in Neanderthals it is lacking. This seems to be a majority point of view among archaeologists at the moment; however, it is certainly not shared by all of them. In 2018 an international team of archaeologists analysed pigments, perforated marine shells and primitive drawings found in a cave along the Spanish south-west coast, and demonstrated, by dating the carbonate crusts in the cave, that this material was 115,000 years old. This would bring the use of symbolic expression within the reach of Neanderthals, a view that contrasts with the *'language late'* view. This isn't the last word on the evolution of language.

Living in groups: altruistic behaviour

Life in groups has been crucial for the evolution of humans. The abiotic and biotic environment was not the only template moulding the life cycle and all complex behaviours of humans; the presence of other people in the direct vicinity also formed an important evolutionary pressure. Human beings are utterly social creatures.

One of many social relationships people have with other people is characterised by *altruism*: behaviour that is good for the other but disadvantageous for yourself. The classical example of altruism is to be found in the Bible. In the parable of the *Good Samaritan* (Luke 10:30-37), Jesus tells about a man who was on his way from Jerusalem to Jericho and was mugged by robbers who left him half dead by the side of the road. A passing Samaritan (the Samaritans were an ethnic group that was not very popular with Jews at the time) took care of the victim, nursed him and took him to an inn. The next morning, before he left, he gave the innkeeper two denarii to make sure that the wounded man would be treated well.

Altruism appears to have a neurological basis. Three nuclei are involved; the *anterior cingulate cortex*, the *insula anterior* and the *ventral striatum* (fig. 6.9). These nuclei typically communicate with each other if a person performs an altruistic act or observes one. But it looks like the signalling is different for different motivations. Altruism can be applied out of pity (empathy; 'I want to be good to him because I feel sorry for him'), or it can be shown out of mutuality (reciprocity; 'I want to give something back, because he has been kind to me'). Functional MRI studies on test persons, who were guided into these situations, showed that the mutual communication between the three nuclei depends on motivation (empathy or reciprocity) (fig. 6.9). The fact that there is such an obvious basis for altruism in our brain supports the idea that it can be considered an evolved behaviour.

In terms of evolution, altruism means that your own fitness reduces while that of others increases. This is seemingly difficult to justify, because why would one do that? Every peer who doesn't engage in altruism has an evolutionary advantage over you. Still, altruistic behaviour is quite common among human beings and a widespread phenomenon; examples of this are friendship, idealistic acts and philanthropy. From the perspective of evolutionary biology, there are three explanations:

- *Reciprocal altruism*. Behaviour that is beneficial to the other is done while expecting that the other will be kind to you as well someday ('you scratch my back today, I'll scratch yours tomorrow').
- Altruism by *kin selection*. In many cases the peers in a group will be related by kinship. This means that you can promote the spreading of your own genes by helping genetically related peers.
- *Group selection*. The altruistic behaviour spreads because it is beneficial to the group. Groups of which the members are altruistic towards each other will overcome groups that are unfamiliar with such behaviour.

The term 'kin selection' expresses that in the fitness of an individual also the fitness of his or her related peers should be included. The altruistic deeds would then be proportional to the degree of kinship, taking costs into account. This is expressed in *Hamilton's rule*:

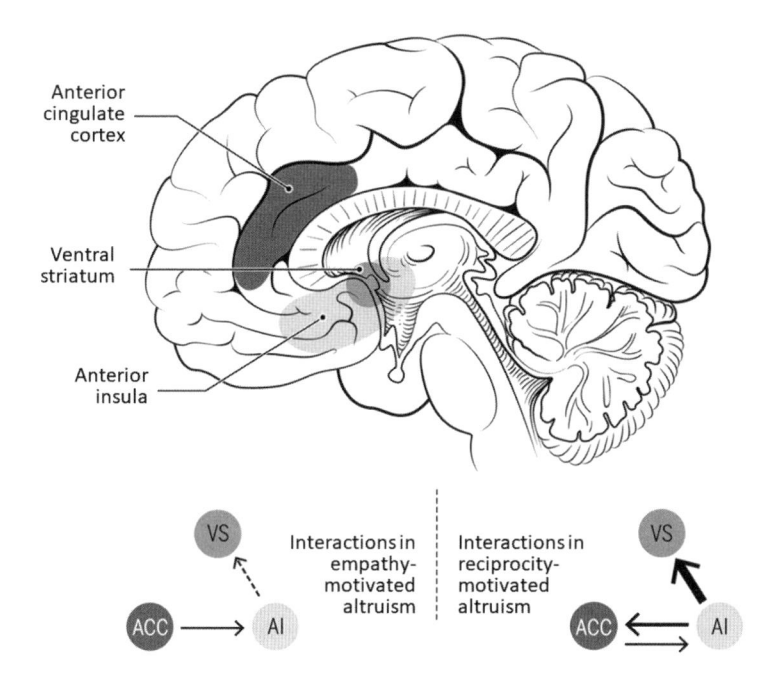

Fig. 6.9: According to functional MRI scans, three parts of the brain transmit signals to each other when a person is looking at or carrying out altruistic actions. Depending on the motivation, the mutual functional relationship between the brain areas differs. Altruism may derive from feelings of empathy or from a tendency to reciprocity.

>Any gene that promotes a certain altruistic behaviour will increase in frequency if r B > C,

where *r* is the genetic relationship between donor and acceptor, *B* is the advantage for the acceptor and *C* the costs to be incurred by the donor. Benefits as well as costs must be expressed in terms of fitness. That is why this is called '*inclusive fitness*'. The degree to which relatives are taken into account is specified by the *inbreeding coefficient r*, a parameter that is technically defined as the chance that two randomly drawn alleles of a locus are identical because they derive from the same ancestor. The coefficient is 1/2 between siblings and between parents and their children. It is 1/4 between grandparents and grandchildren and between uncle and aunt and their nephews and nieces, and it is 1/8 between nephews and nieces.

The rule derived by Hamilton derives from the theory of quantitative genetics, developed as part of the 'Modern Synthesis', discussed in chapter 4. The English geneticist John Maynard Smith suggested the term '*kin selection'* in 1964. J.B.S. Haldane (1892-1964) popularised Hamilton's rule in an unforgettable way by stating that he would be willing to give his life for two brothers, four nephews/nieces or eight cousins without that deed affecting his *inclusive fitness* (fig. 6.10). (Interestingly, the English biologist Richard Dawkins cites a Haldane paper from 1955 in which he remarks, as an afterthought, 'on the two occasions where I have pulled possibly drowning people out of the water (at an infinitesimal risk to myself) I had no time to make such calculations'.)

Undoubtedly the best-known popular version of the kin selection theory is due to the book *The Selfish Gene* by Dawkins, published in 1976. Dawkins assumes that much human behaviour can be understood from the evolutionary advantages that it offers. Affectionate behaviour towards relatives is seen as a way to improve the spreading of your own genes. Although the behaviour is altruistic, the genes causing that behaviour are selfish.

There are few books on biology that have had more impact outside the discipline, than Dawkins' book. Particularly in evolutionary psychology its impact has been enormous. Dawkins can be viewed as an exponent of

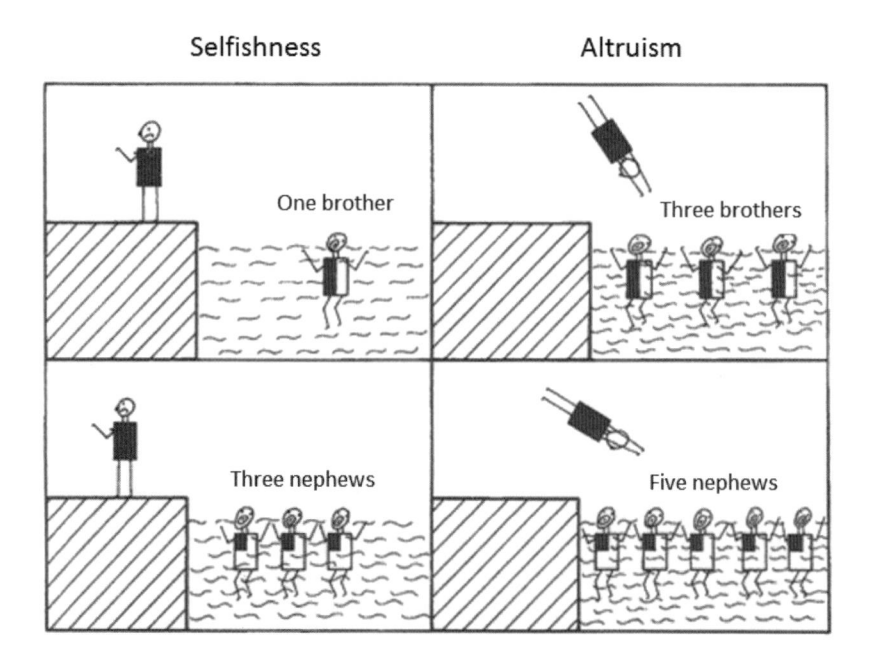

Fig. 6.10: Cartoon illustrating how J.B.S. Haldane explained Hamilton's rule: if, by jumping into the water, you can save multiple relatives while drowning yourself, in an evolutionary sense that would be beneficial, if you could save more than two brothers or more than four nieces or nephews.

sociobiology, a branch of evolutionary biology that was introduced by the American ecologist E.O. Wilson, with his famous book *Sociobiology: The New Synthesis*, which was published in 1975. Wilson stated that the social sciences that try to explain human behaviour can learn from evolutionary biology: many components of human behaviour are biologically determined and explicable in terms of evolution.

The third explanation for the evolution of social behaviour we mentioned above is based on group selection, in which groups are the entities on which selection acts. The argument goes that groups that are doing well while competing with other groups have significant internal cohesion. Altruistic behaviour towards peers is encouraged because it contributes to the strength of the group.

The action of group selection was specifically stressed by the British ornithologist V.C. Wynne Edwards. He claimed that many higher animal species are able to regulate their populations, for example, because clutch sizes are deliberately adjusted to the available amount of food. Populations that adopt this habit prevent collapse in times of food scarcity. Socially valued behaviour, such as helping each other, standing on the lookout to warn others, tempering reproduction in times of rapid population increase, *etc.*, evolve because they are beneficial to the group, according to the group selection argument.

The problem is that group selection does not act according to classical Darwinian selection and is not likely to be effective. Unlike individuals, groups do not reproduce frequently. Exchange between groups reduces selection on the group, and altruistic behaviour in the group is sensitive to cheaters. Numerous evolutionary biologists, including George C. Williams, who worked at the State University of New York at Stony Brook, have demonstrated that group selection is ineffective under natural circumstances. John Maynard Smith showed through game-theoretical models that a non-altruistic cheater intruding on an altruistic population always has an advantage; eventually his descendants will dominate the group.

At the beginning of this century, the discussion about group selection was given a new impulse with the publication of an article (in 2007) by the anthropologist David Sloan Wilson and the previously mentioned E.O. Wilson. The two Wilsons argued that group selection was unjustly criticised in the last century. It definitely does exist, as part of what they call '*multilevel selection*'. Selection can function within the group and between groups at the same time. Also these views raised strong opposition, for instance, from the British ecologist Stuart West from the University of Oxford. West showed that the group selection concept of the two Wilsons wasn't genuine group selection, but individual selection in a population structured by groups (fig. 6.11).

Currently the prevailing sentiment among evolutionary biologists is still anti-group selection. The idea that group selection seems to occur in small groups rather than in large groups, suggests that it is factually kin selection. However, it cannot be ruled out that human beings, specifically, are an example of a species that makes group selection possible, due to their hyper-social behaviour and evolutionary past of living in small groups.

Research on the group size of primates has shown that life in a group has both advantages and disadvantages, and that the size of the group is a reflection thereof. Fig. 6.12 provides a summary of the different determinants of group size, studied over a large number of primate species. This overview, drawn up by British primatologists under the direction of Robin Dunbar of the University of Liverpool, shows that predation risk has a positive, whereas the neocortex ratio has a negative correlation with group size. It is evident that life in a group protects many primates against predators. But the group should not become too large because the number of required social interactions may become so excessive that a primate with a small

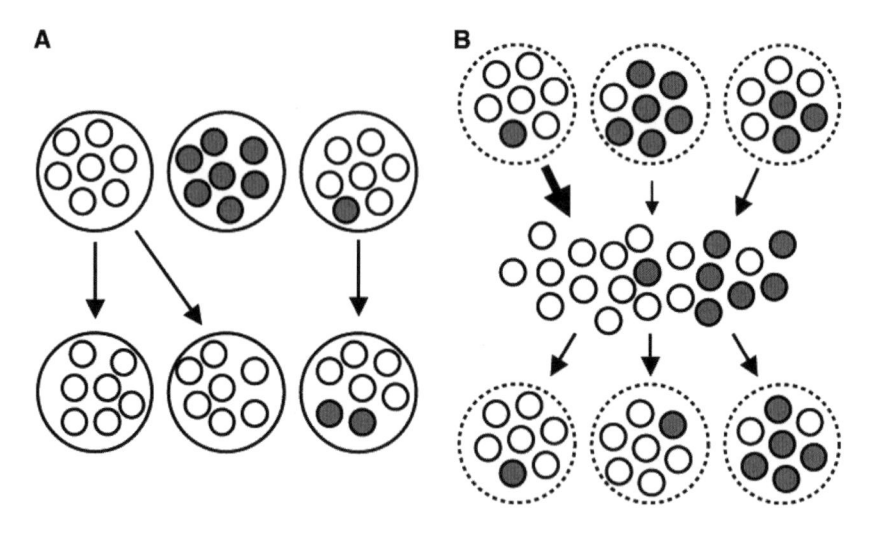

Fig. 6.11: Figure A shows the classical group selection process. There are three groups, each consisting of cooperative (open circles) and selfish (grey circles) individuals. The number of selfish individuals decreases because the groups containing many cooperative individuals are more successful and create multiple new groups. Figure B shows the group selection process as suggested by David Sloan Wilson. In this instance, the groups are more loosely defined and contribute to the offspring in different ways, from which new groups are formed. According to the British evolutionary biologist Stuart West, situation B basically consists of individual or kin selection in a population structured by groups, and is not to be considered true group selection.

neocortex is no longer able to comprehend them. Social interactions among monkeys take place via grooming, whereby the two peers get acquainted and remember each other's individuality. Dunbar is also the source of the so-called *Dunbar's number*, the assumed cognitive limit to group size. Extrapolating from the group sizes of primates, Dunbar argued that the maximum number of individuals with whom a person can maintain stable social relationships is 148.

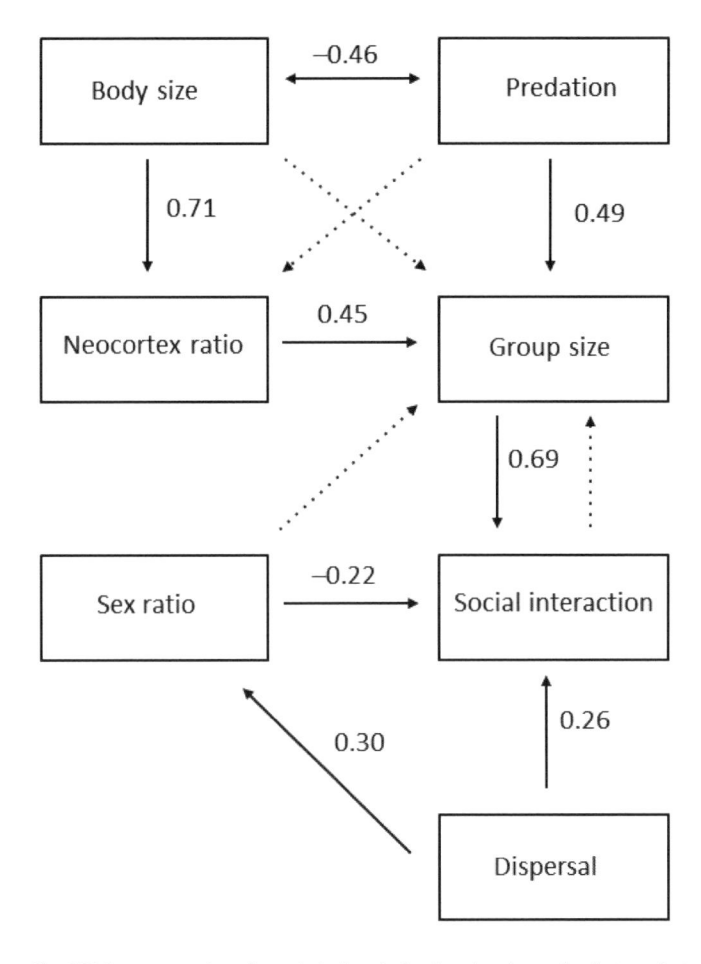

Fig. 6.12: Representation of a statistical analysis, aimed to determine factors that may influence the group size of primates. The arrows show causal relationships, whereas the figures show standardised regression coefficients. The dotted arrows represent less important relationships that were not included in the ultimate model. The figure shows, among others, that species with large brain volumes (neocortex ratio) live in larger groups on the average, just as species that are exposed to high predation risk.

Still, many students of human social behaviour feel that kin selection and group selection cannot explain the full spectrum of human altruistic behaviour. An attempt to explain social interactions in another way was forwarded by the American economist and behavioural scientist Herbert Gintis. He proposed the concept of *'strong reciprocity'*, an extension of the simple reciprocal altruism seen in primates (the first alternative mentioned above). Simple reciprocity is driven by the expectation that somebody else will grant a favour in response to a favour given. However, Gintis argued that humans take it one step further: many people are willing to take costs on themselves or to punish others even if there is no immediate benefit in doing so. The motivation for such behaviours would come from basic urges such as shame, anger, guilt and pride, which are natural instincts everybody has. The theory is tested using a variety of games, such as the *prisoner's dilemma*, the *dictator game* or the *public goods game*, in which test persons are asked to allocate certain amounts of money to specific purposes while taking the expected responses of others into account. Others dispute whether the behaviours recorded in such games can be said to result from evolution in real life. Anyway, this is a very lively area of research.

Cultural evolution

The emergence and existence of cultural phenomena is often explained in evolutionary terms. Think, for example, of the numerous cultural customs displayed by various societies: their interaction with the environment, the care for the dead, the belief in a divine power, their rituals regarding music and dance, *etc*. Richard Dawkins called such behaviours *'memes'* and characterised them as the *'new replicators'*. The term 'meme' is comparable with gene: it can emerge via mutation (inventions, innovations), it can have a high fitness (a high psychological attractiveness) and thus increase in frequency (for example, the 'going viral' of a video on the internet) and it can die out (customs are phased out or are forgotten). Memes can even be subject to *cultural drift*, comparable with genetic drift in a small population (see chapter 4). An example of this is the custom among the Fore people of New Guinea, to eat the brains of deceased fellow villagers, a custom that was maintained due to the small, enclosed population, which led to a high prevalence of the prion disease kuru. Table 6.5 provides an overview of the similarities and differences between cultural evolution of memes and biological evolution of genes.

Culture is defined as behaviour conducted by most members of a group for a prolonged period of time, which is passed on through imitation and

Table 6.5: Overview of analogies and differences between cultural evolution of memes and biological evolution of genes

	Memes	Genes
Origin	Via cultural innovation, often focused on solving a problem	Via mutation, an undirected process, not targeted to a solution
Variation	Due to insufficient reporting or flawed copying	Allelic variation is created by mutation
Retention, spreading	By psychological attraction and acceptance in a group	By selection of the variants that provide the highest fitness
Selection	Adaptivity of a behaviour is of minor importance; coincidence and cultural drift play a major role	Only adaptive behaviour is capable of spreading
Transfer	Horizontal transmission (learning, mimicking)	Vertical transmission (reproduction)
Type of evolution	Lamarckian	Darwinian

social learning. Culture can only develop if the peers are aware of each other's inner world, are able to mimic each other, are sensitive to encouragement or disapproval and can communicate about cultural expressions, for instance, in a language. These conditions explain why culture appears practically among human beings only. Still some behaviours of monkeys have been described in ways that also meet these conditions. An often-cited example is that of a group Japanese macaques (*Macaca fuscata*) on the island of Koshima, which in 1958 discovered that the sweet potatoes that were thrown at them tasted better after they had been washed in seawater. One individual discovered this and the young animals copied the behaviour, until eventually the entire group showed it, and still does to date, a clear example of a cultural tradition (fig. 6.13).

Culture also shows a clear interaction with biological evolution. Due to biological evolution the social environment changes, as a result of which the evolution of culture also changes. Conversely, biological evolution can be given a new direction through certain cultural phenomena. These effects can be illustrated on the basis of a Neanderthal fossil, discovered near La Chapelle-aux-Saints, in France. This fossil is the remnant of an old man who barely had any teeth and molars left. The question is how such an old man managed to stay alive in the population; he was undoubtedly helpless. Apparently, in the Neanderthal group, a cultural habit ruled to take care of the elderly. This influenced the biological evolution of the Neanderthal: elderly were part of the population. Examples in modern society are the various implements for social communication (*e.g.* smart phones) that are the products of human creativity but also act as a selective force upon human behaviours. We have also seen that cooking could be considered a behavioural trait that influenced the evolution of the gut (chapter 3).

Regarding animals, learned behaviour is subject to genetic assimilation, a concept that we discussed in chapter 4. There we presented the idea, due to Conrad Waddington, that phenotypic plasticity, if it is costly, tends to become canalised and genetically fixed. Applied to behaviours this would imply that learned behaviours with a great positive fitness effect should tend to become instinctive. Learning is a costly matter and also success is not always assured. If a certain learned behaviour is applied for a prolonged period of time, any mutation that fixes this behaviour in the genome is quite beneficial. According to this line of reasoning evolution leads to learned behaviour being anchored in the brain, via genetic assimilation, over time.

However, humans seem to have escaped this effect. Almost all of our behaviours must be learned. The only instinctive behaviours, things that

Fig. 6.13: Japanese macaque (*Macaca fuscata*) eating a sweet potato after having washed it in the sea, on the island of Kojima, Japan. This behaviour started to develop in 1958 and has been culturally transmitted ever since.

Table 6.6:   Evolutionary explanations for religion, given by different authors

Author	Year	Brief summary of the explanation
C. Darwin	1871	Religion is a projection of human characteristics on nature. Due to his high cognitive capacities, mankind felt the necessity to understand natural phenomena; the allocation of a soul or a spirit proved to be helpful.
C.H. Waddington	1960	Religious conduct can be seen as a consequence of the human tendency to accept or tolerate authority. That trait was strongly selected, because it enlarged the cultural transmission of parent to child.
K. Lorenz	1966	Religion is result of the human tendency to hold on to rituals. The retention of fixed rituals has an adaptive value, as these are habits that have proven to be effective and safe.
E.O. Wilson	1976	The tendency to stick to group rules is beneficial because, as a member of the group you are being protected.
R.D. Alexander	1987	Religion is an institutionalisation of the moral. If all members of a group subject to shared rules, it is beneficial for the group.
M. Strickberger	2000	Religion is a way in which humans can control the misunderstood world of natural phenomena. If nature is given human characteristics, it can be understood better.
P. Boyer	2001	Religion is a psychologically very attractive concept as it provides support and comfort when trying to understand the environment.
D.S. Wilson	2002	Religion is beneficial for the group as an organic entity, because it makes collective actions possible that a single individual could never perform.
D.C. Dennett	2006	Religion consists of 'domesticated' folk wisdom, originating from nature. By granting a supernatural power to such wisdom its spreading and acceptation was enlarged.

don't require 'showing' by others, have to do with basic urges, such as eating, reproduction and aggression. All our cultural behaviour being learned also means that it could be lost in case the social environment disappears. A sad example of this are the original inhabitants of Tasmania (Parlevar) who, due to their continuously declining population, in the end even forgot how to make fire and how to make use of marine food sources. After contact with Western colonizers, subsequently, the Parlevar culture completely vanished.

One of the perhaps most widely discussed cultural phenomena is the practice of *religion*. Religion is so common, seen in all peoples worldwide, that undoubtedly it must have a biological – evolutionary – basis, we tend to think. It turns out that indeed there is a genetic basis for religious behaviour. Twin studies show that approximately 45% of the differences between people's religious inclination, is caused by additive genetic differences.

This mainly concerns the psychological aspects, such as the tendency of *self-transcendence*, as opposed to religious behaviour imposed by the environment, such as church visits. So, is religion, in that respect, an evolved behaviour? Did it provide certain benefits in the environment of the first humans in the African savannah? A large number of authors have addressed this question. Table 6.6 gives an impression of the range of ideas that have been proposed on this topic. It is remarkable that many arguments ultimately depend on the action of group selection, to a greater or lesser extent.

Similar questions are asked regarding numerous other typical human behavioural traits, such as: Why do women like pink more than men? What is the meaning of musicality? What are the characteristics of a natural leader? All these questions are often discussed in the context of extensive evolutionary explanations, which assume that the behaviour provides a certain benefit and could therefore have evolved in the community of primitive humans on the African savannah. The environment in which

Fig. 6.14: Artist impression of the 'environment of evolutionary adaptedness' (EEA). The conditions in the African savannah, where the human species lived during the major part of its evolution, according to the English psychologist John Bowlby, have favoured adaptations that still determine our behaviour to date. A group of *Homo ergaster* is depicted, trying to scare away a sabre-toothed cat from an antelope's carcass.

this is assumed to occur, from 320,000 to 10,000 years ago, covering 97% of *H. sapiens* evolution, is indicated as the *'environment of evolutionary adaptedness'*, abbreviated EEA, a term that was introduced by the British psychologist John Bowlby (1907-1990). With this he highlighted to totality of environmental and social conditions that ruled during the largest part of human evolution and which were the template for our evolution. An artist's impression of the African savannah as our EEA is reproduced in fig. 6.14. That much of our behaviour evolved in the EEA and some of it was fixed a long time ago, makes it extremely difficult at the same time, to test hypotheses about the evolution of our current behaviour.

To give just one example, we refer to the discussion on why girls like pink. Experiments with test persons have clearly shown this to be the case, both in UK societies and among Chinese people. The explanation often is that it benefitted women more than men if they were attracted to red colours on a green background because that would help them in berry picking. Such an explanation could be true or it couldn't, but more importantly it usually remains untested. Other explanations could be given as well, *e.g.* it would benefit women to be able to clearly distinguish blushes in men when asked about their seeing other women. Such explanations are considered *'just-so stories'* because they are always true and therefore contribute little to scientific progress in our understanding of the evolutionary determinants of human behaviour.

7. Do humans still evolve?

In the sciences traditionally devoted to the study of humans, such as medicine, anthropology, sociology and psychology, until recently little attention was paid to evolutionary biology. Medical research is concerned with defects and diseases; the research is focused on treatments to cure a patient. Anthropological and sociological research focuses on the social relationships between people, and the society they live in, not on their biology or their evolutionary history. Even in evolutionary biology itself there has always been scarce attention to humans as a study object. This is caused in particular by the difficulty of obtaining experimental data on multiple generations. Evolutionary biological research was almost exclusively performed on animals, such as birds and insects.

Currently this is changing, partly because health is gradually regarded more as an interaction between the genetic background of a person and his or her environment, including their lifestyle. The extent to which people react to parasites (like malaria) or to toxins (such as alcohol) is often partly genetically determined, which means that a pathogen or a stress factor can cause evolutionary changes in a human population. But such changes also have a strong environmental component, which in turn is influenced by the behaviour of the individual. These are the exact topics that evolutionary biologists are concerned with. Partly thanks to the availability of human genomes, it is now also possible to link genetic variation to phenotypic differences in disease profiles and pathological defects. As a result, the possibility of answering the question of whether humans still evolve is coming within reach.

Quantitative characters and heritability

The characteristics in which human scientists are interested often have a quantitative nature. They are called metric or *biometric* features: statistics that can be measured in an individual and expressed in a figure, such as body length, number of children, age at the beginning of the menopause, partner preference, intelligence quotient, religious behaviour, *etc*. If such characters are subject to evolution, they should have at least a partly genetic basis. The study of the heredity of metric characteristics is a separate part of evolutionary biology: *quantitative genetics*.

Biometric characters usually follow a normal distribution. This can be expected theoretically on the basis of the *central limit theorem* of statistics: the sum of a large number of independently varying variables approximately follows a normal distribution. In biological terms, you expect a normal distribution for traits that have a polygenic determination, in which each gene has a minor effect, while the positive and negative effects of all genes can be added together. The fact that biometric characters are distributed normally means that they can be conveniently analysed using the classical statistical techniques of analysis of variance, correlation and regression.

By way of example, fig. 7.1 shows a frequency distribution of IQ scores of English school children. The frequency distribution is practically normal, but when studying it closely, two minor deviations appear: the distribution is somewhat curved to the left and a bit narrower than a normal distribution (leptokurtic, in the words of the statistician). The consequence is that the fraction of children with a low IQ is larger than would be expected on the basis of the average and the standard deviation of the best-fitting normal distribution. Another distribution (Pearson type IV) describes the data more accurately. The difference from the normal distribution can only be

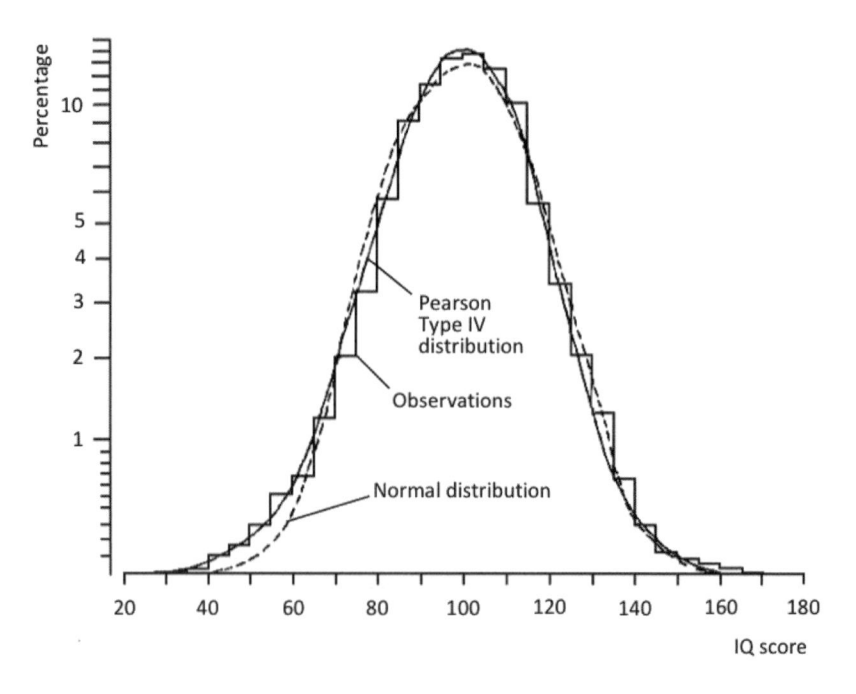

Fig. 7.1: Biometric characters are quantifiable traits of individuals that usually follow a normal distribution; their inheritance can be examined by means of quantitative genetics. Here, a frequency distribution is shown of IQ scores for 4,665 English school children. A Pearson type IV distribution describes the data slightly better than a normal distribution.

demonstrated though in the case of very large data records, such as here, on IQ scores.

A key concept in quantitative genetics is *heritability*, defined as the additive-genetic variance as a fraction of total phenotypic variance. Stating that the IQ in a certain population group has a heritability of 70%, we mean to say that 70% of the differences in IQ score between people can be attributed to additive-genetic variance between those people. The rest can be attributed to variable influences from the environment (upbringing, school), to non-additive genetic effects and to gene-environment interaction.

There are different ways in which heritability in a population can be measured. A very insightful approach is a regression between parents and offspring (fig. 7.2). If you plot the average value of a certain trait among offspring against the average value of their parents, the slope of the best-fitting line through those points is an estimate of heritability, indicated as h^2. Its value varies between 0 (no heredity at all, all variation is environmentally determined) and 1 (all variation is due to differences in genotype).

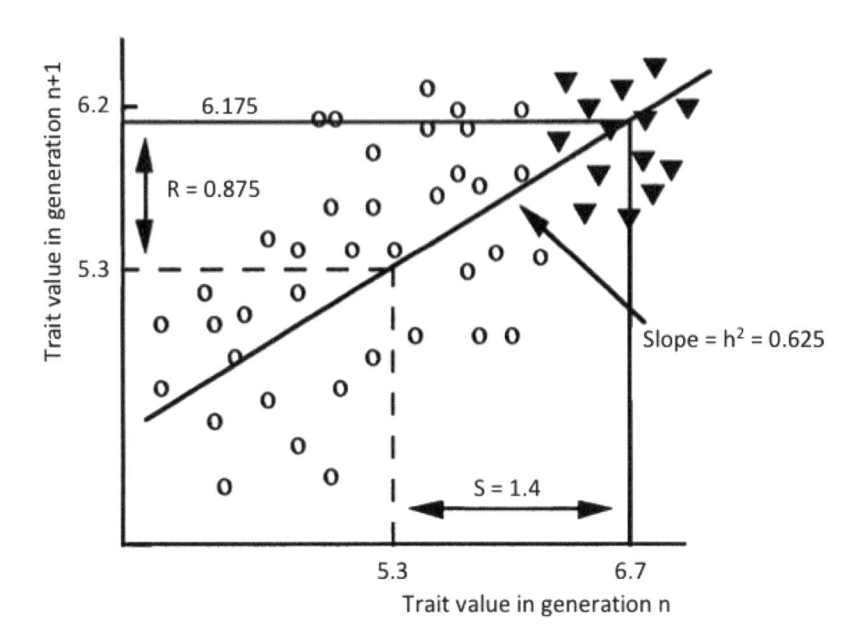

Fig. 7.2: Illustration of the relationship between response to selection and heritability of a trait in a parent-offspring regression. If in this population the individuals with a high score (indicated by triangles) with average 6.7 are selected, the selection differential S will equal 1.4. In case of a heritability (h^2) of 0.625, the response to selection R will equal 0.625 x 1.4 = 0.875. With this evolutionary pressure and heritability, the character will increase by 0.875 per generation.

For humans, parent-offspring regression is not practical, due to the long generation time. Estimates of h^2 for humans are usually derived from twin studies. According to the theory of quantitative genetics, h^2 can be estimated as

$$h^2 = 2(r_{MZ} - r_{DZ})$$

where r_{MZ} is the correlation between monozygotic twins and r_{DZ} is the correlation between dizygotic twins. Both correlations are high, but if heredity plays a role, the monozygotic correlation is larger than the dizygotic. In the case of traits that do not have any heritable basis, identical twins resemble each other as much as dizygotic twins, with the result that $r_{MZ} = r_{DZ}$ and $h^2 = 0$. Also, from other pedigrees, such as correlations between half-brothers and half-sisters, in comparison with full siblings, estimates of h^2 can be derived.

If the heritability of a trait is known, we are in a position to predict how it will react to selection. For that purpose the following formula is used:

$$R = h^2 S$$

where *R* is the response to selection, h^2 is heritability and *S* is the '*selection differential*'. The latter term refers to the difference between the average trait value of the group that is selected, and the average value of the parents (fig. 7.2). An equivalent term is '*selection gradient*', which is defined as the change in fitness with increasing trait value. Selection gradients are popular with evolutionary ecologists because they can be measured in field populations as a regression of lifetime reproductive success on trait value.

Despite the attractiveness of the term 'heritability', as previously described, it has some significant disadvantages. Generally, it is too simplistic a view of the underlying genetics. Many genes do not operate additively towards each other, but via *epistasis*, meaning that the product of one gene has an effect on the phenotypic expression of another gene. The effect of a certain gene therefore depends on other genes in the genome; this is what we call the *genetic background*. The gene may contribute to a metric character in one individual, whereas it doesn't in another. For example, specific mutations in the genes *BRCA1* and *BRCA2* are a risk factor for breast cancer, but practice teaches us that not all women carrying these mutations actually develop breast cancer, because the effect of mutated *BRCA* genes also depends on other genes, with an epistatic effect on the breast cancer genes.

A second non-additive effect is *pleiotropy*. This means that one gene influences multiple, seemingly unrelated, phenotypic characters. As an example we mention the appearance of ribs on the seventh vertebra, a phenomenon that was studied by biologist Frietson Galis from Leiden University. All mammals have seven cervical vertebrae; the first vertebra to support a rib, is the eighth. But in 0.6 to 0.8% of births it happens that the seventh vertebra of the baby supports a rib or rib bud; this is caused by overexpression of *HoxC* genes (see chapter 2). Galis' research shows that occurrence of neck vertebrae is correlated with a high incidence of miscarriages, stillbirths, and also cancer. Apparently, the improperly adjusted *Hox* gene expression causes negative effects elsewhere in embryonic development, an obvious case of pleiotropy.

A third problem with the simple concept of heritability is that interactions with the environment are the rule rather than the exception. A gene can have a determining effect on a metric character in one environment, whereas in another environment it hasn't. If these types of genotype-environment interactions play an important role, the concept of heritability loses much of its value, because the response to the same selective factor will then depend on the environment that the population is situated in.

Also we should realise that heritability is a relative measure. If environmental variation changes, heritability will also increase or decrease, while the underlying genetics remains the same. As an example, the heritability of the IQ score of children in the Western world has significantly risen over the last fifty years, due to the fact that the influence of the environment was minimalised. While in the past the environment a child was born into was important, now children in the Western world generally have the opportunity to get good educations, so the only noticeable differences between children in the modern school system are the inborn differences in talent, while h^2 tends towards 100%.

Furthermore, it is important to take account of changes in h^2 during the selection process. As a consequence of directional selection (table 4.4), additive-genetic variation will decrease and, along with it, the value for h^2. Eventually, after prolonged selection, genetic variation will be exhausted and further selection will no longer have an effect. This occurs in agricultural crops and livestock with prolonged selection for an increasingly high production. To seek further optimisation, crosses with breeds from other countries are applied.

Despite all these warnings, heritability is still a useful term in evolutionary analyses. As we will notice in the following paragraph, it is possible to reconstruct the evolutionary process through long-term studies on biomedical variables or life cycle features. From a regression of the measured variable on fitness, the selection gradient follows. With an estimate for heritability, the response on selection can be predicted. Conversely, if the response on selection is measured, heritability can be calculated. This estimate is called *'realised heritability'*, which is often lower than the estimate from parent-offspring regression or from twin studies.

The complex relationship between genotype and phenotype, with additive effects in addition to pleiotropy, epistasis and environmental interaction, could perhaps explain the fact that complex traits with high heritability are still difficult to trace back to the genome. For example, many studies report a heritability of 70 to 80% for the IQ of adults, but an unequivocal collection of genes has never been identified that predicts with some degree of accuracy a person's intelligence. Even the number of *DUF1220* domains, one of the strongest genetic determinants of intelligence, explains but a part of the variation (see chapter 3). That applies to almost all cognitive and behavioural features. Many researchers cannot resist the temptation, when they have found a gene related to some complex trait, to present that gene as the sole explanation for the complex behaviour, for example, the 'God gene' *VMAT2*, the 'adventure gene' *DRD4*, the 'intelligence gene' *NBPF*, and the 'gene for impulsivity' *HTR2B*, but time after time (after further study) things turn out to be much more complicated than was originally suggested. The probable cause is that epistasis in the genome is a highly dominant phenomenon. Instead of relating phenotypic characters to a few genes, a network approach seems to be a better option.

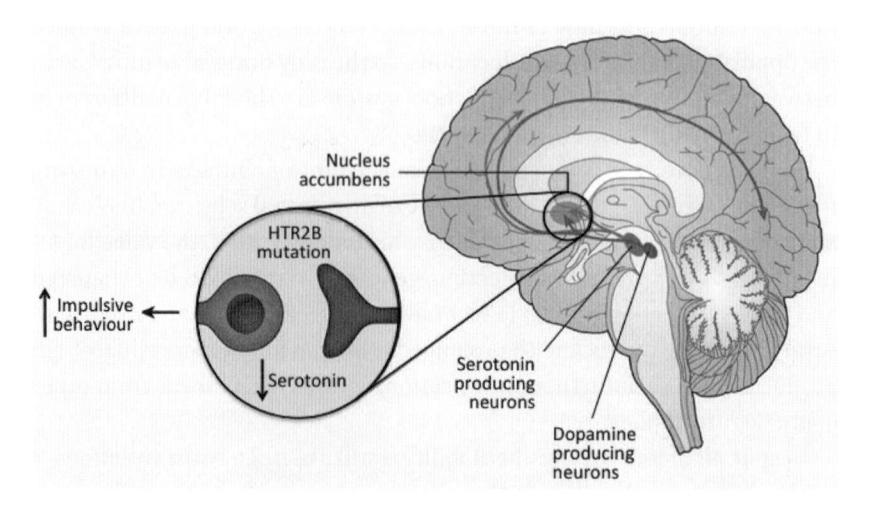

Fig. 7.3: A Finnish study shows that a mutation in the serotonin receptor HTR2B occurs three times more often among men who have been convicted for seriously violent crimes than among the entire population. In the nucleus accumbens, the signals of dopamine and serotonin secreting neurons are integrated; in case of non-functional HTR2B, this leads to reduced release of serotonin, which promotes unrestrained behaviour.

Here is an example of a complex phenotype that is rather strongly associated with one gene. In fig. 7.3 a schematic clarification is given of the functioning of the 'impulsivity gene', *HTR2B*, a *serotonin receptor* that is expressed in the brain. The receptor is involved with the integration of signals originating from serotonin and dopamine producing neurons in the nucleus accumbens. Mutation of the receptor will lead to a low concentration of the neurohormone serotonin, which normally has an inhibiting effect on the prefrontal cortex. A shortage of serotonin causes impulsive and unrestrained behaviour and is also associated with depression and *ejaculatio praecox*. A genotyping survey in Finnish prisons showed that the mutation in *HTR2B* occurred three times more among men who were convicted for serious crimes of violence than among the general population. Just like *DUF1220* this is a rather strong link between one gene and a complex phenotype, but also here the relationship in a quantitative sense is quite limited.

Ecogeographic variation in human body form

A classical example of recent evolutionary adaptation in the human species is provided by the so-called '*ecogeographic rules*' of Bergmann and Allen. In 1847, the German biologist Carl Bergmann was the first to document the trend that within a group of comparable homoeothermic animals, body size tends to increase with increasing latitude (distance from the equator). Bergman's initial observations were on moose in Sweden but the principle was soon applied to other animals and investigated in humans. The American zoologist Joel A. Allen developed a similar rule in 1877, which stated that within a group of related species, the ones living in cold climates tend to have shorter extremities and a more stocky body form.

Both Bergmann's rule and Allen's rule find their logic in surface volume considerations: bigger objects of the same shape have relatively smaller surfaces, and shorter extremities reduce the surface, also. Heat loss being proportional to surface and heat generation (metabolism) related to volume (weight), there is a seemingly sound physiological foundation for the two laws. Both Bergmann's rule and Allen's rule can be seen as reflecting an adaptive response and were recognised in the consolidation of evolutionary theory in the 1940s (the 'Modern Synthesis') as evidence that evolution acted not only in nature but also upon the human species.

In support of Allen's rule we see a clear relationship between the '*crural index*' of people and the average annual temperature at the place where they

Table 7.1: Crural and brachial indices for some human populations and Neanderthals

Populations	Crural index (%)			Brachial index (%)		
	Mean	SD	n	Mean	SD	n
Neanderthal	78.7	1.6	4	73.2	2.5	5
Koniag	80.5	2.3	20	75.3	2.6	20
European	82.9	2.4	243	75.1	2.5	240
Northern African	85.0	2.3	133	78.6	2.4	136
Southern African	86.1	2.2	66	79.6	2.5	67

Crural index = tibia length relative to femur, brachial index = radius length relative to humerus, SD = standard deviation, n = number of samples

live. The crural index is a measure for the length of the leg; since leg length varies more in the lower leg (tibia and fibula) than in the upper leg (femur) the crural index is expressed as tibia length relative to femur length. In a similar manner, the 'brachial index' is defined as radius length relative to humerus length. Table 7.1 lists some values of both indices. Although the absolute differences are small, they are significant and show an obvious relationship with latitude, confirming Allen's rule. There are similar trends for the hand (the first metacarpals of people living in cold environments are shorter than those in tropical climates), for the diameter of the caput femoris (wider in cold climate), and for hip width (wider). It is also evident from the data in table 7.1 that Neanderthals have a crural index most close to humans living in arctic environments (Koniag), which confirms that Neanderthals should really be considered to be cold-adapted humans, as may be expected from their living during the ice ages in Europe and Siberia.

Interestingly, the association of crural and brachial indices with latitude also holds for immature individuals. The indices remain constant over the course of growth. The relative length of the limbs thus seems to be something that is canalised in development (presumably in response to selection in the past) and does not show a lot of phenotypic plasticity towards the climate. An interpretation of Allen's rule in terms of the molecular mechanisms for limb formation that we met in chapter 2, is, however, not (yet) available.

Do the trends in bone dimensions indeed reflect altered surface-volume relationships? Yes, there is good evidence for this. It is not easy to reliably estimate the surface of the human body. One way to do it is to simulate the body with a number of cylinders and truncated cones, a model called 'Tin Man' (fig. 7.4), calibrated on real bone dimensions, from which the volume and the surface follow easily. This approach, applied to data from

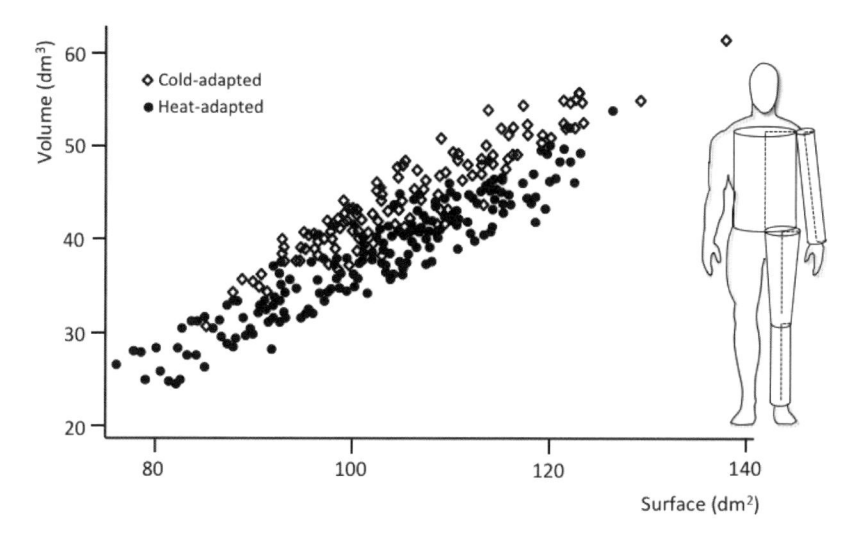

Fig. 7.4: Relationship between surface and volume of the human body for a number of people living in many different areas of the world. The populations have been divided according to whether they are cold- or heat-adapted, the latter showing a larger surface at the same volume. On the right the 'Tin Man' model is depicted, which was used to estimate volume and surface from skeletal elements.

thousands of individuals out of 46 different populations all over the world, show a quite consistent difference between 'cold-adapted populations' and 'heat-adapted populations', the latter having about 9% more body surface at the same body volume (fig. 7.4).

We should realise though that the dimensions of the various bones of the human body are not independent from each other. When the correlations among bone lengths are taken into account it appears that humerus length does not decrease but increases with increasing latitude, which seems to be a compensatory though nonadaptive response to the decreasing radius; a similar though less pronounced response is seen in the femur relative to the tibia. In addition, any comparison of traits between populations should be corrected for genetic relatedness. If two populations have similar body proportions, this may not only be due to their living in a comparable climate but also to a historical relationship between them. By and large, two populations are often similar to each other just because they live close by, as we saw in chapter 4 (the IBD principle). When a correction is made for genetic distance the Allen's rule effect diminishes, but is still present for distal limb lengths and hip width, but not for femoral head width. So an explanation for Allen's rule that is only based on natural selection overstates the issue: there are multiple evolutionary forces shaping the various morphologies of the human body and in some features the latitudinal trend may be due to a serial founder effect rather than to natural selection.

Evolution of biomedical traits

The most important mortality factors in prehistoric times are assumed to be external and biotic, such as food shortage, infections through pathogenic microorganisms and parasites, and injuries inflicted by fellow humans. Humans avail of different mechanisms to minimise the damage caused by these factors, such as the adaptive and native immune systems, and responses like blood clotting and wound healing. As these mechanisms have a direct and large effect on fitness, it is expected that they be under strong evolutionary pressure. To answer the question whether defence mechanisms do indeed evolve, we must be able to monitor a population across multiple generations. Only then can the variation of advantageous traits be estimated and their inheritance and changes in frequency studied. In this manner, it becomes possible to establish whether natural selection influences these inherited phenotypic features and we can answer the question whether humans still evolve.

The *Framingham Heart Study* is one of the most famous examples of such a cohort study across multiple generations. The study was a project of the National Heart, Lung, and Blood Institute, in collaboration with Boston University (since 1948), and it is still ongoing. Originally, it was intended to study the progress of cardiovascular disorders across multiple generations, involving over 5,000 Americans of European origin. Ultimately, it was decided to monitor seventy biomedical characteristics in this cohort, including blood pressure, cholesterol content, and glucose level in the blood. Every two years, these traits are scored and currently data are available for three generations. The fitness of the individuals is quantified by noting the total number of children of a person during his or her entire life (*life-time reproductive success*, LRS).

A short while ago the first analysis was published, in which seventy biomedical characters across three generations were correlated with fitness. There was a significant correlation for six of them. The characters 'body weight' and 'age at the beginning of the menopause' showed a positive correlation with fitness, which implies that there is positive selection on these traits. Four of the characters showed a significant negative correlation with reproductive success; apparently, these were selected against. Those were cholesterol content, total body length, blood pressure and the age of women upon the birth of their first child.

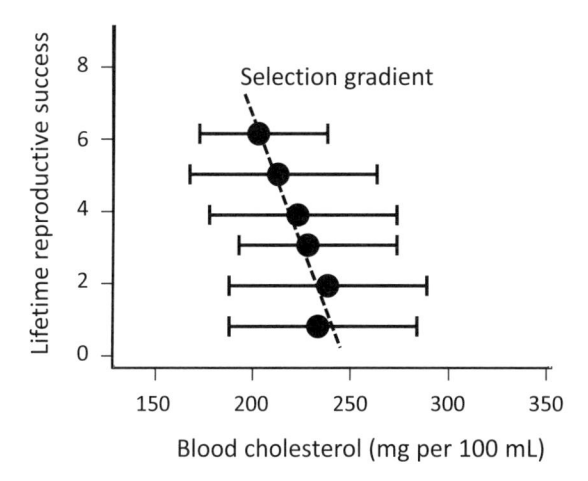

Fig. 7.5: Data concerning three generations of 2,227 women in the Framingham Heart Study show that women with higher reproductive success on the average have a lower total blood cholesterol level. This implies that in modern society there is negative selection on blood cholesterol.

In fig. 7.5 the results are shown for one of the variables, cholesterol blood concentration. Although the variability is large, the results show a significant trend: women with a lower cholesterol content on the average have a higher reproductive success. The slope of the line is called the *selection gradient*; this gradient indicates the degree by which average cholesterol content decreases, if natural selection promotes an increase of LRS. The selection gradient is equivalent to the selection differential in fig. 7.2; multiplied by the heritability, it provides the response to selection.

A group of American scientists, headed by the evolutionary ecologist Stephen Stearns, calculated that selection will cause noticeable changes in health over the next ten generations. Cholesterol concentration will decline by 4% and blood pressure by 2%. Furthermore, the average age of menarche (first occurrence of menstruation) will reduce by one year while the age of the menopause will increase by a year; people will be two centimetres shorter and weigh one kilogram more. Beyond doubt the analysis of the Framingham Heart Study data shows that modern humans still evolve in terms of biomedical traits.

It is important though to note that the results are context-dependent: the data are based on Americans with a European background. Comparable studies with other population groups may produce other results. Still, they can also provide useful information, because they can provide insights into which population groups are more or less exposed to health risks.

A second example involves the use of a new carbohydrate source in our diet: *lactose* in milk. Lactose is the most important source of nutrition for newborn mammals; during the first months of their life newborns do not get any other nutrition than mother's milk and lactose is the only carbohydrate in milk. It is known that newborns express the enzyme *lactase* (*LCT*) in their gut, to convert the disaccharide lactose into the monosaccharides glucose and galactose, which can be readily absorbed. *LCT* expression is shut down after the young is weaned and switches to a diet of plants or meat. Until the development of agriculture, this was also the case in human populations. Since adults do no longer produce the enzyme lactase, lactose in food has a negative influence on the digestion; it may cause diarrhoea. Lactose intolerance in adults can be regarded as the plesiomorphic condition.

But when humans started to live with farm animals that produce milk, a new food source emerged. A 'mutant' version of *LCT*, in which the gene is not switched off after childhood, gained a great advantage, because people with this mutation could consume milk without any problems. Actually the mutation is not in the *LCT* gene itself but in an upstream element that regulates *LCT* expression. And to make it more complicated: there are in total six different mutations that cause the lactase-persistence phenotype. Two of them are genetically linked and are common in Eurasia, North Africa and Central Africa, while four others mainly occur in the Middle East and East Africa. The best-investigated allele is *-13.910*T*, which is a mutation (C substituted for T) in an intron of a gene called *MCM6*, 14 kb upstream of *LCT*. *MCM6* encodes an enhancer protein that regulates *LCT* expression. Because of the large genomic distance between *LCT* and *MCM6* it took a long time before molecular biologists discovered the reason for this regulatory change in *LCT* expression. The mutation creates a new binding site for another regulatory protein, causing the regulatory complex on the *LCT* promoter to remain active and stimulate *LCT* expression. Another mutation in *MCM6* designated *-22.018*A* is in genetic linkage with *-13.910*T* and most likely acts as an epistatic factor, as it cannot induce the lactase persistence phenotype by itself. The four other mutations also cluster in *MCM6*, but whether they act in the same way is not known.

The lactase persistence allele *-13.910*T*, here abbreviated *LCTp*, inherits as an autosomal dominant trait, which simplifies the evolutionary analysis. A U.S. team of medical geneticists was the first, in 2004, to show that *LCTp* must have been under strong positive selection, because it was located on an unusually long haplotype block (> 1 Mb), and it showed clear evidence of a *'selective sweep'* (see the case of G6PD discussed in chapter 4). The team was also able to estimate selection coefficients for *LCTp* from its change in frequency, assuming it increased after a single mutation in an ancestral population. Such estimates of selection coefficients have now been published

Table 7.2: Estimates of selection coefficients for the -13.910*T allele of LCT, causing lactase persistence, and estimated timing of selection, in a number of human populations

Population	Selection coefficient	Start of selection (years before present)
European American	0.082*	11,419*
Finnish/Swedish	0.140*	2,407*
European American	0.039	9,323
European American	0.048	5,575
Western Finnish	0.043	5,200
European	0.095	7,441
European	0.012	n.e.
Finnish	0.025	11,200

*Average of range given by authors; n.e. = not estimated

by a number of other studies and are summarised in table 7.2. The data are consistent with a model in which *LCTp* arose early in the Neolithic and was subjected to strong directional selection. The selection coefficients are among the highest reported for loci in human populations.

The data reconstructed from present frequencies of lactase persistence are confirmed by ancient DNA, which show a similar pattern. Interestingly, *LCTp* is lacking from any ancient DNA older than about 5,000 years, so it indeed represents a very recent mutation. What may have caused such strong selection for it? The most common explanation is the switch to agriculture in the Neolithic revolution, which brought an advantage to people who could digest lactose during their whole life, since this allowed them to consume milk from livestock. This adaptive explanation is supported by a correlation between pastoralism and lactase persistence across Old World populations, and it is often cited as a classical case of evolution acting upon the human species.

The adaptive explanation is, however, plagued by many exceptions and inconsistencies. The current distribution of *LCTp* is quite irregular (fig. 7.6); several Asiatic populations (Mongols, Kazakhs) are pastoral but have a very low frequency of lactase persistence. And why is the mutation not found in populations older than 5,000 years ago, while agriculture was developing for thousands of years earlier in several places? How is selection influenced by the common practice of milk fermentation (to make yoghurt), which removes most of the lactose? And how strong is the negative effect of lactose, anyway? (Many lactase-deficient people can tolerate a moderate amount of lactose without physical consequences.) Most likely the story of lactase

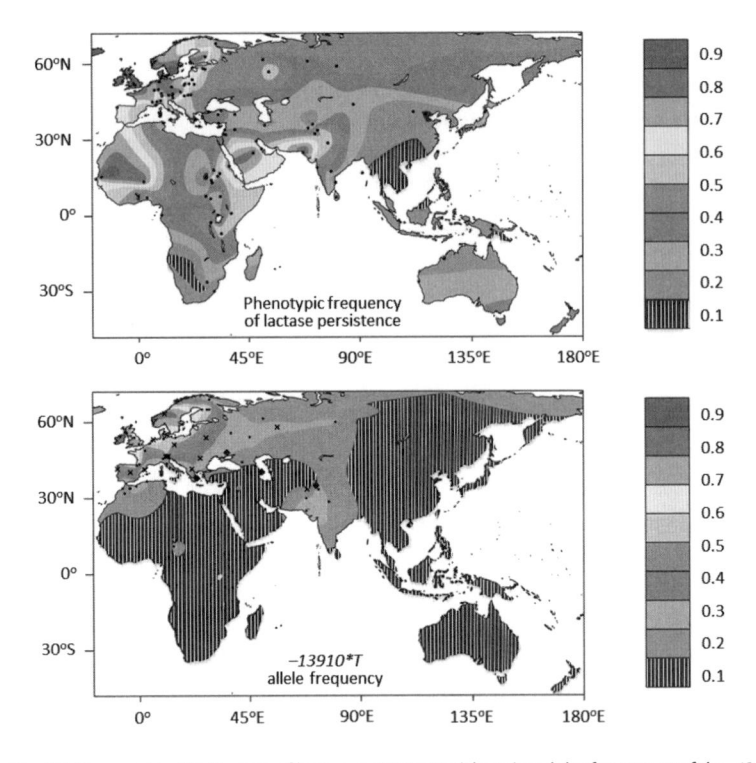

Fig. 7.6: Geographic distribution of lactase persistence (above) and the frequency of the *-13910*T* allele of lactase (below). The mutant allele promotes a phenotype that does not show deactivation of lactase expression after childhood, which makes lifelong digestion of lactose possible. The swift spreading of the new allele is interpreted from a selective advantage associated with raising dairy cattle, however, present geographical matching between lactase persistence phenotypes and *-13910*T* allele frequencies is meagre.

Table 7.3: Traits in today's human populations demonstrated to be under selection, with a listing of loci involved

Traits	Genes involved
Brain size, cognitive capacity	NPBF, MAOA, DRD4
Light skin colour	SLC24A5, MC1R, OCA2, KITLG
Increased fertility	H2 haplotype
Protection against hypertension	AGT, CYP3A
Protection against malaria	Duffy antigen, Hemoglobin C, G6PD, TNFS5N
Protection against smallpox and AIDS	CCR5
Milk drinking, lactase persistence	LCT
Enhanced starch digestion	AMY1
Bitter tasting	TAS2R38 (PTC)
Living at high altitudes	EPAS1
Sustained diving, thyroid and spleen size	BDKRB2, PDE10A

persistence involves natural selection as well as migration, founder effects and maybe correlations with other phenotypes.

To complete this short overview of evidence for recent human evolution in response to biomedical factors we present in table 7.3 a list of currently known and well-supported cases of ongoing evolution in present society. A number of these cases we have already met in other parts of this book (DRD4, G6PD, LCT) – others we leave to the reader for further study.

Evolution of the life cycle

The human life cycle may be subdivided into six stages: infant, child, juvenile, adolescent, adult and elderly. The relative time periods of these stages have been subject to significant change during the evolution of the hominins. Due to continued developmental delay (neoteny), the beginning of the adult stage was extended, which made the adolescent period much more important than it is in other species. Women begin reproducing at a later age, but wean their infants at younger age and have a relatively short inter-birth interval. Finally, the protection from infectious diseases provided by modern medicine caused a marked increase of life expectancy, with the consequence that a significant part of the human population now consists of post-reproductive persons and elderly. Teeth-eruption patterns in fossils demonstrate that the new life cycle started with *H. erectus* and was likely completed in the ancestor we share with Neanderthals, since growth patterns reconstructed from Neanderthal skeletons show that most parts of their body developed at the same rate as they do in humans.

One of the more mysterious human features is the menopause. The fertility of women decreases with age, until infertility is reached, after the menopause, around the age of fifty. Oocytes seem to have a fixed lifespan, independent of the life cycle of the female. The fact that women have a postmenopausal life with apparent zero fitness is difficult to explain from a purely selective point of view. In wild chimpanzees and other great apes, the menopause does not occur; it only does if females reach a high age in captivity. It may be sheer coincidence that the biological production of eggs ceases after 35-40 years. In that case the menopause would be a physiological necessity. But there are also elegant evolutionary explanations.

The most appealing theory is known as the *'grandmother hypothesis'*. This hypothesis states that grandmothers, after the menopause, can increase their fitness by helping to raise their grandchildren. Those grandchildren on the average possess a quarter of their grandmother's DNA (compare Hamilton's

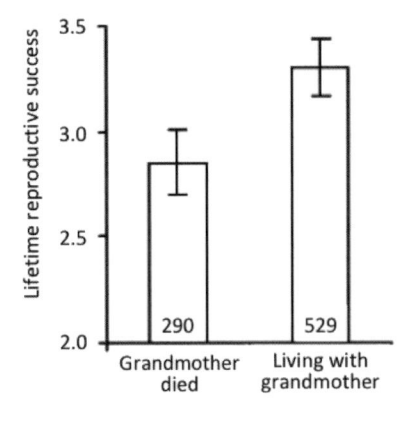

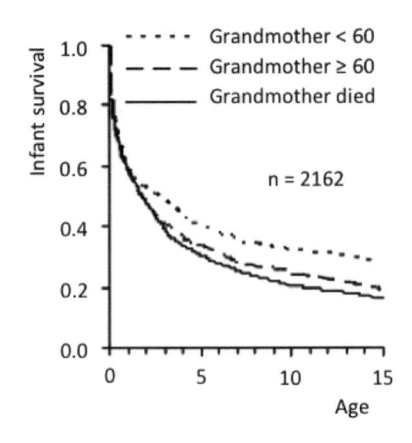

Fig. 7.7: Demographic data on farming families from the nineteenth century support the grandmother hypothesis: a woman's reproductive success increases when the grandmother is living with her (left) and the survival of the children increases in the presence of the grandmother, especially if she is below sixty (right). This shows that the non-reproductive postmenopausal period, which is exceptionally long for humans, has an adaptive significance.

rule in chapter 6). So enabling her daughters to have more children can enlarge a grandmother's share in the gene pool of the next generation. This hypothesis sounds far-fetched, yet it is supported by several studies. Detailed demographic research on Canadian and Finnish population groups in the nineteenth century show that the presence of a grandmother in her daughter's family has a positive effect on the survival of the grandchildren (fig. 7.7). The age at the first pregnancy of the daughter is also lower on average. Finally, it appeared that mothers, being supported by grandmothers, had their first three children with shorter intervals. These data confirm the hypothesis that lengthening of the life expectancy for women after the menopause could be an adaptation to enhance their own (inclusive) fitness.

After the industrial revolution, living conditions changed drastically and human society became much more complex. The details of pre-industrial communities, as depicted in fig. 7.7, are no longer valid in modern society, so whether the grandmother effect is still functional, is uncertain. It now seems that the age at which the menopause starts is increasing by selection (compare the Framingham Heart Study).

Partner choice and sexual selection

Many animals produce volatile chemical compounds that influence the sexual behaviour of potential partners. These substances are called *pheromones*. The function of pheromones is best understood in *Drosophila*. Fertile males produce the organic esterified compound 11-cis-vaccenyl acetate (*cVA*), present in the epidermis and the ejaculate. This substance acts as an *aphrodisiac* on females; it arouses the sex drift, due to which they become receptive and accept copulation by males. Regarding males, on the other hand, *cVA* acts as an anti-aphrodisiac and may even cause aggression. Females detect the pheromone via an olfactory receptor in the antennas, which, in the case of binding by *cVA* send an electrical signal to the central nervous system. The question is why the same *cVA* has a different effect on males and females. This is due to the neural circuit that is activated by the *cVA* signal. Up to the protocerebrum, the neural pathway is identical, but afterwards there is a strong differentiation between the two sexes, as a result of which the signals reach different neurons in males and females. Therefore the same signal on the antennas may result in sexual willingness in females and sexual restraint in males.

Pheromones also play an important role in vertebrates. A well-known example is the reaction of a sow to the smell of a boar. As a consequence of the boar's smell, a sow instinctively assumes a copulation position, allowing the boar to mount her. This effect is attributed to the pheromone *androstenone*, a substance related to the male sex hormone testosterone. Androstenone is detected by olfactory receptors in the *vomeronasal organ* (also called Jacobson's organ), which is situated in the top of the nasal cavity of vertebrates. This organ has over 400 olfactory receptors and can detect a very large number of volatile compounds.

In humans, as discussed in chapter 4, hundreds of olfactory receptors were lost. Also the vomeronasal organ of humans is reduced, although traces of an olfactory nerve can still be found in the nose cavity epithelium. It is therefore possible that human beings could also be sensitive to androstenone-like chemical signals emitted by the other sex. Such signals could play a role in partner choice.

The fact that human partner choice is subject to unexpected and unconscious signals was first suggested in experiments conducted by the Swiss researcher Claus Wedekind in 1995. In the famous '*sweaty T-shirt*' test male test persons were asked to wear a T-shirt for two days without using a deodorant and refrain from eating strongly smelling food items. Afterwards female test persons were invited to smell the T-shirts, without knowing which male they belonged to, and indicate to which T-shirt they felt most attracted. Simultaneously, all test persons were genotyped for genes of the immune system (*MHC II*). Upon analysis of the results, it turned out that the females showed a preference for partners with immune genes different

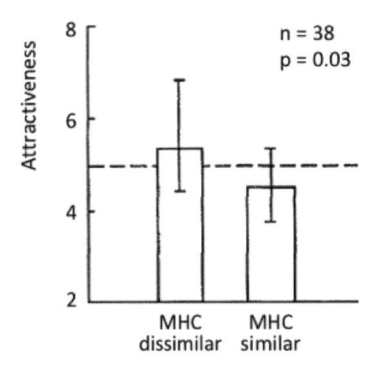

Fig. 7.8: Females prefer T-shirts worn by males with an MHC genotype different from their own. This classical experiment by Klaus Wedekind shows that females unintentionally promote MHC heterozygosity of their offspring, which is evolutionary beneficial.

from themselves (fig. 7.8). The evolutionary advantage of this is obvious: by making this choice they increased the chance of their children being heterozygous. *MHC II* genes are codominant and heterozygotes are more resistant against different infection diseases than homozygotes. Hence, there is a strong selection for women to enhance the chance of healthy children through partner preference.

The Wedekind experiment has been repeated many times and proved to be largely consistent. Also, the students attending the human evolution course given by the authors of this book show immune system-dependent partner preference. The outcome of the experiment is influenced, though, by the menstruation cycle of the woman and the use of contraceptives. The probable mechanism behind the effect is that the *MHC II* genes are in strong linkage disequilibrium. Over a distance of several Mb the recombination frequency is quite low. Coincidentally, in this area there are also a large number of olfactory receptors, which are genetically linked to the immune genes. However, it is not known whether there is any causal connection between olfactory receptors and the response in the 'sweaty T-shirt' test.

Obviously males and females do not choose their partners merely on the basis of chemical signals. Chemical communication is unconscious and perhaps only plays a role when other matters are equal. If males and females are asked what factors are of personal importance in partner choice, the obvious factors will come out, such as amiability, educational level and cultural background. The preferences of men and women differ when they are asked about the physical beauty of the potential partner (on average, men find this more important) and income level (which women on the average find more important). It is striking that these gender differences are manifest in almost every culture. A standard list of eighteen questions was drawn up and used in thirty-three countries around the world, resulting in the same stereotypes every time. This might have an evolutionary background: men have a large interest in a fertile woman (indicated by beauty) and women, due to the enormous investment they make through their pregnancy and nursing period, have an interest in a partner who will ensure sufficient food and other resources.

In principle, partner choice can lead to *sexual selection*, which means that characters in one sex evolve in response to certain preferences by the other sex (*intersexual selection*) or as a reaction to the competition between members of the same sex (*intrasexual selection*). In the animal kingdom these processes are often excessive and lead to extravagant attire as well as large differences in body size and 'armament'. Take, for instance, the colourful feathers of birds, the exaggerated antlers of deer or the ridiculous eye stalks of stalk-eyed flies (Diopsidae). Darwin recognised the importance of sexual selection in his book *The Descent of Man* (1871).

Sexual selection is also a meaningful process in human beings, and humans can even be considered a suitable model species for this type of research, since preferences can be observed easily and the research is not invasive. We saw in chapter 1 that during evolution of the hominins, sexual dimorphism in body size has decreased, indicating reduced intrasexual selection (among men). Sexual selection is most obvious in intersexual preferences for certain visible body features. A large number of experiments have been conducted on this. Regarding the preference of men for women, two traits emerge as important, the ratio between waist and hip width (*waist-to-hip ratio*, WHR) and the characteristics of the face (symmetry in particular, as well as gentle facial features).

Studies show that men consistently prefer a waist-to-hip ratio of 0.7. This was demonstrated, for example, with manipulated photos of women, whereby the brain activity of the man was studied. If men must choose between women with a WHR of 0.9 and 0.7 they obviously prefer the latter (fig. 7.9). This choice has an evolutionary background; the waist width of a woman is correlated with her fertility. Due to their unconscious preference for a WHR of 0.7, men unconsciously choose the most fertile women.

Facial symmetry and 'gentle traits' aren't really correlated with fertility. The fact that men still have a strong preference for such characters may relate to gentle traits being an indicator of young age. By preferring 'baby face'-like traits males in fact choose a partner that may be expected to be fertile. Something similar probably holds for breast size. Despite the male

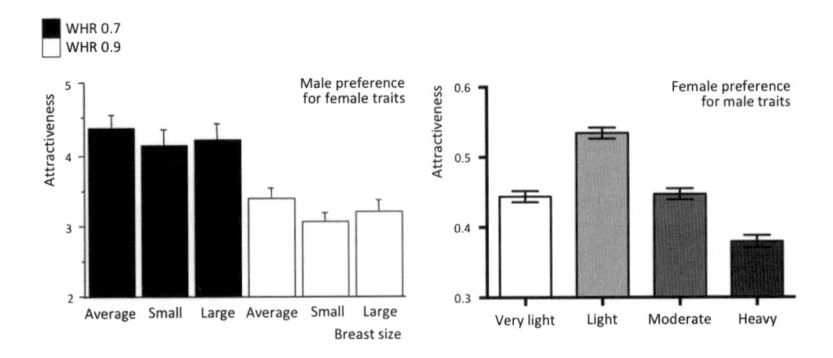

Fig. 7.9: Illustration of sexual selection among humans. Left: men prefer women with a waist-to-hip ratio of 0.7, independent of the breast size of the woman in question. Right: women prefer men with moderate body hair. In both cases it is likely that partner choice is driven by 'good genes'.

fascination for this part of the female body, breast size is not strongly correlated with fertility, although in some cultures large breasts (and buttocks) are considered a sign of health. Research, however, shows that WHR in the male's eye is considered more important than breast volume (fig. 7.9).

Physical characteristics are often assumed to play a smaller role in the preference of females for men, than vice versa, but women do actually use physical characteristics to assess men. If females show partner preference, they pay attention to indicators that reflect the testosterone levels in the male's blood. Most familiar are the jawline and the body hair. In fig. 7.9 the results of Australian research are shown, which prove that women fancy light body hair and are less attracted to potential partners with heavy body hair. These results are consistent with the sexual selection hypothesis for the nakedness of *H. sapiens*, as discussed in chapter 3.

There has been some speculation about the question whether also penis size would be subject to sexual selection. Humans have a relatively large penis, larger than that of great apes. In a recent study on the preference of seventy-five American female students who were shown 3D models of penises, it turned out that both for a one-night stand and for a long-term relationship females preferred an erection slightly above the American average. Although this type of research is highly context-dependent and never includes random sampling of test persons from a specified population, it is not unimaginable that penis size of human beings has been subject to sexual selection. While in mammals, the scrotum and the testicles are assumed to have a sexual signalling function (*cf.* fig. 3.17), in humans this role may have been played by the penis and the evolution of larger penises could have been driven by female mate choice.

Evolutionary medicine

Knowledge of evolution may provide refreshing insights in biomedical research. A well-known example is the emergence of *jaundice*. Jaundice is caused by accumulation of *bilirubin* in the body. Bilirubin is a degradation product of haemoglobin that is released as red blood cells are degraded (fig. 7.10). Normally, the poorly soluble bilirubin is made soluble in the liver and disposed of via the bile and the faeces. In the case of jaundice, this process is disrupted in different ways. Some patients break down too much haemoglobin, while others show a defect in the liver due to which bilirubin cannot be disposed of. Finally, the bile duct may get clogged, as a result of which bilirubin cannot be excreted. In the case of *Gilbert's syndrome*, an enzyme that conjugates bilirubin with glucuronic acid (UDPGT, uridine 5'-diphospho-glucuronosyltransferase) is mutated. Normally, the conjugation with glucuronic acid solubilises bilirubin, allowing it to be excreted via the bile. In the event of a mutated UDPGT, bilirubin starts accumulating in the blood (fig. 7.10).

The odd thing is that haemoglobin is converted into bilirubin via a well-soluble intermediate form, *biliverdin*. One might suggest that it would

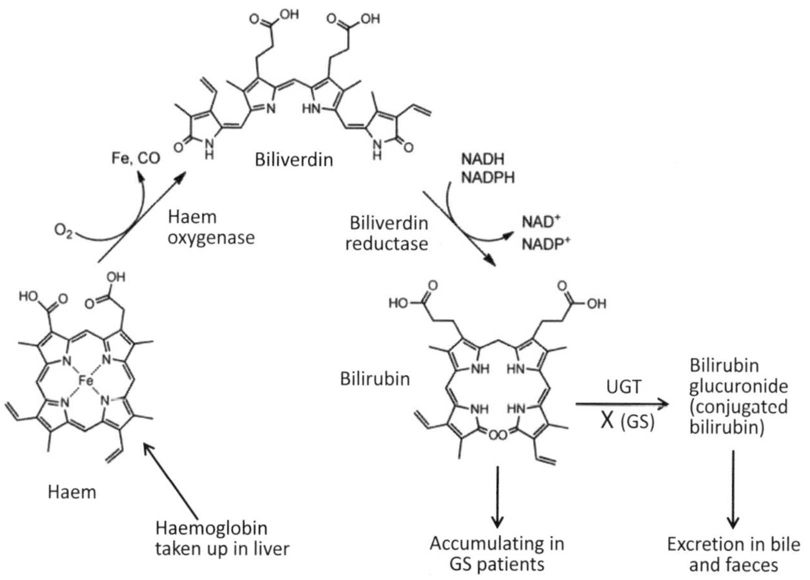

Fig. 7.10: Diagram of haem metabolism in liver. The end product is conjugated bilirubin, which is excreted via the bile. In patients with Gilbert's syndrome (GS), the enzyme uridine 5'-diphospho-glucuronosyltransferase (UGT) is mutated, due to which biliverdin accumulates and reaches a high concentration in the blood, which leads to jaundice, but also provides protection against heart failure.

be much more convenient to directly excrete the water-soluble and non-toxic biliverdin. The fact that this doesn't happen can be attributed to the antioxidant effect of bilirubin; it offers protection against oxidative stress. Apparently an increased (yet not too high) concentration of bilirubin is beneficial for the body. Epidemiological studies show indeed that people with a relatively high bilirubin content, suffering from chronic jaundice, have a five times lower risk of cardiovascular disease. So bilirubin also has a positive effect; natural selection has promoted the degradation pathway of haemoglobin to bilirubin, or at least maintained it. Evolutionary biology is therefore able to explain why bilirubin is produced still, although the substance has a pathological effect when present in high concentrations. This insight is easily overlooked in medical studies focused on the adverse effects of jaundice.

A second example in which evolutionary thinking provides a new insight in biomedical research is the in vitro fertilisation through *intracytoplasmic sperm-injection* (ICSI). This technique is applied when males wish to have children, but are unable to fertilise their partner in a natural way, due to insufficient production of active sperm. In many cases this is caused by a genetic disorder: such men lack part of the Y chromosome which carries genes that encode so-called azoospermia factors, proteins that are essential for the formation of a healthy sperm cell. In a natural situation these men would experience zero fitness; after all, they cannot produce any offspring. As a consequence, there is strong selection against this Y chromosomal disorder. This is what we call *purifying selection* (*cf.* table 4.4).

In ICSI, a man with bad semen quality is still able to beget children; with the consequence that medical technology is used to intervene in human evolution. All boys born via ICSI-fertilised eggs are 100% sterile: the defective Y chromosome after all, is passed on to his sons. As a result, we're facing an ethical dilemma. On the one hand we agree that every human being has the right to have children, on the other hand, we run the risk that, due to medical intervention, genetic defects continue to exist in the population or will even increase. In the case of ICSI, the group that is treated is very small and will hardly cause the genetic defect in the Y chromosome to increase in the total population. Other medical techniques, for example, gene therapy, will cause similar effects, though. A gene-therapeutic method being is developed to treat cystic fibrosis, which will reduce the purifying selection on mutant alleles of the *CFTR* gene, which causes the disease.

These medical interventions, as a result of which we take our evolution into our own hands, are expected to become increasingly more important in the future. For example, currently it is possible to check human embryos for genetic defects after an in vitro fertilisation and to repair these before the embryos are placed back inside the womb. A revolutionary new molecular technology known as *CRISPR-Cas9* has become available which allows '*genome editing*' without leaving traces. In this technique genetic defaults may be repaired in the embryo and thus removed from the germ line, since the children will not show the genetic defect anymore. In principle, this technology could also be used to enhance desired characteristics (body length, hair colour, muscular development *et al.*). It would go too far to discuss this technology here. In December 2015 a group of scientists called for a moratorium on such experiments until it is clear what the risks are and how we deal with the ethical implications.

A third example of ongoing evolution in modern society concerns the battle against infectious diseases. Until the early twentieth century, diseases caused by bacteria and viruses were one of the most important causes of death in human populations. The construction of sewers turned out to be an enormous improvement. The fact that the Scottish physician Alexander Fleming discovered penicillin in 1928, subsequently ensured a further reduction of bacterial infections. Antibiotics such as penicillin proved to be very effective against all kinds of gram-positive pathogenic bacteria. Nowadays they are widely used, not only in hospitals, but also in livestock farming, to contain infections on farm animals in the stables.

Excessive use of antibiotics is a strong selective force that induces evolution of resistance in bacteria. The medical and economic consequences of the emergence of resistant pathogens are substantial. A study in the United States showed that resistance is responsible for approximately 450,000 more hospitalisations, 5,000 fatalities and over 7 million extra days of hospitalisation, with an economic loss amounting 75 billion dollars.

Resistant bacteria emerge predominantly in and around *intensive care* units, where a lot of antibiotics are applied, due to the high risk of infection for weakened patients. But every introduction of a new antibiotic in the market is unfortunately soon followed by the discovery of pathogenic bacteria that do no longer react to it and are, as such, resistant to the new medicine. In that sense we can say that there is a new arms race going on, between the development of new antibiotics on the one side and evolution of bacterial resistance against that new drug on the other side.

The emergence of antibiotic resistance is a purely evolutionary process. A well-documented example of this is the resistance against the drug *macrolide*. The mode of action of this substance is based on binding to 28S rRNA in the ribosomes of the target bacteria. Protein translation is inhibited and the bacterium is killed. Macrolide recognises only a small area in the

28S rRNA molecule, the sequence GAAGA. Resistance occurs when the nucleotides in this area change due to a mutation. Since the nucleotide sequence GAAGA is not essential to the functioning of the 28S rRNA during protein synthesis, a mutation that negates the action of macrolide emerges relatively easily.

Evolutionary biologists have given a lot of thought to solutions to restrain the resistance problem in hospitals. One possible solution is to change the types of antibiotics over time. After completion of the first treatment, a second one follows with a different antibiotic. This method is called *cycling*; it works until double resistance has been developed against both antibiotics. Also treatments were developed on the basis of evolutionary models that may provide a more stable resistance. This approach consists of alternating antibiotics with different modes of action within the period of the same treatment. In addition, it is useful to switch the antibiotics per patient during the treatment at random moments. As such, a mosaic develops of different antibiotic applications in space and time within a hospital, as a result of which the emergence of resistance is permanently reduced (fig. 7.11).

It is not only medical care that influences human evolution. Society is changing so fast that evolutionary adaptation to new conditions is hardly

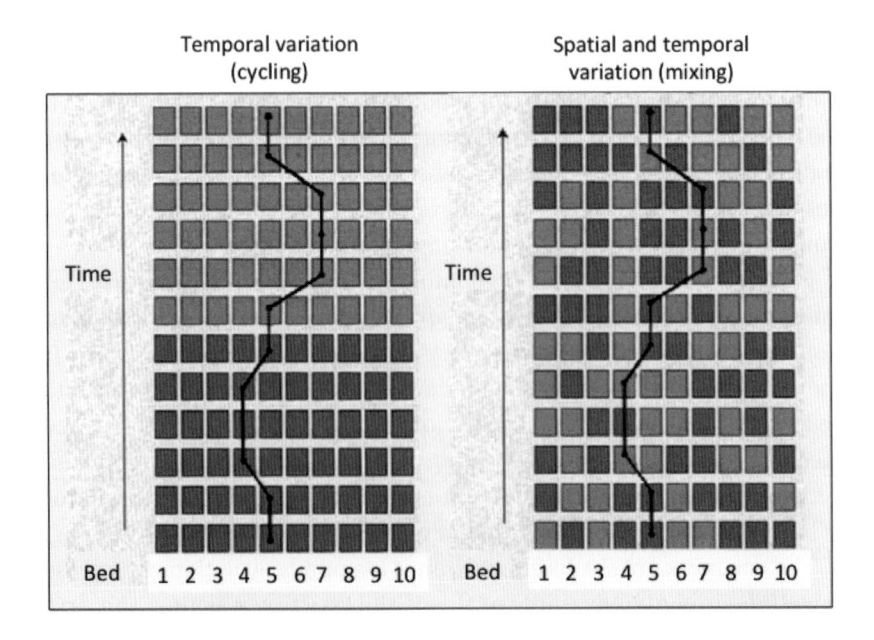

Fig. 7.11: Strategies to prevent antibiotic resistance in a hospital. By using different antibiotics in successive treatments (*cycling*) or by haphazardly varying the use in space and time (*mixing*), evolution of resistance can be restricted. The line displays a possible course of jumping infections.

possible, as evolution requires a large number of generations. So in modern society, a *'mismatch'* is occurring between the changing environment and our adaptations, which largely date from the *'environment of evolutionary adaptedness'* (EEA) (see fig. 6.14 and fig. 7.12). In a recent book by Ronald Giphart and Mark Van Vugt, *How Our Stone Age Brain Deceives Us Every Day*, numerous examples of mismatches are discussed.

A widely suggested mismatch is the discrepancy between current food availability and our physiological use of food components for maintenance, activity, growth and reproduction. Since the emergence of the first hominins there has always been a strong selection on survival in the event of food scarcity. Humans in the EEA had to feed themselves by collecting fruits and seeds. Later, meat was added as a protein source. Over time, the diet became more diverse and food intake gradually improved through the development of cooking techniques. This also had consequences for the anatomy of our intestines, as we have seen in chapter 3. But since then the human body has hardly had time to adapt to an excess of food. Our body is still equipped to make very efficient use of carbohydrates, fats and proteins, whereby surpluses are stored as reserves in fat tissue. As a result, overweight has become an increasing problem in modern society. Obesity has become the most abundant medical disorder of the twenty-first century. The World Health Organization recently made an inventory of diabetes prevalence around the world. It was remarkable to find that in quickly developing countries, such as India and China, the number of diabetes patient has grown exponentially, to more than 8% of the population.

The health risks of being overweight are related to the *'body mass index'* (BMI), the body weight in kilograms divided by the square of the body length in metres. Corpulence is determined by many different genes, each of which has limited influence. A lot of research focuses on so-called *'thrifty genes'*, which are involved in fat storage in adipocytes. The transcription

Fig. 7.12: Human behaviour with respect to food acquisition has evolved over a period of almost 2 million years. Compared to this history, modern society is only very recent, which explains why our behaviour still includes adaptations to an environment that no longer exists.

factor *peroxisome proliferator-activated receptor gamma 2* (PPAR) seems a good candidate. Within the European population a PPAR variant was discovered that provides protection against type-2 diabetes. Persons with this variant are more sensitive to the influence of insulin on blood sugar, ensuring better regulation. Unfortunately, the ancestral variant of PPAR, associated with a higher risk of type-2 diabetes, is still the most abundant allele in the world population – so it seems there is, indeed, a mismatch. Evolution by selection for the favourable PPAR allele has not been able to increase its frequency to a significant extent.

While this example is widely cited as evidence supporting the *mismatch* hypothesis, there also are publications that argue against it. In a recent genome-wide association study, 115 genes were found that are associated with a high BMI. Nine of them show a form of positive selection, but five are selected for the allele that protects against obesity; those genes advance a slim body. There were only three genes for which positive selection could be demonstrated for an allele that increases the chance of obesity. The retention of energy and the accumulation of fat in order to overcome periods of famine apparently weren't always beneficial for primitive humans. However logical the mismatch hypothesis sounds as an explanation for the obesity epidemic, the genetic proof is currently by no means convincing.

Do humans still evolve? This chapter provided various examples to show that this is certainly true. Evolution influences our body shape, our life cycle, our partner preference, our ability to digest lactose and countless medically relevant traits and behavioural syndromes. Knowledge of evolutionary mechanisms will not only help to better understand our own body, but also to improve health care.

Epilogue

In this book we have advocated an integrative approach of human evolution, in which mutation, development, selection and drift are studied in conjunction with each other. In such an approach, not only is there room for natural selection, the traditional principle of Darwin, but also for the limitations and constraints resulting from the body plan and for the influence of mutation and molecular tinkering on the development of a phenotype. We have emphasised that we cannot understand evolution by limiting the discussion to natural selection; we also have to study the origin of new characters. Molecular turbulence in the genome is the source of novel phenotypes which subsequently spread and fix by selection but sometimes also through neutral processes, such as colonisation and drift.

The definition of evolution according to the paradigm of the 'Modern Synthesis', cited in chapter 4, states: 'due to mutation new alleles emerge of which the frequency changes through selection and drift'. This definition is still at the core of evolutionary theory. However, over the years a growing group of evolutionary biologists have argued that the Modern Synthesis is too restrictive a framework and have pointed at the incompleteness of the current paradigm.

Since 2009, the centenary of Darwin's birth, a new evolutionary concept has been developed in which the Modern Synthesis definition is extended in the light of progress in developmental biology, genetics and genomics of the twenty-first century. An important exponent thereof is the American philosopher and theoretical biologist Massimo Pigliucci. Together with Gerd Müller, professor of zoology at the University of Vienna, he edited *Evolution: The Extended Synthesis* (2010), a volume in which a number of leading biologists convey their vision on the new developments in evolutionary biology. In the figure below, a number of components of the new theory are schematically shown. Our treatment of human evolution fits well with this new paradigm.

At the end of *On the Origin of Species* Charles Darwin used the image of a *tangled bank* (in the first edition it says 'entangled bank'), a model for the environment in which many plants grow in an interconnected manner and animals crawl around in mutually complex interactions, in brief, an environment in which everything is interrelated and each member of the community must fight for its position. Darwin may have been inspired by Alexander Von Humboldt's ideas on the unity of organic life, as he had a copy of Humboldt's book on board the Beagle. The image of the tangled bank has inspired also many a modern biologist. Carl Zimmer named his evolution book after it and Michael Eisen wrote a poem about it. Graham Bell used the image in his explanation for the evolution of sexual reproduction.

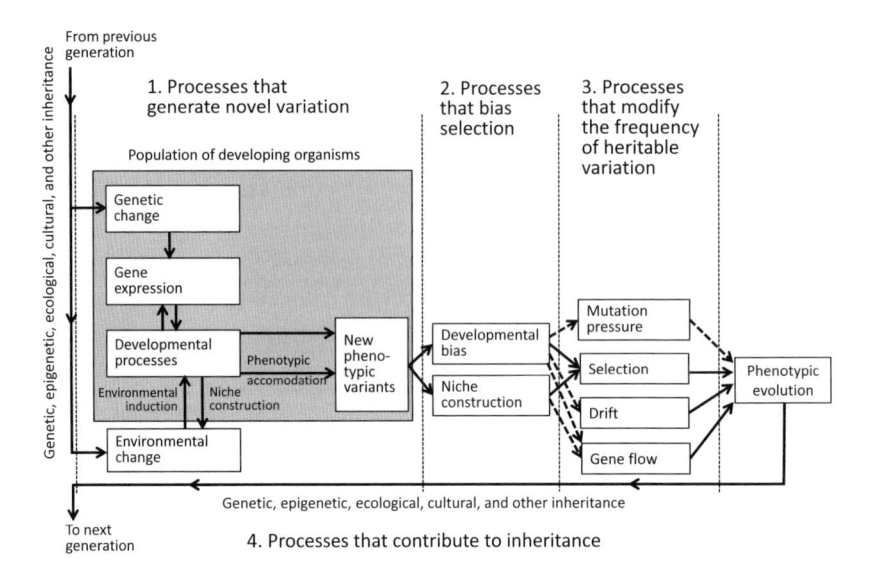

The 'tangled bank' of Darwin stresses the selective function of the ecology, of external factors, abiotic and biotic. But if there is one thing that has become clear in our tour across human evolution, it is that a second entanglement exists, the internal environment, which is at least as tangled, and moreover is full of developmental constraints, complicated connectedness, inimitable tinkering and autonomous, environment-independent drivers. The British geneticist Gabriel Dover evoked the image of an *'internal tangled bank'*, emphasising that evolution is a process not only guided by external factors (selection, adaptation, drift, extinction), but also by internal causes (mutation, recombination, development, tinkering). Evolution stems from the confrontation of the internal tangled bank with the external tangled bank.

Human evolution is a good illustration of what we call the *'dual tangled bank'* principle. A lot is known about our own species; humans make a very good evolutionary model, fossil data can be integrated with genetics, development, physiology, ecology and behaviour. Furthermore, our knowledge on human evolution is advancing at incredibly rapid pace. Each year new fossils are discovered that shed a different light on our ancestry. The new insights obtained from sequencing ancient DNA are astonishing and raise new questions every time. Many of the discoveries are published and widely reported in newspapers and magazines. We hope that the reader,

Quotation from Charles Darwin, *On the Origin of Species* (1859), p. 489, showing his amazement about the 'entangled bank'

It is interesting to contemplate an entangled bank, clothed with many plants of many kinds, with birds singing on the bushes, with various insects flitting about, and with worms crawling through the damp earth, and to reflect that these elaborately constructed forms, so different from each other, and dependent on each other in so complex a manner, have all been produced by laws acting around us.

after studying our elementary introduction, is sufficiently equipped to judge future discoveries in their proper perspective.

This book is published at the moment that a new revolution seems to be occurring in the life sciences. By the application of 'genome editing' using CRISPR/Cas9 and other systems, biologists are able to reconstruct the traits of evolutionary ancestors *in vivo*. A group working under Neill Shubin (University of Chicago) has managed to modify the fins of zebra fish, making them look rather like legs. A group led by Bhart-Anjan Bhullar (Yale University) has mutated the development of a chicken beak such that it displays the characteristics of a dinosaur snout. With the new biotechnology, it is possible to avoid any gratuitous just-so stories and subject hypotheses about the traits of ancestral species, including the evolutionary trajectory to the present, to rigorous experimental testing. Whether this will also enhance our knowledge on human evolution is currently not known, but it is certainly not altogether impossible.

Further reading

We provide here a list of textbooks, popular books and classical texts to guide the reader who wants to study the subject further.

Aiello, L. & C. Dean (2006), *An Introduction to Human Evolutionary Anatomy*. Elsevier Scientific Press, Amsterdam.
Albers, P.C.H. & J. de Vos (2010), *Through Eugène Dubois' Eyes: Stills of a Turbulent Life*. Brill, Leiden.
Alberts, B., A. Johnson, J. Lewis, M. Raff, K. Roberts & P. Walter (2014), *Molecular Biology of the Cell*, 5th ed. Garland Science, New York.
Barrett, L., R. Dunbar & J. Lycett (2002), *Human Evolutionary Psychology*. Palgrave Macmillan, London.
Behe, M.J. (1996), *Darwin's Black Box*. The Free Press, New York.
Bell, G. (1982), *The Masterpiece of Nature: The Evolution and Genetics of Sexuality*. University of California Press, Berkeley.
Benton, M.J. & D.A.T. Harper (2009), *Introduction to Paleobiology and the Fossil Record*. John Wiley & Sons, Chichester.
Boughner, J.C. & C. Rolian, eds. (2016), *Developmental Approaches to Human Evolution*. John Wiley & Sons, Hoboken, NJ.
Boyd, R. & J.B. Silk (2012), *How Humans Evolved*, 6th ed. W.W. Norton & Company, New York.
Boyer, P. (2001), *Religion Explained: The Evolutionary Origins of Religious Thought*. Basic Books, New York.
Carroll, S.B. (2005), *Endless Forms Most Beautiful: The New Science of Evo Devo*. W.W. Norton & Company, New York.
Carroll, S.B., J.K. Grenier & S.D. Weatherbee (2005), *From DNA to Diversity*. Blackwell Publishing, Malden.
Cela-Conde, C.J. & F.J. Ayala (2007), *Human Evolution: Trails from the Past*. Oxford University Press, Oxford.
Conroy, G.C. (2005), *Reconstructing Human Origins*, 2nd ed. W.W. Norton & Company, Inc., New York.
Conway Morris, S. (2003), *Life's Solution*. Cambridge University Press, Cambridge.
Corbey, R. & W. Roebroeks, eds. (2001), *Studying Human Origins: Disciplinary History and Epistemology*. Amsterdam University Press, Amsterdam.
Darwin, C. (1859), *On the Origin of Species by Means of Natural Selection, or the Preservation of Favoured Races in the Struggle for Life*. John Murray, London.
Darwin, C. (1871), *The Descent of Man and Selection in Relation to Sex*. John Murray, London.
Dawkins, R. (1976), *The Selfish Gene*. Granada Publishing Ltd., Frogmore.
Dawkins, R. (1986), *The Blind Watchmaker*. Norton & Company Inc., New York.
Dawkins, R. (2004), *The Ancestor's Tale: A Pilgrimage to the Dawn of Life*. Phoenix, London.
De Waal, F.B.M. (1997), *Good Natured: The Origins of Right and Wrong in Humans and Other Animals*. Harvard University Press, Cambridge, MA.
De Waal, F., R. Wright, C.M. Korsgaard, P. Kitcher & P. Singer (2006), *Primates and Philosophers: How Morality Evolved*. Princeton University Press, Princeton.
Dennett, D.C. (1995), *Darwin's Dangerous Idea: Evolution and the Meanings of Life*. Simon & Schuster, New York.
Dennett, D.C. (2006), *Breaking the Spell: Religion as a Natural Phenomenon*. Penguin Group (USA), New York.

Desalle, R. & I. Tattersall (2012), *Human Origins: What Bones and Genomes Tell Us about Ourselves*. Texas A&M University Press, College Station.

Diamond, J. (1998), *Guns, Germs and Steel: A Short History of Everybody for the Last 13,000 Years*. The Random House Group Ltd., London.

Donald, M. (2001), *A Mind So Rare: The Evolution of Human Consciousness*. W.W. Norton & Company, New York.

Dover, G. (2000), *Dear Mr. Darwin: Letters on the Evolution of Life and Human Nature*. University of California Press, Berkeley.

Falconer, D.S. (1981), *Introduction to Quantitative Genetics*. Longman, New York.

Flint, J., R.J. Greenspan & K.S. Kendler (2010), *How Genes Influence Behavior*. Oxford University Press, Oxford.

Fontdevila, A. (2011), *The Dynamic Genome: A Darwinian Approach*. Oxford University Press, Oxford.

Freeman, S. & J.C. Herron (1998), *Evolutionary Analysis*. Prentice Hall, Upper Saddle River.

Futuyma, D.J. (2009), *Evolution*, 2nd ed. Sinauer Associates, Inc., Sunderland.

Gärdenfors, P. (2006), *How Homo Became Sapiens: On the Evolution of Thinking*. Oxford University Press, Oxford.

Gilbert, S.F. & D. Epel (2009), *Ecological Developmental Biology: Integrating Epigenetics, Medicine and Evolution*. Sinauer Associates, Inc., Sunderland.

Giphart, R. & M. Van Vugt (2018), *Mismatch: How Our Stone Age Brain Deceives Us Every Day (and What We Can Do about It)*. Little Brown Book Group, London.

Goffman, E. (1956), *The Presentation of Self in Everyday Life*. Monograph no. 2. University of Edinburgh Social Science Research Centre, Edinburgh.

Gould, S.J. (1977), *Ontogeny and Phylogeny*. The Belknap Press of Harvard University Press, Cambridge, MA.

Gould, S.J. (1977), Ever Since Darwin. Norton G. Company, Inc. New York.

Graur, D. & W.-H. Li, (2000), *Fundamentals of Molecular Evolution*, 2nd ed. Sinauer Associates, Inc., Sunderland.

Gundling, T. (2005), *First in Line: Tracing Our Ape Ancestry*. Yale University Press, New Haven.

Haeckel, E. (1866), *Generelle Morphologie der Organismen. Erster Band: Allgemeine Anatomie der Organismen*. Verlag von Georg Reimer, Berlin.

Hall, B.K. (1999), *Evolutionary Developmental Biology*, 2nd ed. Kluwer Academic Publishers, Dordrecht.

Jablonka, E. & M.J. Lamb (2005), *Evolution in Four Dimensions*. The MIT Press, Cambridge, USA.

Jacob, F. (1981), *Le jeu des possibles. Essai sur la diversité du vivant*. Librairie Arthème Fayard, Paris.

Kaas, J.H., ed. (2009), *Evolutionary Neuroscience*. Academic Press, Amsterdam.

Kimura, M. (1983), *The Neutral Theory of Molecular Evolution*. Cambridge University Press, Cambridge.

Klein, J. & N. Takahata (2002), *Where Do We Come From? The Molecular Evidence for Human Descent*. Springer, Berlin.

Konner, M. (2002), *The Tangled Wing*, 2nd ed. Henry Holt & Co., New York.

Koonin, E.V. (2012), *The Logic of Chance*. Pearson Education, Upper Saddle River.

LaFrenière, P. (2010), *Adaptive Origins: Evolution and Human Development*. Psychology Press, New York.

Lamarck, J.B.P.A. (1809), *Philosophie zoologique*. J.B. Ballièrre, Paris.

Larsen, C.S. (2008), *Our Origins: Discovering Physical Anthropology*. W.W. Norton & Company, London.

Lee, S.-H. & S.-Y. Yoon (2015), *Close Encounters with Humankind: A Palaeoanthropologist Investigates Our Evolving Species*. W.W. Norton & Company, New York.

LeGros Clark, W.E. (1964), *The Fossil Evidence for Human Evolution*. Rev. ed. University of Chicago Press, Chicago.
Lever, J. (1958), *Creatie en evolutie*. N.V. Gebr. Zomer & Keunings Uitgeversmaatschappij, Wageningen.
Lewin, R. & R.A. Foley (2004), *Principles of Human Evolution*, 2nd ed. Blackwell Science, Malden.
Li, W.-H. (1997), *Molecular Evolution*. Sinauer Associates, Inc., Sunderland.
Lynch, M. (2007), *The Origins of Genome Architecture*. Sinauer Associates, Inc, Sunderland.
Maynard Smith, J. (1972), *On Evolution*. Edinburgh University Press, Edinburgh.
McKee, J.K., F.E. Poirier & W.S. McGraw (2005), *Understanding Human Evolution*, 5th ed. Pearson Education, Inc., Upper Saddle River.
Minugh-Purvis, N. & K.J. McNamara, eds. (2002), *Human Evolution through Developmental Change*. The John Hopkins University Press, Baltimore.
Molnar, S. (1998), *Human Variation: Races, Types, and Ethnic Groups*. Prentice Hall, Upper Saddle River.
Morris, D. (1969), *The Naked Ape*. Jonathan Cape Ltd., London.
Nei, M. (2013), *Mutation-Driven Evolution*. Oxford University Press, Oxford.
Nei, M. & S. Kumar (2000), *Molecular Evolution and Phylogenetics*. Oxford University Press, Oxford.
Netter, F.H. (2014), *Atlas of Human Anatomy*, 6th ed. Elsevier Saunders, Philadelphia.
Nielsen, C. (1995), *Animal Evolution: Interrelationships of the Living Phyla*. Oxford University Press, Oxford.
Paley, W. (1802), *Natural Theology: or, Evidences of the Existence and Attributes of the Deity*. J. Faulder, London.
Pigliucci, M. & G.B. Müller, eds. (2010), *Evolution: The Extended Synthesis*. The MIT Press, Cambridge, MA.
Plomin, R., J.C. DeFries, V.S. Knopik & J.M. Neiderhiser (2013), *Behavioral Genetics*, 6th ed. Worth Publishers, New York.
Portmann, A. (1969), *Einführung in die vergleichende Morphologie der Wirbeltiere. Vierte, überarbeitete und ergänzte Auflage*. Schwabe & Co. Verlag, Basel/Stuttgart.
Reich, D. (2018), *Who We Are and How We Got Here*. Oxford University Press, Oxford.
Reid, R.G.B. (2007), *Biological Emergences: Evolution by Natural Experiment*. The MIT Press, Cambridge, MA.
Roberts, A. (2014), *The Incredible Unlikeliness of Being: Evolution and the Making of Us*. Heron Books, London.
Russell, P.J. (2002), *iGenetics*. Pearson Education/Benjamin Cummings, San Francisco.
Schoenwolf, G.C., S.B. Bleyl, P.R. Brauer & P.H. Francis-West (2015), *Larsen's Human Embryology*, 5th ed. Elsevier Churchill Livingstone, Philadelphia.
Shubin, N. (2008), *Your Inner Fish: A Journey into the 3.5-Billion-Year History of the Human Body*. Allen Lane, London.
Slack, J.M.W. (2006), *Essential Developmental Biology*. Blackwell Publishing, Malden.
Stinson, S., B. Bogin, R. Huss-Ashmore & D. O'Rourke, eds. (2000), *Human Biology: An Evolutionary and Biocultural Perspective*. Wiley-Liss, Inc., New York.
Stone, L. & P.F. Lurquin (2007), *Genes, Culture, and Human Evolution*. Blackwell Publishing, Malden.
Strickberger, M.W. (2000), *Evolution*, 3rd ed. Jones and Bartlett Publishers, Sudbury.
Stringer, C. (2012), *Lone Survivors: How We Came to Be the Only Humans on Earth*. Times Books/Henry Holt and Company, New York.
Swaab, D. (2010), *Wij zijn ons brein. Van baarmoeder tot Alzheimer*. Uitgeverij Contact, Amsterdam.
Sykes, B. (1999), *The Human Inheritance: Genes, Language and Evolution*. Oxford University Press, Oxford.

Tattersall, I. (1998), *Becoming Human: Evolution and Human Uniqueness*. Oxford University Press, Oxford.

Tattersall, I. (2015), *The Strange Case of the Rickety Cossack and Other Cautionary Tales from Human Evolution*. Palgrave Macmillan, New York.

Thompson, D'A.W. (1961), *On Growth and Form*, abridged ed. Cambridge University Press, Cambridge.

Valentine, J.W. (2004), *On the Origin of Phyla*. University of Chicago Press, Chicago.

Van Straalen, N.M. & D. Roelofs (2012), *An Introduction to Ecological Genomics*, 2nd ed. Oxford University Press, Oxford.

Van Straalen, N.M. & D. Roelofs (2017), *Evolueren wij nog?* Amsterdam University Press, Amsterdam.

Waddington, C.H. (1957), *The Strategy of the Genes: A Discussion of Some Aspects of Theoretical Biology*. George Allen & Unwin Ltd., London.

Wade, N. (2006), *Before the Dawn: Recovering the Lost History of Our Ancestors*. Penguin Books, New York.

West-Eberhard, M.J. (2003), *Developmental Plasticity and Evolution*. Oxford University Press, Oxford.

Wilson, E.O. (1975), *Sociobiology: The New Synthesis*. Belknap, Cambridge, USA.

Wilson, E.O. (1978), *On Human Nature*. Harvard University Press, Cambridge, USA.

Wilson, D.S. (2002), *Darwin's Cathedral: Evolution, Religion, and the Nature of Society*. University of Chicago Press, Chicago.

Wolpert, L., R. Beddington, T. Jessel, P. Lawrence, E. Meyerowitz & J.L. Smith (2002), *Principles of Development*, 2nd ed. Oxford University Press, Oxford.

Wood, B. (2005), *Human Evolution: A Very Short Introduction*. Oxford University Press, Oxford.

Wright, A. (2004), *A Short History of Progress*. House of Anansi Press, Toronto.

Wulf, A. (2015) *The Invention of Nature. The Adventures of Alexander Von Humboldt. The Lost Hero of Science*. John Murray (Publishers), London.

Wynne, C.D.L. (2004), *Do Animals Think?* Princeton University Press, Princeton.

Zimmer, C. (2013), *The Tangled Bank: An Introduction to Evolution*, 2nd ed. W.H. Freeman & Co., New York.

Primary literature

We provide here, for each section, the most important primary sources on which we based our arguments, in addition to textbooks and general works.

The story of our ancestors

The revolutionary innovation: walking upright
Richmond, B.G., D.R. Begun & D.S. Strait (2001), Origin of human bipedalism: the knuckle-walking hypothesis revisited. *Yearbook of Physical Anthropology* 44: 70-105.
Richmond, B.G. & D.S. Strait (2000), Evidence that human evolved from a knuckle-walking ancestor. *Nature* 404: 382-385.
Wong, K. (2003), Stranger in a new land. *Scientific American* November: 74-83.
Zhu, Z., R. Dennell, W. Huang, Y. Wu, S. Qiu, S. Yang, Z. Rao, Y. Hou, J. Xie, J. Han & T. Quyang (2018), Hominin occupation of the Chinese Loess Plateau since about 2.1 million years ago. *Nature* 559: 608-612.

How old is that fossil?
Deino, A.L., P.R. Renne & C.C. Swisher III (1998), ^{40}Ar/^{39}Ar dating in paleoanthropology and archeology. *Evolutionary Anthropology* 6: 63-75.
Grün, R. (2006), Direct dating of human fossils. *Yearbook of Physical Anthropology* 49: 2-48.
Ungar, P.S. & M. Sponheimer (2011), The diet of early hominins. *Science* 334: 190-193.

The hominin tree
González-José, R., I. Escapa, W.A. Neves, R. Cúneo & H.M. Pucciarelli (2008), Cladistic analysis of continuous modularized traits provides phylogenetic signals in *Homo* evolution. *Nature* 453: 775-778.
Gunz, P. (2012), Evolutionary relationships among robust and gracile australopiths: an 'evo-devo' perspective. *Evolutionary Biology* 39: 472-487.
Skelton, R.R. & H.M. McHenry (1992), Evolutionary relationships among early hominids. *Journal of Human Evolution* 23: 309-349.
Strait, D.S. & F.E. Grine (2004), Inferring hominoid and early hominid phylogeny using craniodental characters: the role of fossil taxa. *Journal of Human Evolution* 47: 399-452.
Strait, D.S., F.E. Grine & M.A. Moniz (1997), A reappraisal of early hominid phylogeny. *Journal of Human Evolution* 32: 17-82.
White, T.D., G. WoldeGabriel, B. Asfaw, S. Ambrose, Y. Beyene, R.L. Bernor, J.-R. Boisserie, B. Currie, H. Gilbert, Y. Haile-Selassie *et al.* (2006), Asa Issie, Aramis and the origin of *Australopithecus*. *Nature* 440: 883-889.
Wood, B. & B.G. Richmond (2000), Human evolution: taxonomy and paleobiology. *Journal of Anatomy* 196: 19-60.

The earliest hominins

Brunet, M., F. Guy, D. Pilbeam, H.T. Mackaye, A. Likius, D. Ahounta, A. Beauvillain, C. Blondel, H. Bocherens, J.-R. Boisserie et al. (2002), A new hominid from the upper Miocene of Chad, Central Africa. *Nature* 418: 145-151, erratum 801.

Lovejoy, C.O. (2009), Reexamining human origins in light of *Ardipithecus ramidus*. *Science* 326: 74.

Lovejoy, C.O., B. Latimer, G. Suwa, B. Asfaw & T.D. White (2009), Combining prehension and propulsion: the foot of *Ardipithecus ramidus*. *Science* 326: 72.

Senut, B., M. Pickford, D. Gommery, P. Mein, K. Cheboi & Y. Coppens (2001), First hominid from the Miocene (Lukeino formation, Kenya). *Comptes Rendues d'Academie des Sciences de Paris, Sciences de la terre et des planètes* 332: 137-144.

White, T.D., B. Asfaw, Y. Beyene, Y. Haile-Selassie, C.O. Lovejoy, G. Suwa & G. WoldeGabriel (2009), *Ardipithecus ramidus* and the paleobiology of early hominids. *Science* 326: 75-86.

Wolpoff, M.H., B. Senut, M. Pickford & J. Hawks (2002), *Sahelanthropus* or *Sahelpithecus*? *Nature* 419: 581-582.

Wolpoff, M.H., J. Hawks, B. Senut, M. Pickford & J. Ahern (2006), An ape or *the* ape: is the Toumaï cranium TM 266 a hominid? *PaleoAnthropology* 2006: 36-50.

Wood, B. & T. Harrison (2011), The evolutionary context of the first hominins. *Nature* 470: 347-352.

Zollikofer, C.P.E., M.S. Ponce de Leon, D.E. Lieberman, F. Guy, D. Pilbeam, A. Likius, H.T. Mackaye, P. Vignaud & M. Brunet (2005), Virtual cranial reconstruction of *Sahelanthropus tchadensis*. *Nature* 434: 755-759.

The heyday of the ape-men

Alemseged, Z., F. Spoor, W.H. Kimbel, R. Bobe, D. Geraads, D. Reed & J.G. Wynn (2006), A juvenile early hominin skeleton from Dikika, Ethiopia. *Nature* 443: 296-301.

Berger, L.R. (2013), The mosaic nature of *Australopithecus sediba*. *Science* 340: 163-165.

Berger, L.R., D.J. de Ruiter, S.E. Churchill, P. Schmid, K.J. Carlson, P.H.G.M. Dirks & J.M. Kibii (2010), *Australopithecus sediba*: a new species of *Homo*-like Australopith from South Africa. *Science* 328: 195-204.

Haile-Selassie, Y., L. Gibert, S.M. Melillo, T.M. Ryan, M. Alene, A. Deino, N.E. Levin, G. Scott & B.Z. Saylor (2015), New species from Ethiopia further expands Middle Pliocene hominin diversity. *Nature* 521: 483-488.

Henry, A.G., P.S. Ungar, B.H. Passey, M. Sponheimer, L. Rossouw, M. Bamford, P. Sandberg, D.J. de Ruiter & L. Berger (2012), The diet of *Australopithecus sediba*. *Nature* 487: 90-93.

Johanson, D.C., T.D. White & Y. Coppens (1978), A new species of the genus *Australopithecus* (Primates: Hominidae) from the Pliocene of Eastern Africa. *Kirtlandia* 28: 1-14.

Leakey, M.G., F. Spoor, F.H. Brown, P.N. Gathogo, C. Kiarie, L.N. Leakey & I. McDougall (2001), New hominin genus from eastern Africa shows diverse middle Pliocene lineages. *Nature* 410: 433-440.

Ungar, P.S. & M. Sponheimer (2011), The diet of early hominins. *Science* 334: 190-193.

The first Homo

Antón, S.C., R. Potts & L.C. Aiello (2014), Evolution of early *Homo*: an integrated biological perspective. *Science* 345: 45.

Argue, D., M.J. Morwood, T. Sutikna, Jatmiko & E.W. Saptomo (2009), *Homo floresiensis*: a cladistic analysis. *Journal of Human Evolution* 57: 623-639.

Berger, L.R., J. Hawks, D.J. de Ruiter, S.E. Churchill, P. Schmid, L. K. Delezene, T. L. Kivell, H.M. Garvin, S.A. Williams, J.M. DeSilva *et al.* (2015), *Homo naledi*, a new species of the genus *Homo* from the Dinaledi Chamber, South Africa. *eLIFE* 4: e09560.

Berger, L., J. Hawks, P.H.G.M. Dirks, M. Elliott & E.M. Roberts (2017), *Homo naledi* and Pleistocene hominin evolution in subequatorial Africa. *eLIFE* 6: e24234.

Brown, P., T. Sutikna, M.J. Morwood, R.P. Soejono, Jatmiko, E. Wahyhu Saptomo & R. Awe Due (2004), A new small-bodied hominin from the late Pleistocene of Flores, Indonesia. *Nature* 431: 1055-1061.

Brumm, A., G.D. van den Bergh, M. Storey, I. Kurniawan, B.V. Alloway, R. Setiawan, E. Setiyabudi, R. Grün, M.W. Moore, D. Yurnaldi *et al.* (2016), Age and context of the oldest known hominin fossils from Flores. *Nature* 534: 249-252.

Dennell, R. & W. Roebroeks (2005), An Asian perspective on early human dispersal from Africa. *Nature* 438: 1099-1104.

Dirks, P.H.G.M., E.M. Roberts, H. Hilbert-Wolf, J.D. Kramers, M. Elliot, M. Evans, R. Grün, J. Hellstrom, A.I.R. Herries, R. Joannes-Boyau *et al.* (2017), The age of *Homo naledi* and associated sediments in the Rising Star Cave, South Africa. *eLIFE* 6: e24231.

Falk, D., C. Hildebolt, K. Smith, M.J. Morwood, T. Sutikna, P. Brown, Jatmiko, E. Wayhu Saptomo, B. Brunsden & F. Prior (2005), The brain of LB1, *Homo floresiensis*. *Science* 308: 242-245.

Hawks, J., M. Elliot, P. Schmid, S.E. Churchill, D.J. De Ruiter, E.M. Roberts, H. Hilbert-Wolf, H.M. Garvin, S.A. Williams, L.K. Delezene *et al.* (2017), New fossil remains of *Homo naledi* from the Lesedi Chamber, South Africa. *eLIFE* 6: e24232.

Jacob, T., E. Indriati, R.P. Soejono, K. Hsü, D.W. Frayer, R.B. Eckhardt, A.J. Kuperavage, A. Thorne & M. Henneberg (2006), Pygmoid Australomelanesian *Homo sapiens* skeletal remains from Liang Bua, Flores: population affinities and pathological abnormalities. *Proceedings of the National Academy of Sciences of the United States of America* 103: 13421-13426.

Leakey, M.G., F. Spoor, M.C. Dean, C.S. Feibel, S.C. Antón, C. Kiarie & L.N. Leakey (2012), New fossils from Koobi Fora in northern Kenya confirm taxonomic diversity in early *Homo*. *Nature* 488: 201-204.

Lordkipanidze, D., M.S. Ponce de Leon, A. Margvelashvili, Y. Rak, G.P. Rightmire, A. Vekua & C.P.E. Zollikofer (2013), A complete skull from Dmanisi, Georgia, and the evolutionary biology of early *Homo*. *Science* 342: 326-331.

Morwood, M.J., P. Brown, Jatmiko, T. Sutikna, E. Wahyhu Saptomo, K.E. Westaway, R. Awe Due, R.G. Roberts, T. Maeda, S. Wasisto & T. Djubiantono (2005), Further evidence for small-bodied hominins from the late Pleistocene of Flores, Indonesia. *Nature* 437: 1012-1017.

Schwartz, J.H., I. Tattersall & Z. Chi (2014), Comments on 'A complete skull from Dmanisi, Georgia, and the evolutionary biology of early *Homo*'. *Science* 344: 360-a.

Spoor, F., M.G. Leakey, P.N. Gathogo, F.H. Brown, S.C. Antón, I. McDougall, C. Kiarie, F.K. Manthi & L.N. Leakey (2007), Implications of new early *Homo* fossils from Ileret, east of Lake Turkana, Kenya. *Nature* 448: 688-691.

Storm, P., F. Aziz, J. de Vos, D. Kosasih, S. Baskoro, Ngaliman & L.W. van den Hoek Ostende (2005), Late Pleistocene *Homo sapiens* in a tropical rainforest in East Java. *Journal of Human Evolution* 49: 536-545.

Tucci, S., S.H. Vohr, R.C. McCoy, B. Vernot, M.R. Robinson, C. Barbieri, B.J. Nelson, W. Fu, G.A. Purnomo, H. Sudoyo *et al.* (2018), Evolutionary history and adaptation of a human pygmy population of Flores Island, Indonesia. *Science* 361, 511-516.

Van den Bergh, G.D., Y. Kaifu, I. Kurniawan, R.T. Kono, A. Brumm, E. Setiyabudi, F. Aziz & M.J. Morwood (2016), *Homo floresiensis*-like fossils from the early Middle Pleistocene of Flores. *Nature* 534: 245-248.

Van Heteren, A.H. & J. de Vos (2007), Heterochrony as a typical island adaptation in *Homo floresiensis*. In: *Recent Advances on Southeast Asian Paleoanthropology and Archaeology*.

Yogyakarta, Indonesia, Laboratory of Bioanthropology and Paleoanthropology, Faculty of Medicine Gadja Mada University, Yogyakarta, pp. 95-106.
Villmoare, B., W.H. Kimbel, C. Seyoum, C.J. Campisano, E.N. DiMaggio, J. Rowan, D.R. Braun, J.R. Arrowsmith & K.E. Reed (2015), Early *Homo* at 2.8 Ma from Ledi-Geraru, Afar, Ethiopia. *Science* 347: 1352-1355.
Wood, B. (2014), Fifty years after *Homo habilis*. *Nature* 508: 31-33.
Wood, B. & M. Collard (1999), The human genus. *Science* 284: 65-71.

Towards modern times

Arsuaga, J.-L., I. Martinez, L.J. Arnold, A. Aranburu, A. Gracia-Téllez, W.D. Sharp, R.M. Quam, C. Falguères, A. Pantoja-Pérez, J. Bischoff *et al.* (2014), Neandertal roots: cranial and chronological evidence from Sima de los Huesos. *Science* 344: 1358-1363.
Benazzi, S., K. Doula, C. Fornai, C.C. Bauer, O. Kullmer, J. Svoboda, I. Pap, F. Malegni, P. Bayle, M. Coquerelle *et al.* (2011), Early dispersal of modern humans in Europe and implications for Neanderthal behaviour. *Nature* 479: 525-528.
De Vos, J. (2009), Receiving an ancestor in the phylogenetic tree: Neanderthal Man, *Homo erectus* and *Homo floresiensis*: *l'histoire se répète*. *Journal of the History of Biology* 42: 361-379.
Excoffier, L. (2006), Neanderthal genetic diversity: a fresh look from old samples. *Current Biology* 16: R650-R652.
Galway-Witham, J. & C. Stringer (2018), How did *Homo sapiens* evolve? Genetic and fossil evidence challenges current models of modern human evolution. *Science* 360: 1296-1298.
Green, R.E., J. Krause, A.W. Briggs, T. Maricic, U. Stenzel, M. Kircher, N. Patterson, H. Li, W. Zhai, M. Hsi-Yang Fritz *et al.* (2010), A draft sequence of the Neandertal genome. *Science* 328: 710-722.
Hublin, J.-J. (2009), The origin of Neandertals. *Proceedings of the National Academy of Sciences of the United States of America* 106: 16022-16027.
Hublin, J.-J., A. Ben-Neer, S.E. Balley, S.E. Freidline, S. Neubauer, M.M. Skinner, I. Bergmann, A. La Cabec, S. Benazzi, K. Harvati & P. Gunz (2017), New fossils from Jebel Irhoud, Marocco and the pan-African origin of *Homo sapiens*. *Nature* 546: 289-292.
Krause, J., Q. Fu, J.M. Good, B. Viola, M.V. Shunkov, A.P. Derevianko & S. Pääbo (2010), The complete mitochondrial DNA genome of an unknown hominin from southern Siberia. *Nature* 464: 894-897.
Lalueza-Fox, C., J. Krause, D. Caramelli, G. Catalano, L. Milani, M. L. Sampietro, F. Calafell, C. Martinez-Maza, M. Bastir, A. Garcia-Tabernero *et al.* (2006), Mitochondrial DNA of an Iberian Neandertal suggests a population affinity with other European Neandertals. *Current Biology* 16: R629-R630.
Langbroek, M. (2014), Ice age mentalists: debating neurological and behavioural perspectives on the Neandertal and modern mind. *Journal of Anthropological Sciences* 92: 285-289.
Li, Z.-Y., X.-J. Wu, L.-P. Zhou, W. Liu, X. Gao, X.-M. Nian & E. Trinkaus (2017), Late Pleistocene archaic human crania from Xuchang, China. *Science* 355: 969-972.
Neubauer, S., J.-J. Hublin & P. Gunz (2018), The evolution of modern human brain shape. *Science Advances* 4: eaao5961.
Roebroeks, W. & P. Villa (2011), On the earliest evidence for habitual use of fire in Europe. *Proceedings of the National Academy of Sciences of the United States of America* 108: 5209-5214.
Serre, D., A. Langaney, M. Chech, M. Teschler-Nicola, M. Paunovic, P. Mennecier, M. Hofreiter, G. Possner & S. Pääbo (2004), No evidence of Neandertal mtDNA contribution to early modern humans. *PLoS Biology* 2: 0313-0317.
Shimelmitz, R., S.L. Kuhn, A.J. Jelinek, A. Ronen, A.E. Clark & M. Weinstein-Evron (2014), 'Fire at will': the emergence of habitual fire use 350,000 years ago. *Journal of Human Evolution* 77: 196-203.

Stringer, C. (2012), The status of *Homo heidelbergensis* (Schoetensack 1908). *Evolutionary Anthropology* 21: 101-107.
Toussaint, M. & D. Bonjean, Eds. (2014), *The Scladina 1-4A Juvenile Neandertal*. Études et Recherches Archéologiques de l'Université de Liège, Andenne.
Trinkaus, E. (2005), Early modern humans. *Annual Review of Anthropology* 34: 207-230.
White, T.D., B. Asfaw, D. DeGusta, H. Gilbert, G.D. Richards, G. Suwa & F.C. Howell (2003), Pleistocene *Homo sapiens* from middle Awash, Ethiopia. *Nature* 423: 742-747.

From ovum to human

Heterochrony and Haeckel's law
Cañestro, C., H. Yokai & J.H. Postlewaith (2007), Evolutionary developmental biology and genomics. *Nature Reviews Genetics* 8: 932-942.
Duboule, D. (1994), Temporal colinearity and the phylotypic progression: a basis for the stability of a vertebrate Bauplan and the evolution of morphologies through heterochrony. *Development*, Supplement: 135-142.
McNulty, K.P. (2012), Evolutionary development in *Australopithecus africanus*. *Evolutionary Biology* 39: 488-498.
Richardson, M.K., J. Hanken, M.L. Gooneratne, C. Pieau, A. Raynaud, L. Selwood & G.M. Wright (1997), There is no highly conserved embryonic stage in the vertebrates: implications for current theories of evolution and development. *Anatomy and Embryology* 196: 91-106.
Slack, J.M., P.W.H. Holland & C.F. Graham (1993), The zootype and the phylotypic stage. *Science* 361: 490-492.

Cleavages and germ layers
Barron Abitua, P., E. Wagner, I.A. Navarrete & M. Levine (2012), Identification of a rudimentary neural crest in a non-vertebrate chordate. *Nature* 492: 104-107.
Horie, R., A. Hazbun, K. Chen, C. Cao, M. Levine, & T. Horie (2018), Shared evolutionary origin of vertebrate neural crest and cranial placodes. *Nature* 560: 228-232.
Lauri, A., T. Brunet, M. Handberg-Thorsager, A.H.L. Fischer, O. Simakov, P.R.H. Steinmetz, R. Tomer, P.J. Keller & D. Arendt (2014), Development of the annelid axochord: insights into notochord evolution. *Science* 345: 1365-1368.
Lavialle, C., G. Cornelis, A. Dupressoir, C. Esnault, O. Heidemann, C. Vernochet & T. Heidmann (2013), Paleovirology of 'syncytins', retroviral *env* genes exapted for a role in placentation. *Philosophical Transactions of the Royal Society of London. B. Biological Sciences* 368: 20120507.
Roberts, R.M., J.A. Green & L.C. Schulz (2016), The evolution of the placenta. *Reproduction* 152: R179-R189.

Axes to provide direction
Arbeitman, M.N., E.E.M. Furlong, F. Imam, E. Johnson, B.H. Null, B.S. Baker, M.A. Krasnow, M.P. Scott, R.W. Davis & K.P. White (2002), Gene expression during the life cycle of *Drosophila melanogaster*. *Science* 297: 2270-2275.
Hejnol, A. (2010), A twist in time – the evolution of spiral cleavage in the light of animal phylogeny. *Integrative and Comparative Biology* 50: 695-706.
Hill, A.A., C.P. Hunter, B.T. Tsung, G. Tucker-Kellogg & E.L. Brown (2000), Genomic analysis of gene expression in *C. elegans*. *Science* 290: 809-812.

Lambert, J.D. (2008), Mesoderm in spiralians: the organizer and the 4d cell. *Journal of Experimental Zoology* (*Mol. Dev. Evol.*) 310B: 15-23.

Martin-Durán, J.M., Y.J. Passamaneck, M.Q. Martindale & A. Hejnol (2016), The developmental basis for the recurrent evolution of deuterostomy and protostomy. *Nature Ecology & Evolution* 1: 0005.

Mousseau, T.A. & C.H. Fox (1998), The adaptive significance of maternal effects. *Trends in Ecology and Evolution* 13: 403-407.

Model animals in developmental biology

C. elegans Sequencing Consortium (1998), Genome sequence of the nematode *C. elegans*: a platform for investigating biology. *Science* 282: 2012-2018.

Darling, J.A., A.R. Reitzel, P.M. Burton, M.E. Mazza, J.F. Ryan, J.C. Sullivan & J.R. Finnerty (2005), Rising starlet: the starlet sea anemone, *Nematostella vectensis*. *BioEssays* 27.2: 211.

Hellsten, U., R.M. Harland, M.J. Gilchrist, D. Hendrix, J. Jurka, V.V. Kapitonov, I. Ovcharenko, N.H. Putnam, S. Shu, L. Taher et al. (2010), The genome of the western clawed frog *Xenopus tropicalis*. *Science* 328: 633-636.

Leopold, P. & N. Perrimon (2007), Drosophila and the genetics of the internal milieu. *Nature* 450: 186-188.

Rubin, G.M. & E.B. Lewis (2000), A brief history of Drosophila's contributions to genome research. *Science* 287: 2216-2218.

Sea Urchin Sequencing Consortium (2006), The genome of the sea urchin *Strongylocentrotus purpuratus*. *Science* 314: 941-952.

Sommer, R.J. (2009), The future of evo-devo: model systems and evolutionary theory. *Nature Reviews Genetics* 10: 416-422.

Tribolium Genome Sequencing Consortium (2008), The genome of the model beetle and pest *Tribolium castaneum*. *Nature* 452: 949-955.

Wood, W.B. (1988), Introduction to *C. elegans* biology. In: *The Nematode Caenorhabditis elegans* (Wood, W.B., ed.). Cold Spring Harbor Laboratory Press, New York, pp. 1-16.

The molecular tool kit for development

Amores, A., A. Force, Y.-L. Yan, L. Joly, C. Amemiya, A. Fritz, R.K. Ho, J. Langeland, V. Prince, Y.-L. Wang, M. Westerfield, M. Ekker & J.H. Postlewaith (1998), Zebrafish *hox* clusters and vertebrate genome evolution. *Science* 282: 1711-1714.

Duboule, D. (2007), The rise and fall of Hox gene clusters. *Development* 134: 2549-2560.

Hoekstra, H.E. & J.A. Coyne (2007), The locus of evolution: evo devo and the genetics of adaptation. *Evolution* 61: 995-1016.

Holland, P. W. H. (2013), Evolution of homeobox genes. *WIREs Developmental Biology* 2: 31-45.

Patel, N.H. (2004), Time, space and genomes. *Nature* 431: 28-29.

Ryan, J.F., P.M. Burton, M.E. Mazza, G.K. Kwong, J.C. Mullikin & J.R. Finnerty (2006), The cnidarian-bilaterian ancestor possessed at least 56 homeoboxes: evidence from the starlet sea anemone, *Nematostella vectensis*. *Genome Biology* 7: R64.

Wellik, D.M. & M.R. Capecchi (2003), *Hox10* and *Hox11* genes are required to globally pattern the mammalian skeleton. *Science* 301: 363-367.

New axes for limbs

Lovejoy, C.O., M.A. McCollum, P.L. Reno & B.A. Rosenman (2003), Developmental biology and human evolution. *Annual Review of Anthropology* 32: 85-109.

Panganiban, G., S.M. Irvine, C. Lowe, H. Roehl, L.S. Corley, B. Sherbon, J.K. Grenier, J.F. Fallon, J. Kimble, M. Walker et al. (1997), The origin and evolution of animal appendages. *Proceedings of the National Academy of Sciences of the United States of America* 94: 5162-5166.

Our tinkered body

Tinkers, watchmakers and a Boeing 747
Gould, S.J. & E.S. Vrba (1982), Exaptation – a missing term in the science of form. *Paleobiology* 8: 4-15.
Jacob, F. (1977), Evolution and tinkering. *Science* 196: 1161-1166.
Liu, R. & H. Ochman (2007), Stepwise formation of the bacterial flagellar system. *Proceedings of the National Academy of Sciences of the United States of America* 104: 7116-7121.

The naked human
Falk, D. (1990), Brain evolution in *Homo*: The "radiator" theory. *Behavioral and Brain Sciences* 13: 333-344.
Kittler, R., M. Kayser & M. Stoneking (2003), Molecular evolution of *Pediculus humanus* and the origin of clothing. *Current Biology* 13: 1414-1417.
Reed, D.L., V.R. Smith, S.L. Hammond, A.R. Rogers & D.H. Clayton (2004), Genetic analysis of lice supports direct contact between modern and archaic humans. *PLoS Biology* 2: 1972-1983.
Reed, D.L., J.E. Light, J.M. Allen & J.J. Kirchmann (2004), Pair of lice lost or parasites regained: the evolutionary history of anthropoid primate lice. *BMC Biology* 5: 7.
Van Straalen, N.M. (2018), The naked ape as an evolutionary model, 50 years later. *Animal Biology* 68: 227-246.
Verhaegen, M. (2013), The aquatic ape evolves: common misconceptions and unproven assumptions about the so-called aquatic ape hypothesis. *Human Evolution* 28: 237-266.
Weiss, R.A. (2009), Apes, lice and prehistory. *Journal of Biology* 8: 20.
Wheeler, P.E. (1984), The evolution of bipedality and loss of functional body hair in hominids. *Journal of Human Evolution* 13: 91-98.
Winter, H., L. Langbein, M. Krawczak, D.N. Cooper, L.F. Jave-Suarez, M.A. Rogers, S. Praetzel, P.J. Heidt & J. Schweizer (2001), Human type I hair keratin pseudogene $\psi hHaA$ has functional orthologs in the chimpanzee and gorilla: evidence for recent inactivation of the human gene after the *Pan-Homo* divergence. *Human Genetics* 108: 37-42.
Wu, D.D., D.M. Irwin & Y.-P. Zhang (2008), Molecular evolution of the keratin associated protein gene family in mammals, role in the evolution of mammalian hair. *BMC Evolutionary Biology* 8: 241.

Adaptations to bipedalism in the locomotor apparatus
Bennett, M.R., J.W.K. Harris, B.G. Richmond, D.R. Braun, E. Mbua, P. Kiura, D. Olago, M. Kibunjia, C. Omuombo, A.K. Behrensmeyer et al. (2009), Early hominin foot morphology based on 1.5-million-year-old footprints from Ileret, Kenya. *Science* 323: 1197-1201.
DeSilva, J.M., K.G. Holt, S.E. Churchill, K.J. Carlson, C.S. Walker, B. Zipfel & L.R. Berger (2013), The lower limb and mechanics of walking in *Australopithecus sediba*. *Science* 340: 163-165.
Gruss, L.T. & D. Schmitt (2015), The evolution of the human pelvis: changing adaptations to bipedalism, obstetrics and thermoregulation. *Philosophical Transactions of the Royal Society of London. B. Biological Sciences* 370: 20140063.

Harcourt-Smith, W.E. & A. Aiello (2004), Fossils, feet and the evolution of human bipedal locomotion. *Journal of Anatomy* 204: 403-416.

Kibii, J.M., S.E. Churchill, P. Schmid, K.J. Carlson, N.D. Reed, D.J. De Ruiter & L.R. Berger (2011), A partial pelvis of *Australopithecus sediba*. *Science* 333: 1407-1411.

Pickford, M., B. Senut, D. Glommery & J. Treil (2002), Bipedalism in *Orrorin tugenensis* revealed by its femora. *Comptes Rendus Palevol* 1: 191-203.

Ponce de León, M.S., L. Golovanova, V. Doronichev, G. Romanova, T. Akazawa, O. Kondo, H. Ishida & P.E. Zollikofer (2008), Neanderthal brain size at birth provides insights into the evolution of human life history. *Proceedings of the National Academy of Sciences of the United States of America* 105: 13764-13768.

Pontzer, H., C. Rolian, G.P. Rightmire, T. Jashashvili, M.S. Ponce de Leon, D. Lordkipanidze & C. Zollikofer (2010), Locomotor anatomy and biomechanics of the Dmanisi hominins. *Journal of Human Evolution* 58: 492-504.

Savell, K.R.R., B.M. Auerbach & C.C. Roseman (2016), Constraint, natural selection, and the evolution of human body form. *Proceedings of the National Academy of Sciences of the United States of America* 113: 9492-9497.

Simpson, S.W., J. Quade, N.E. Levin, R. Butler, G. Dupont-Nivet, M. Everett & S. Semaw (2008), A female *Homo erectus* pelvis from Gona, Ethiopia. *Science* 322: 1089-1092.

Williams, S.A., K.R. Ostrofsky, N. Frater, S.E. Churchill, P. Schmid & L.R. Berger (2013), The vertebral column of *Australopithecus sediba*. *Science* 340: 163-165.

Zipfel, B., J.M. DeSilva, R.S. Kidd, K.J. Carlson, S.E. Churchill & L.R. Berger (2011), The foot and ankle of *Australopithecus sediba*. *Science* 333: 1417-1420.

Gill slits, larynx and middle ear

Gould, S.J. (1990), An earful of jaw. *Natural History* 3/90: 12-23.

Luo, Z.-X. (2007), Transformation and diversification in early mammal evolution. *Nature* 450: 1011-1019.

Meng, J., Y. Wang & C. Li (2011), Transitional mammalian middle ear from a new Cretaceous Jehol eutricinodont. *Nature* 472: 181-185.

Rich, T.H., J.A. Hopson, A.M. Musser, T.F. Flannery & P. Vickers-Rich (2005), Independent origins of middle ear bones in monotremes and therians. *Science* 307: 910-914.

The intestines and the lung

Aiello, A. & P. Wheeler (1995), The expensive tissue hypothesis. the brain and the digestive system in human and primate evolution. *Current Anthropology* 36: 199-221.

Beasley, D.E., A.M. Koltz, J.E. Lambert, N. Fierer & R.R. Dunn (2015), The evolution of stomach acidity and its relevance to the human microbiome. *PLoS One* 10: e0134116.

Carmody, R.N. & R.W. Wrangham (2009), The energetic significance of cooking. *Journal of Human Evolution* 57: 379-391.

Furness, J.B., J.J. Cottrell & D.M. Bravo (2015), Comparative physiology of digestion. *Journal of Animal Science* 93: 485-491.

Girard-Madoux, M.J.H., M. Gomez de Agüero, S.C. Ganal-Vornarburg, C. Mooser, G.T. Belz, A.J. Macpherson & E. Vivier (2018), The immunological functions of the appendix: an example of redundancy? *Seminars in Immunology* 36: 31-44.

Hoffman, B.U. & E.A. Lumpkin (2018), A gut feeling. *Science* 361: 1203-1204.

Ingicco, T., G.D. van den Bergh, C. Jago-on, J.-J. Bahain, M.G. Chacón, N. Aamano, H. Forestier, C. King, K. Manalo, S. Nomade *et al.* (2018), Earliest known hominin activity in the Philippines by 709 thousand years ago. *Nature* 557: 233-237.

Smith, H.F., R.E. Fisher, M.L. Everett, A.D. Thomas, R.R. Bollinger & W. Parker (2009) Comparative anatomy and phylogenetic distribution of the mammalian cecal appendix. *Journal of Evolutionary Biology* 22: 1984-1999.

Standen, E.M., T.Y. Du & H.C.E. Larsson (2014), Developmental plasticity and the origin of tetrapods. *Nature* 513: 54-58.

Stevens, C.E. & I.D. Hume (1998), Contributions of microbes in vertebrate gastrointestinal tract to production and conservation of nutrients. *Physiological Reviews* 78: 393-427.

Wrangham, R. & N. Conklin-Brittain (2003), Cooking as a biological trait. *Comparative Biochemistry and Physiology Part A* 136: 35-46.

Heart and urogenital system

Bishopric, N.H. (2005), Evolution of the heart from bacteria to man. *Annals of the New York Academy of Sciences* 1047: 13-29.

Kleisner, K., R. Ivell & J. Flegr (2010), The evolutionary history of testicular externalization and the origin of the scrotum. *Journal of Biosciences* 35: 27-37.

Van Praagh, R. (2011), The evolution of the human heart and its relevance to congenital heart disease. *Kardiochirurgia i Torakochirurgia Polska* 8: 427-431.

Van Praagh, R. (2011), The cardiovascular keys to air-breathing and permanent land-living in vertebrates: the normal human embryonic aortic switch procedure produced by complete right-left asymmetry in the development of the subarterial conal free walls, and the evolution of the right ventricular sinus. *Kardiochirurgia i Torakochirurgia Polska* 8: 1-22.

Evolution of the brain

Bae, B.-I., D. Jayaraman & C.A. Walsh (2015), Genetic changes shaping the human brain. *Developmental Cell* 32: 423-434.

Boyd, J.L., S.L. Skove, J.P. Rouanet, L.-J. Pilaz, T. Bepler, R. Gôrdan, G.A. Wray & D.L. Silver (2015), Human-chimpanzee differences in a *FZD8* enhancers alter cell cycle dynamics in the developing neocortex. *Current Biology* 25: 772-779.

Davis, J.M., V.B. Searles, N. Anderson, J. Keeney, A. Raznahan, L.J. Horwood, D.M. Fergusson, M.A. Kennedy, J. Giedd & J.M. Sikela (2015), DUF1220 copy number is linearly associated with increased cognitive function as measured by total IQ and mathematical aptitude scores. *Human Genetics* 134: 67-75.

Doan, R.N., B.-I. Bae, B. Cubelos, C. Chang, A.A. Hossain, S. Al-Saad, N.M. Mukaddes, O. Oner, M. Al-Saffar, S. Balkhy *et al.* (2016), Mutations in human accelerated regions disrupt cognition and social behavior. *Cell* 167: 341-354.

Enard, W., P. Khaitovich, J. Klose, S. Zöllner, F. Heissig, P. Giavalisco, K. Nieselt-Struwe, E. Muchmore, A. Varki, R. Ravid, G.M. Doxiadis, R.E. Bontrop & S. Pääbo (2002), Intra- and interspecific variation in primate gene expression patterns. *Science* 296: 340-343.

Fiddes, I., G.A. Lodewijk, M. Mooring, C.M. Bosworth, A.D. Ewing, G.L. Mantalas, A.M. Novak, A. van den Bout, A. Bishara, J.L. Rosenkrantz *et al.* (2018), Human-specific NOTCH2NL genes affect Notch signaling and cortical neurogenesis. *Cell* 173: 1356-1369.

Florio, M., M. Albert, E. Taverna, T. Namba, H. Brandl, E. Lewitus, C. Haffner, A. Sykes, F.K. Wong, J. Peters *et al.* (2015), Human-specific gene ARHGAP11B promotes basal progenitor amplification and neocortex expansion. *Science* 347: 1465-1470.

Franchini, L.F. & K.S. Pollard (2017), Human evolution: the non-coding revolution. *BMC Biology* 15: 89.

Gilad, Y., A. Oshlack, G.K. Smyth, T.P. Speed & K.P. White (2006), Expression profiling in primates reveals a rapid evolution of human transcription factors. *Nature* 440: 242-245.

González-Forero, M. & A. Gardner (2018), Inference of ecological and social drivers of human brain-size evolution. *Nature* 557: 554-557.

Hofman, M.A. (2001), Evolution and complexity of the human brain: some organizing principles. In: *Brain Evolution and Cognition*, ed. G. Roth & M.F. Wullimann. Wiley and Sons, New York, pp. 501-521.

Hofman, M.A. & D.F. Swaab (1991), Sexual dimorphism in the human brain: myth and reality. *Experimental and Clinical Endocrinology* 98: 161-170.

Keeney, J.G., L. Dumas & J.M. Sikela (2014), The case for DUF1220 domain dosage as a primary contributor to anthropoid brain expansion. *Frontiers in Human Neuroscience* 8: 427.

Keeney, J.G., J.M. Davis, J. Siegenthaler, M.D. Post, B.S. Nielsen, W.D. Hopkins & J.M. Sikela (2015), DUF1220 protein domains drive proliferation in human neural stem cells and are associated with increased cortical volume in anthropoid primates. *Brain Structure and Function* 220: 3053-3060.

King, M.-C. & A.C. Wilson (1975), Evolution at two levels in humans and chimpanzees. *Science* 188: 107-116.

Leigh, S.R. (2012), Brain size growth and life history in human evolution. *Evolutionary Biology* 39: 587-599.

Levchenko, A., A. Kanapin, A. Samsonova & R.R. Gainetdinov (2017), Human accelerated regions and other human-specific sequence variations in the context of evolution and their relevance for brain development. *Genome Biology and Evolution* 10: 166-188.

Maguire, E.A., N. Burgess & J. O'Keefe (1999), Human spatial navigation: cognitive maps, sexual dimorphism and neural substrates. *Current Opinion in Neurobiology* 9: 171-177.

Mitchell, C. & Silver, D.L. (2018), Enhancing our brains: genomic mechanisms underlying cortical evolution. *Seminars in Cell & Developmental Biology* 76: 23-32.

Nuttle, X., G. Giannuzzi, M.H. Duyzend, J.G. Schraiber, I. Narvaiza, P.H. Sudmant, O. Penn, G. Chiatante, M. Malig, J. Huddleston et al. (2016), Emergence of a *Homo sapiens*-specific gene family and chromosome 16p11.2 CNV susceptibility. *Nature* 536: 205-209.

O'Bleness, M.S., C.M. Dickens, L.J. Dumas, H. Kehrer-Sawatzki, G.J. Wyckoff & J.M. Sikela (2012), Evolutionary history and genome organization of DUF1220 protein domains. *G3 Genes|Genomes|Genetics* 2: 977-986.

Perry, G.H., B.C. Verrelli & A.C. Stone (2004), Comparative analyses reveal a complex history of molecular evolution for human MYH16. *Molecular Biology and Evolution* 22: 379-382.

Pollard, K.S., S.R. Salama, N. Lambert, M.-A. Lambot, S. Coppens, J.S. Pedersen, S. Katzman, B. King, C. Onodera, A. Siepel et al. (2006), An RNA gene expressed during cortical development evolved rapidly in humans. *Nature* 443: 167-172.

Preuss, T.M., M. Cáceres, M.C. Oldham & D.H. Geschwind (2004), Human brain evolution: insights from microarrays. *Nature Reviews Genetics* 5: 850-860.

Previc, F.H. (1999), Dopamine and the origins of human intelligence. *Brain and Cognition* 41: 299-350.

Reardon, P.K., J. Seidlitz, S. Vandekar, S. Liu, R. Patel, M.T.M. Park, A. Alexander-Bloch, R.C. Clasen, J.D. Blumentahl, F.M. Lalonde et al. (2018), Normative brain size variation and brain shape diversity in humans. *Science* 360: 1222-1227.

Rushton, J.P. & C.D. Ankney (2009), Whole brain size and general mental ability: a review. *International Journal of Neuroscience* 119: 692-732.

Scott, N., M. Prigge, O. Yizhar & T. Kimchi (2015), A sexually dimorphic hypothalamic circuit controls maternal care and oxytocin secretion. *Nature* 525: 519-522.

Semendeferi, K., A. Lu, N. Schenker & H. Damasio (2002), Humans and great apes share a large frontal cortex. *Nature Neuroscience* 5: 272-276.

Stedman, H.H., B.W. Kozyak, A. Nelson, D.M. Thesier, L.T. Su, D.W. Low, C.R. Bridges, J.B. Shrager, N. Minugh-Purvis & M.A. Mitchell (2004), Myosin gene mutation correlates with anatomical changes in the human lineage. *Nature* 428: 415-418.

Uylings, H.B.M. & C.G. van Eden (1990), Qualitative and quantitative comparison of the prefrontal cortex in rat and primates, including humans. In: *Progress in Brain Research* 85, ed. H.B.M. Uylings, C.G. Van Eden, J.P.C. De Bruin, M.A. Corner & M.G.P. Feenstra. Elsevier Science Publishers B.V., Amsterdam, pp. 31-62.

Van Dongen, P.A.M. (1998), Brain Size in Vertebrates. In: *The Central Nervous System of Vertebrates*, ed. R. Nieuwenhuys, H.J. ten Donkelaar & C. Nicholson. Springer-Verlag, Berlin, pp. 2099-2134.

Zimmer, F. & S.H. Montgomery (2015), Phylogenetic analysis supports a link between DUF1220 domain number and primate brain expansion. *Genome Biology and Evolution* 7: 2083-2088.

There must be differences

Giant leaps, neutral fluctuations or gradual adaptation?

Chouard, T. (2010), Revenge of the hopeful monster. *Nature* 463: 864-867.

Dennett, D.C. (1983), Intentional systems in cognitive ethology: the 'Panglossian paradigm' defended. *Behavioral and Brain Sciences* 6: 343-390.

Eldredge, N. & S.J. Gould (1972), Punctuated equilibria: an alternative to phyletic gradualism. In: *Models in Paleobiology*, ed. T.J.M. Schopf. Freeman, Cooper & Co., San Francisco, pp. 82-115.

Gould, S.J. & N. Eldredge (1977), Punctuated equilibria: the tempo and mode of evolution reconsidered. *Paleobiology* 3: 115-151.

Gould, S.J. & R.C. Lewontin (1979), The spandrels of San Marco and the panglossian paradigm. *Proceedings of the Royal Society of London, Series B* 205: 581-598.

Hurst, L.D. (2009), Genetics and the understanding of selection. *Nature Reviews Genetics* 10: 83-93.

Iltis, H.H. (1983), From teosinte to maize: the catastrophic sexual transmutation. *Science* 222: 886-894.

Koonin, E.V. (2017), Splendor and misery of adaptation, or the importance of neutral null for understanding evolution. *BMC Biology* 14: 114.

Lynch, M. (2007), The frailty of adaptive hypotheses for the origins of organismal complexity. *Proceedings of the National Academy of Sciences of the United States of America* 104, Suppl. 1: 8597-8604.

Nei, M. (2007), The new mutation theory of phenotypic evolution. *Proceedings of the National Academy of Sciences of the United States of America* 104: 12235-12242.

Pigliucci, M. & J. Kaplan (2000), The fall and rise of Dr Pangloss: adaptationism and the *Spandrels* paper 20 years later. *Trends in Ecology and Evolution* 15: 66-70.

The emergence of variation

Charrier, C., K. Joshi, J. Coutinho-Budd, J.-E. Kim, N. Lambert, J. de Marchena, W.-J. Jin, P. Vanderhaeghen, A. Ghosh, T. Sassa & F. Polleux (2012), Inhibition of SRGAP2 function by its human-specific paralogs induces neoteny during spine formation. *Cell* 149: 923-935.

Benirschke, K., L.E. Brownhill & M.M. Beath (1962), Somatic chromosomes of the horse, donkey and their hybrids, the mule and the hinny. *Journal of Reproduction and Fertility* 4: 319-326.

Chimpanzee Sequencing and Analysis Consortium (2005), Initial sequence of the chimpanzee genome and comparison with the human genome. *Nature* 437: 69-87.

Comings, D.E., N. Gonzales, G. Saucier, J.P. Johnson & J.P. MacMurray (2000), The DRD4 gene and the spiritual transcendence scale of the character temperament index. *Psychiatric Genetics* 10: 185-189.

Davis, C.G., M.A. Lehrman, D.W. Russell, R.G.W. Anderson, M.S. Brown & J.L. Goldstein (1986), The J.D. mutation in familial hypercholesterolemia: amino acid substitution in cytoplasmic domain impedes internalization of LDL receptors. *Cell* 45: 15-24.

Dennis, M.Y., X. Nuttle, P.H. Sudmant, F. Antonacci, T.A. Graves, M. Nefedov, J. A. Rosenfield, S. Sajjadian, M. Malig, H. Kotkiewicz *et al.* (2012), Evolution of human-specific neural SRGAP2 genes by incomplete segmental duplication. *Cell* 149: 912-922.

Dimaio, S., N. Grizenko & R. Joober (2003), Dopamine genes and attention deficit hyperactivity disorder: a review. *Journal of Psychiatry and Neuroscience* 28: 27-38.

Ding, Y.-C., H.-C. Chi, D.L. Grady, A. Mirishima, J.R. Kidd, K.K. Kidd, P. Flodman, M.A. Spence, S. Schiuck, J.M. Swanson, Y.-P. Zhang & R.K. Moyzis (2002), Evidence of positive selection acting at the human dopamine receptor D4 gene locus. *Proceedings of the National Academy of Sciences of the United States of America* 99: 309-314.

Gaspar, C., I. Lopes-Cendes, S. Hayes, J. Goto, K. Arvidsson, A. Dias, I. Silveira, P. Maciel, P. Coutinho, M. Lima *et al.* (2001), Ancestral origin of the Machado-Joseph disease mutation: a worldwide haplotype study. *American Journal of Human Genetics* 68: 523-528.

Hughes, G.M., E.C. Teeling & D.G. Higgins (2014), Loss of olfactory receptor function in hominin evolution. *PLoS One* 9: e84714.

Lichter, J.B., C.L. Barr, J.L. Kennedy, H.H.M. van Tol, K.K. Kidd & K.J. Livak (1993), A hypervariable segment in the human dopamine receptor D_4 (DRD4) gene. *Human Molecular Genetics* 2: 767-773.

Lima, M., M.T. Smith, C. Silva, A. Abade, F.M. Mayer & P. Coutinho (2001), Natural selection at the MJD locus: phenotypic diversity, survival and fertility among Machado-Joseph disease patients from the Azores. *Journal of Biosocial Science* 33: 361-373.

Mine, O.M., K. Kedikilwe, R.T. Ndebele & S.J. Nsoso (2000), Sheep-goat hybrid born under natural conditions. *Small Ruminant Research* 37: 141-145.

Ohta, Y. & M. Nishikimi (1999), Random nucleotide substitutions in primate nonfunctional gene for $_L$-gulono-g-lactone oxidase, the missing enzyme in $_L$-ascorbic acid biosynthesis. *Biochimica et Biophysica Acta – General Subjects* 1472: 408-411.

Quintana-Murci, L. (2012), Gene losses in the human genome. *Science* 335: 806-807.

Šipek Jr., A., A. Panczak, R. Mihalová, L. Hrčková, E. Suttrová, V. Sobotka, P. Lonsky, N. Kaspřiková & V. Gregor (2015), Pericentric inversion of human chromosome 9 epidemiology study in Czech males and females. *Folia biologica* 61: 140-146.

Wang, X., W.E. Grus & B. Zhang (2006), Gene losses during human origins. *PLoS Biology* 4: 0366-0377.

Yunis, J.J. & P. Prakash (1982), The origin of man: a chromosomal pictorial legacy. *Science* 215: 1525-1529.

Zhang, Z.D., A. Frankish, T. Hunt, J. Harrow & M. Gerstein (2010), Identification and analysis of unitary pseudogenes: historic and contemporary gene losses in humans and other primates. *Genome Biology* 11: R26.

Zhou, S.-F., J.-P. Liu & B. Chowbay (2009), Polymorphism of human cytochrome P450 enzymes and its clinical impact. *Drug Metabolism Reviews* 41: 89-295.

Equilibrium between allele and genotype frequencies

Hawks, J., E.T. Wang, G.M. Cochran, H.C. Harpending & R.K. Moyzis (2007), Recent acceleration of human adaptive evolution. *Proceedings of the National Academy of Sciences of the United States of America* 104: 20753-20758.

Lewontin, R.C. (1964), The interaction of selection and linkage. I. General considerations; heterotic models. *Genetics* 49: 49-67.
Novembre, J. & A. Di Rienzo (2009), Spatial patterns of variation due to natural selection in humans. *Nature Reviews Genetics* 10: 745-755.
Penn, D.J., K. Damjanovich & W.K. Potts (2002), MHC heterozygosity confers a selective advantage against multiple-strain infections. *Proceedings of the National Academy of Sciences of the United States of America* 99: 11260-11264.
Reich, D.E., M. Cargill, S. Bolk, J. Ireland, P.C. Sabeti, D.J. Richter, T. Lavery, R. Kouyoumjian, S.F. Farhadian, R. Ward & E.S. Lander (2001), Linkage disequilibrium in the human genome. *Nature* 411: 199-204.
Sauermann, U., P. Nürnberg, F.B. Bercovitch, J.D. Berard, A. Trefilov, A. Widdig, M. Kessler, J. Schmidtke & M. Krawczak (2001), Increased reproductive success of MHC class II heterozygous males among free-ranging rhesus macaques. *Human Genetics* 108: 249-254.
Saunders, M.A., M. Slatkin, C. Garner, M.F. Hammer & M.W. Nachman (2005), The extent of linkage disequilibrium caused by selection on *G6PD* in humans. *Genetics* 171: 1219-1229.
Shifman, S., J. Kuypers, M. Kokoris, B. Yakir & A. Darvasi (2003), Linkage disequilibrium patterns of the human genome across populations. *Human Molecular Genetics* 12: 771-776.
Ward, L.D. & M. Kellis (2012), Evidence of abundant purifying selection in humans for recently acquired regulatory functions. *Science* 337: 1675-1678.

Neutral evolution

Caballero, A. (1994), Developments in the prediction of effective population size. *Heredity* 73: 657-679.
Harpending, H.C., M.A. Batzer, M. Gurven, L.B. Jorde, A.R. Rogers & S.T. Sherry (1998), Genetic traces of ancient demography. *Proceedings of the National Academy of Sciences of the United States of America* 95: 1961-1967.
Kimura, M. (1980), Average time until fixation of a mutant allele in a finite population under continued mutation pressure: studies by analytical, numerical, and pseudo-sampling methods. *Proceedings of the National Academy of Sciences of the United States of America* 77: 522-526.
Lynch, M. (2007), The evolution of genetic networks by non-adaptive processes. *Nature Reviews Genetics* 8: 803-813.
Ohta, T. (1992), The nearly neutral theory of molecular evolution. *Annual Review of Ecology and Systematics* 23: 263-286.
Sjödin, P., A.E. Sjöstrand, M. Jakobsson & M.G.B. Blum (2012), Resequencing data provide no evidence for a human bottleneck in Africa during the penultimate glacial period. *Molecular Biology and Evolution* 29: 1851-1860.
Smith, E.I., Z. Jacobs, R. Johnsen, M. Ren, E.C. Fisher, S. Oestmo, J. Wilkins, J.A. Harris, P. Karkanas, S. Fitch *et al.* (2018), Humans thrived in South Africa through the Toba eruption about 74,000 years ago. *Nature* 555: 511-515.
Wagner, A. (2008), Neutralism and selectionism: a network-based reconciliation. *Nature Reviews Genetics* 9: 965-974.

Geographical distance causes genetic differences

Cavalli-Sforza, L.L. & M.W. Feldman (2003), The application of molecular genetic approaches to the study of human evolution. *Nature Genetics* 33 Supplement: 266-275.

Ramachandran, S., O. Deshpande, C.C. Roseman, N.A. Rosenberg, M.W. Feldman & L.L. Cavalli-Sforza (2005), Support from the relationship of genetic and geographic distance in human populations for a serial founder effect originating in Africa. *Proceedings of the National Academy of Sciences of the United States of America* 102: 15942-15947.

On top of genetics

Bollati, V. & A. Baccarelli (2010), Environmental epigenetics. *Heredity* 105: 105-112.

Bossdorf, O., C.L. Richards & M. Pigliucci (2008), Epigenetics for ecologists. *Ecology Letters* 11: 106-115.

Crick, F. (1970), Central dogma of molecular biology. *Nature* 227: 561-563.

Crispo, E. (2007), The Baldwin effect and genetic assimilation: revisiting two mechanisms of evolutionary change mediated by phenotypic plasticity. *Evolution* 61: 2469-2479.

Drake, A.J. & B.R. Walker (2004), The intergenerational effects of fetal programming: non-genomic mechanisms for the inheritance of low birth weight and cardiovascular risk. *Journal of Endocrinology* 180: 1-16.

Holoch, D. & D. Moazed (2015), RNA-mediated epigenetic regulation of gene expression. *Nature Reviews Genetics* 16: 71-84.

Jablonka, E. & G. Raz (2009), Transgenerational epigenetic inheritance: prevalence, mechanisms, and implications for the study of heredity and evolution. *The Quarterly Review of Biology* 84: 131-176.

Morgan, H.D., H.G.E. Sutherland, D.I.K. Martin & E. Whitelaw (1999), Epigenetic inheritance at the agouti locus in the mouse. *Nature Genetics* 23: 314-318.

Rando, O.J. & K.J. Verstrepen (2007), Timescales of genetic and epigenetic inheritance. *Cell* 128: 655-668.

Rice, W.R., U. Friberg & S. Gavrilets (2012), Homosexuality as a consequence of epigenetically canalized sexual development. *The Quarterly Review of Biology* 87: 343-368.

Richards, E.J. (2006), Inherited epigenetic variation – revisiting soft inheritance. *Nature Reviews Genetics* 7: 395-400.

Rünneburger, E. & A. Le Rouzic (2016), Why and how genetic canalization evolves in gene regulatory networks. *BMC Evolutionary Biology* 16: 239.

Smallwood, S.A. & G. Kelsey (2012), *De novo* DNA methylation: a germ cell perspective. *Trends in Genetics* 28: 33-42.

Tobi, W.W., J.J. Goeman, R. Monajemi, H. Gu, H. Putter, Y. Zhang, R.C. Slieker, A.P. Stok, P. Thijssen, F. Müller *et al.* (2014), DNA methylation signatures link prenatal famine exposure to growth and metabolism. *Nature Communications* 5: 5592.

Varki, A., D.H. Geschwind & E.E. Eichler (2008), Explaining human uniqueness: genome interactions with environment, behaviour and culture. *Nature Reviews Genetics* 9: 749-763.

Verhoeven, K.J.F., B.M. Vonholdt & V. Sork (2016), Epigenetics in ecology and evolution: what we know and what we need to know. *Molecular Ecology* 25: 1631-1638.

Waddington, C.H. (1942), Canalization of development and the inheritance of acquired characters. *Nature* 150: 563-565.

West-Eberhard, M.J. (2005), Developmental plasticity and the origin of species differences. *Proceedings of the National Academy of Sciences of the United States of America* 102: 6543-6549.

Whitelaw, E. (2006), Sins of the fathers, and their fathers. *European Journal of Human Genetics* 14: 131-132.

The past in the present

Phylogenetic reconstruction
Altschul, S.F., W. Gish, W. Miller, E.W. Myers & D.J. Lipman (1990), Basic Local Alignment Search Tool. *Journal of Molecular Biology* 215: 403-410.
Dunbar, M.J. (1980), The blunting of Occam's Razor, or to hell with parsimony. *Canadian Journal of Zoology* 58: 123-128.
Maley, L.E. & C.R. Marshall (1998), The coming of age of molecular systematics. *Science* 279: 505-506.
Whelan, S., P. Liò & N. Goldman (2001), Molecular phylogenetics: state-of-the art methods for looking into the past. *Trends in Genetics* 17: 262-272.

The molecular clock
Callaway, E. (2012), Studies slow the human DNA clock. *Nature* 489: 343-344.
Fitch, W.M. & C.J. Langley (1979), Protein evolution and the molecular clock. *Federation Proceedings* 35: 2092-2097.
Fu, Q., A. Mittnik, P.L.F. Johnson, K. Bos, M. Lari, R. Bollongino, C. Sun, L. Giemsch, R. Schmitz, J. Burger et al. (2013), A revised time-scale for human evolution based on ancient mitochondrial genomes. *Current Biology* 23: 553-559.
Green, R.E. & B. Shapiro (2013), Human evolution: turning back the clock. *Current Biology* 23: R286.
Harpending, H.C., M.A. Batzer, M. Gurven, L.B. Jorde, A.R. Rogers & S.T. Sherry (1998), Genetic traces of ancient demography. *Proceedings of the National Academy of Sciences of the United States of America* 95: 1961-1967.
Scally, A. & R. Durbin (2012), Revising the human mutation rate: implications for understanding human evolution. *Nature Reviews Genetics* 13: 745-753.
Takahata, N. (1995), A genetic perspective on the origin and history of humans. *Annual Review of Ecology and Systematics* 26: 343-372.

Out of Africa or multiregional evolution?
Ayala, F.J. (1995), The myth of Eve: molecular biology and human origins. *Science* 270: 1930-1936.
Cann, R.L., M. Stoneking & A.C. Wilson (1987), Mitochondrial DNA and human evolution. *Nature* 325: 31-36.
Eswaran, V., H.C. Harpending & A.R. Rogers (2005), Genomics refutes an exclusively African origin of humans. *Journal of Human Evolution* 49: 1-18.
Harpending, H.C., S.T. Sherry, A.R. Rogers & M. Stoneking (1993), The genetic structure of ancient human populations. *Current Anthropology* 34: 483-496.
Ingman, M., H. Kaessmann, S. Pääbo & U. Gyllensten (2000), Mitochondrial genome variation and the origin of modern humans. *Nature* 408: 708-713.
Lieberman, D.E. (1995), Testing hypotheses about recent human evolution from skulls: integrating morphology, function, development, and phylogeny. *Current Anthropology* 36: 159-197.
Manica, A., W. Amos, F. Balloux & T. Hanihara (2007), The effect of ancient population bottlenecks on human phenotypic variation. *Nature* 448: 346-349.
Schlebusch, C.M., H. Malmström, T. Günther, P. Sjödin, A. Coutinho, H. Edlund, A. R. Munters, M. Vicente, M. Steyn, H. Soodyall, M. Lombard & M. Jakobsson (2017), Southern African ancient genomes estimate modern human divergence to 350,000 to 260,000 years ago. *Science* 358: 652-655.

Stewart, J.R. & C.B. Stringer (2012), Human evolution out of Africa: the role of refugia and climate change. *Science* 335: 1317-1321.
Templeton, A.R. (1997), Out of Africa? What do genes tell us? *Current Opinion in Genetics and Development* 7: 841-847.
Vigilant, L., M. Stoneking, H.C. Harpending, K. Hawkes & A.C. Wilson (1991), African populations and the evolution of human mitochondrial DNA. *Science* 253: 1503-1507.
Wolpoff, M.H., J. Hawks, D.W. Frayer & K. Hunley (2001), Modern human ancestry at the peripheries: a test of the replacement theory. *Science* 291: 293-297.

Migrations in all directions

Bae, C.J., K. Donka & M.D. Petraglia (2017), On the origin of modern humans: Asian perspectives. *Science* 358: eaai9067.
Balter, M. (2011), Tracing the paths of the first Americans. *Science* 333: 1692.
Braje, T.J., T.D. Dillehay, J.M. Erlandson, R.G. Klein & T.C. Rick (2017), Finding the first Americans. *Science* 358: 592-594.
Clarkson, C., Z. Jacobs, B. Marick, R. Fullagar, L. Wallis, M. Smith, R.G. Roberts, E. Hayes, K. Lowe, X. Carah *et al.* (2017), Human occupation of northern Australia by 65,000 years ago. *Nature* 547: 306-309.
Dillehay, T.D., C. Ramírez, M. Pino, M.B. Collins, J. Rossen & J.D. Pino-Navarro (2008), Monte Verde: seaweed, food, medicine and the peopling of South America. *Science* 320: 784-786.
Gallego Llorente, M., E.R. Jones, A. Eriksson, V. Siska, K.W. Arthur, J.W. Arthur, M.C. Curtis, J.T. Stock, M. Coltorti, P. Pieruccini *et al.* (2015), Ancient Ethiopian genome reveals extensive Eurasian admixture throughout the African continent. *Science* 350: 820-822.
Fu, Q., C. Posth, M. Hajdinjak, M. Petr, S. Mallick, D. Fernandes, A. Furtwängler, W. Haak, M. Meyer, A. Mittnik *et al.* (2016), The genetic history of Ice Age Europe. *Nature* 534: 200-205.
Gibbons, A. (2000), Europeans trace ancestry to paleolithic people. *Science* 290: 1080-1081.
Gibbons, A. (2017), The first Australians arrived early. *Science* 357: 238-239.
Hammer, M.F. (1995), A recent common ancestry for human Y chromosomes. *Nature* 378: 376-378.
Hammer, M.F., T. Karafet, A. Rasanayagam, E.T. Wood, T.K. Altheide, T. Jenkins, R.C. Griffiths, A.R. Templeton & S.L. Zegura (1998), Out of Africa and back again: nested cladistic analysis of human Y chromosome variation. *Molecular Biology and Evolution* 15: 427-441.
Hoffecker, J.F., S.A. Elias & D.H. O'Rourke (2014), Out of Beringia? *Science* 343: 979-980.
Liu, W., M. Martinón-Torres, Y.-j. Cai, S. Xing, H.-w. Tong, S.-w. Pei, M.J. Sier, X.-h. Wu, R.L. Edwards, H. Cheng *et al.* (2015), The earliest unequivocally modern humans in southern China. *Nature* 526: 696-699.
Malaspinas, A.-S., M.C. Westaway, C. Muller, V.C. Sousa, O. Lao, I. Alves, A. Bergström, G. Athanasiadis, J.Y. Cheng, J.E. Crawford *et al.* (2016), A genomic history of Aboriginal Australia. *Nature* 538: 207-214.
Marean, C.W. (2017), Early signs of human presence in Australia. *Nature* 547: 285-287.
Moreno-Mayar, I., B.A. Potter, L. Vinner, M. Steinrücken, S. Rasmussen, J. Terhorst, J.A. Kamm, A. Albrechtsen, A.-S. Malaspinas, M. Sikora *et al.* (2018), Terminal Pleistocene Alaskan genome reveals first founding population of Native Americans. *Nature* 553: 203-207.
Nei, M. & A.K. Roychoudhury (1993), Evolutionary relationships of human populations on a global scale. *Molecular Biology and Evolution* 10: 927-943.
Potter, B.A., A.B. Beaudoin, C.V. Haynes, V.T. Holliday, C.E. Holmes, J.W. Ives, R. Kelly, B. Llamas, R. Malhi, S. Miller *et al.* (2018), Arrival routes of first Americans uncertain. *Science* 359: 1224-1225.
Pringle, H. (2014), Welcome to Beringia. *Science* 343: 961-963.

Raghavan, M., P. Skoglund, K. Graf, E., M. Metspalu, A. Albrechtsen, I. Moltke, S. Rasmussen, T.W. Stafford Jr., L. Orlando, E. Metspalu *et al.* (2014), Upper Palaeolithic Siberian genome reveals dual ancestry of Native Americans. *Nature* 505: 87-90.

Rasmussen, M., X. Guo, Y. Wang, K.E. Lohmueller, S. Rasmussen, A. Albrechtsen, L. Skotte, S. Lindgren, M. Metspalu, T. Jombart *et al.* (2011), An Aboriginal Australian genome reveals separate human dispersals into Asia. *Science* 334: 94-98.

Rasmussen, M., S.L. Anzick, M.R. Waters, P. Skoglund, M. DeGiorgio, T.W. Stafford Jr., S. Rasmussen, I. Moltke, A. Albrechtsen, S.M. Doyle *et al.* (2014), The genome of a Late Pleistocene human from a Clovis burial site in western Montana. *Nature* 506: 225-228.

Rasmussen, M., M. Sikora, A. Albrechtsen, T. Sand Korneliussen, J.V. Moreno-Mayar, G.D. Poznik, C.P.E. Zollikofer, M.S. Ponce de Leon, M.E. Allentoft, I. Moltke *et al.* (2015), The ancestry and affiliations of Kennewick Man. *Nature* 523: 455-458.

Reich, D., N. Patterson, D. Campbell, A. Tandon, S. Mazieres, N. Ray, M.V. Parra, W. Rojas, C. Duque, N. Mesa *et al.* (2012), Reconstructing Native American population history. *Nature* 488: 370-374.

Scheib, C.L., H. Li, T. Desai, V. Link, C. Kendall, G. Dewar, P.W. Griffith, A. Mörseburg, J.R. Johnson, A. Potter *et al.* (2018), Ancient human parallel lineages within North America contributed to a coastal expansion. *Science* 360: 1024-1027.

Skoglund, P., S. Mallick, M.C. Bortolini, N. Chennagiri, T. Hünemeier, M.L. Petzl-Erler, F.M. Salzano, N. Patterson & D. Reich (2015), Genetic evidence for two founding populations of the Americas. *Nature* 525: 104-108.

Skoglund, P., H. Malmström, M. Raghavan, J. Storå, P. Hall, E. Willerslev, M.T.P. Gilbert, A. Götherström & M. Jakobsson (2012), Origins and genetic legacy of neolithic farmers and hunter-gatherers in Europe. *Science* 336: 466-469.

Stringer, C. & J. Galway-Witham (2018), When did modern humans leave Africa? *Science* 359: 389-390.

Sykes, B. (1999), Using genes to map population structure and origins. In: *The Human Inheritance: Genes, Language, and Evolution*, ed. B. Sykes. Oxford University Press, Oxford, pp. 93-117.

The Y chromosome Consortium (2002), A nomenclature system of the tree of human Y-chromosomal binary haplogroups. *Genome Research* 12: 339-348.

Tobler, R., A. Rohrlach, J. Soubrier, P. Bover, B. Llamas, J. Tuke, N. Bean, A. Abdullah-Highfold, S. Agius, I. O'Donoghue *et al.* (2017), Aboriginal mitogenomes reveal 50,000 years of regionalism in Australia. *Nature* 544: 180-184.

Yang, Z., H. Zhong, J. Chen, X. Zhang, H. Zhang, X. Luo, S. Xu, H. Chen, D. Lu, Y. Han *et al.* (2016), A genetic mechanism for convergent skin lightening during recent human evolution. *Molecular Biology and Evolution* 33: 1177-1187.

Waters, M.R., S.L. Forman, T.A. Jennings, L.C. Nordt, S.G. Driese, J.M. Feinberg, J.L. Keene, J. Halligan, A. Lindquist, J. Pierson *et al.* (2011), The Buttermilk Creek complex and the origins of Clovis at the Debra L. Friedkin site, Texas. *Science* 331: 1599-1603.

Hybridisations between ancient hominins

Abi-Rached, L., M.J. Jobin, S. Kulkarni, A. McWhinnie, K. Dalva, L. Gragert, F. Babrzadeh, B. Gharizadeh, M. Luo, F.A. Plummer *et al.* (2011), The shaping of modern human immune systems by multiregional admixture with archaic humans. *Science* 334: 89-94.

Bustamante, C.D. & B.M. Henn (2010), Shadows of early migrations. *Nature* 468: 1044-1045.

Castellano, S., G. Parra, F.A. Sánchez-Quinto, F. Racimo, M. Kuhlwilm, M. Kircher, S. Sawyer, Q. Fu, A. Heinze, B. Nickel *et al.* (2014), Patterns of coding variation in the complete exomes of three Neandertals. *Proceedings of the National Academy of Sciences of the United States of America* 111: 6666-6671.

Cheviron, Z.A. & R.T. Brumfield (2012), Genomic insights into adaptation to high-altitude environments. *Heredity* 108: 354-361.

Ding, Q., Y. Hu, S. Xu, J. Wang & L. Jin (2013), Neanderthal introgression at chromosome 3p21.31 was under positive selection in East Asians. *Molecular Biology and Evolution* 31: 683-695.

Eriksson, A. & A. Manica (2012), Effect of ancient population structure on the degree of polymorphism shared between modern human populations and ancient humans. *Proceedings of the National Academy of Sciences of the United States of America* 109: 13956-13960.

Fu, Q., M. Hajdinjak, O.T. Moldovan, S. Constantin, S. Mallick, P. Skoglund, N. Patterson, N. Rohland, I. Lazaradis, B. Nickel *et al.* (2015), An early modern human from Romania with a recent Neanderthal ancestor. *Nature* 524: 216-219.

Fu, Q., H. Li, P. Moorjani, F. Jay, S.M. Slepchenko, A.A. Bondarev, P.J.F. Johnson, A. Aximu-Petri, K. Prüfer, C. de Filippo *et al.* (2015), Genome sequence of a 45,000-year-old modern human from western Siberia. *Nature* 514: 445-449.

Gibbons, A. (2016), Five matings for moderns, Neandertals. *Science* 351: 1250-1251.

Gittelman, R.M., J.G. Schraiber, B. Vernot, C. Mikacenic, M.M. Wurfel & J.M. Akey (2016), Archaic hominin admixture facilitated adaptation to out-of-Africa environments. *Current Biology* 26: 3375-3382.

Green, R.E., J. Krause, S.E. Ptak, A.W. Briggs, M.T. Ronan, J.F. Simons, L. Du, M. Egholm, J.M. Rothberg, M. Paunovic & S. Pääbo (2006), Analysis of one million base pairs of Neanderthal DNA. *Nature* 444: 330-336.

Hajdinjak, M., A. Hübner, M. Petr, F. Mafessoni, S. Grote, P. Skoglund, V. Narasimham, H. Rougier, I. Crevecoeur, P. Semal *et al.* (2018), Reconstructing the genetic history of late Neanderthals. *Nature* 555: 652-656.

Hammer, M.F., A.E. Woerner, F.L. Mendez, J.C. Watkins & J.D. Wall (2011), Genetic evidence for archaic admixture in Africa. *Proceedings of the National Academy of Sciences of the United States of America* 108: 15123-15128.

Huerta-Sánchez, E., X. Jin, Asan, Z. Bianba, B.M. Peter, N. Vinckenbosch, Y. Liang, X. Yi, M. He, M. Somel *et al.* (2014), Altitude adaptation in Tibetans caused by introgression of Denisovan-like DNA. *Nature* 512: 194-197.

Kuhlwilm, M., I. Gronau, M.J. Hubisz, C. De Filippo, J. Prado-Martinez, M. Kircher, Q. Fu, H.A. Burbano, C. Lalueza-Fox, M. De la Rasilla *et al.* (2016), Ancient gene flow from early modern humans into Eastern Neanderthals. *Nature* 530: 429-433.

Lalueza-Fox, C. & M.T.P. Gilbert (2011), Paleogenomics of archaic hominins. *Current Biology* 21: R1002-R1009.

Li, Z.-Y., X.-J. Wu, L.-P. Zhou, W. Liu, X. Gao, X.-M. Nian & E. Trinkaus (2017), Late Pleistocene archaic human crania from Xuchang, China. *Science* 355: 969-972.

Mendez, F.L., J.C. Watkins & M.F. Hammer (2012), Global genetic variation at *OAS1* provides evidence of archaic admixture in Melanasian populations. *Molecular Biology and Evolution* 29: 1513-1520.

Mendez, F.L., T. Krahn, B. Schrack, A.-M. Krahn, K.R. Veeramah, A.E. Woerner, F.L.M. Fomine, N. Bradman, M.G. Thomas, T. Karafet & M.F. Hammer (2013), An African American paternal lineage adds an extremely ancient root to the human Y chromosome phylogenetic tree. *The American Journal of Human Genetics* 92: 454-459.

Meyer, M., J.-L. Arsuaga, C. De Filippo, S. Nagel, A. Aximu-Petri, B. Nickel, I. Martinez, A. Gracia, J.M. Bermúdez de Castro, E. Carbonell *et al.* (2016), Nuclear DNA sequences from the Middle Pleistocene Sima de los Huesos hominins. *Nature* 531: 594-507.

Meyer, M., Q. Fu, A. Aximu-Petri, I. Glocke, B. Nickel, J.-L. Arsuaga, I. Martínez, A. Gracia, J.M. Bermúdez de Castro, E. Carbonell & S. Pääbo (2013), A mitochondrial genome sequence of a hominin from Sima de los Huesos. *Nature* 505: 403-406.

Meyer, M., M. Kircher, M.T. Gansauge, H. Li, F. Racimo, S. Mallick, J.G. Schraiber, F. Jay, K. Prüfer, C. de Filippo *et al.* (2012), A high-coverage genome sequence from an archaic Denisovan individual. *Science* 338: 222-226.

Posth, C., C. Wißing, K. Kitagawa, L. Pagani, L. Van Holstein, F. Racimo, K. Wehrberger, N.J. Conard, C.J. Kind, H. Bocherens & J. Krause (2017), Deeply divergent archaic mitochondrial genome provide lower time boundary for African gene flow into Neanderthals. *Nature Communications* 8: 16046.

Prüfer, K., F. Racimo, N. Patterson, F. Jay, S. Sankararaman, S. Sawyer, A. Heinze, G. Renaud, P.H. Sudmant, C. de Filippo *et al.* (2014), The complete genome sequence of a Neanderthal from the Altai Mountains. *Nature* 505: 43-49.

Reich, D., R.E. Green, M. Kircher, M. Krause, N. Patterson, E.Y. Durand, B. Viola, A.W. Briggs, U. Stenzel, P.L.F. Johnson *et al.* (2010), Genetic history of an archaic hominin group from Denisova Cave in Siberia. *Nature* 468: 1053-1060.

Sankararaman, S., S. Mallick, M. Dannemann, K. Prüfer, J. Kelso, S. Pääbo, N. Patterson & D. Reich (2014), The genomic landscape of Neanderthal ancestry in present-day humans. *Nature* 507: 354-357.

Sankararaman, S., N. Patterson, H. Li, S. Pääbo & D. Reich (2012), The date of interbreeding between Neandertals and modern humans. *PLoS Genetics* 8: e1002947.

Simonti, C.N., B. Vernot, L. Bastarache, E. Bottinger, D.S. Carrell, R.L. Chisholm, D.R. Crosslin, S.J. Hebbring, G.P. Jarvik, I.J. Kullo *et al.* (2016), The phenotypic legacy of admixture between modern humans and Neandertals. *Science* 351: 737-741.

Slon, V., C. Hopfe, C.L. Weiß, F. Mafessoni, M. De la Rasilla, C. Laluela-Fox, A. Rosas, M. Soressi, M.V. Knul, R.G. Miller *et al.* (2017), Neandertal and Denisovan DNA from Pleistocene sediments. *Science* 356: 605-608.

Slon, V., F. Mafessoni, B. Vernot, C. de Filippo, S. Grote, B. Viola, M. Hajdinjak, M, S. Peyrégne, S. Nagel, S. Brown *et al.* (2018), The genome of the offspring of a Neanderthal mother and a Denisovan father. *Nature* 561: 113-116.

Vernot, B., S. Tucci, J. Kelso, J.G. Schraiber, A.B. Wolf, R.M. Gittelman, M. Dannemann, S. Grote, R.V. McCoy, H. Norton *et al.* (2016), Excavating Neandertal and Denisovan DNA from the genomes of Melanasian individuals. *Science* 352: 235-239.

Yang, M.A., A.-S. Malaspinas, E.Y. Durand & M. Slatkin (2012), Ancient structure in Africa unlikely to explain Neanderthal and non-African genetic similarity. *Molecular Biology and Evolution* 29: 2987-2995.

The cultural human

Prehistoric tools and cave drawings

Akhilesh, K., S. Pappu, H.M. Rajapara, Y. Gunnell, A.D. Shukla & A.K. Singhvi (2018), Early Middle Palaeolithic culture in India around 385-172 ka reframes Out of Africa models. *Nature* 554: 97-101.

Almécija, S., I.J. Wallace, S. Judea, D.M. Alba & S. Moyà-Solà (2015), Comment on 'Human-like hand use in *Australopithecus africanus*'. *Science* 348: 1101-a.

Aubert, M., A. Brumm, M. Ramli, T. Sutikna, E.W. Saptomo, B. Hakim, M.J. Morwood, G.D. van den Bergh, L. Kinsley & A. Dosseto (2014), Pleistocene cave art from Sulawesi, Indonesia. *Nature* 514: 223-227.

Bar-Yosef, O. & J.-G. Bordes (2010), Who were the makers of the Châtelperronian culture? *Journal of Human Evolution* 59: 586-593.

Conard, N.J., P.M. Grootes & F.H. Smith (2004), Unexpected recent dates for human remains from Vogelherd. *Nature* 430: 198-201.

Harmand, S., J.E. Lewis, C.S. Feibel, C.J. Lepre, S. Prat, A. Lenoble, X. Boës, R.L. Quinn, M. Brenet, A. Arroyo *et al.* (2015), 3.3-million-year-old stone tools from Lomekwi 3, West Turkana, Kenya. *Nature* 521: 310-315.

Higham, T., T. Compton, C. Stringer, R. Jacobi, B. Shapiro, E. Trinkaus, B. Chandler, F. Gröning, C. Collins, S. Hillson *et al.* (2011), The earliest evidence for anatomically modern humans in northwestern Europe. *Nature* 479: 521-524.

Holden, C. (2004), Oldest beads suggest early symbolic behavior. *Science* 304: 369.

Joordens, J.C.A., F. D'Errico, F.P. Wesselingh, S. Munro, J. De Vos, J. Wallinga, C. Ankjaergaard, T. Reimann, J.R. Wijbrans, K.F. Kuiper *et al.* (2014), *Homo erectus* at Trinil on Java used shells for tool production and engraving. *Nature* 518: 228-231.

Kivell, T.L., J.M. Kibii, S.E. Churchill, P. Schmid & L.R. Berger (2011), *Australopithecus sediba* hand demonstrates mosaic evolution of locomotor and manipulative abilities. *Science* 333: 1411-1417.

Linde-Medina, M. (2011), Adaptation or exaptation? The case of the human hand. *Journal of Biosciences* 36: 575-585.

McPherron, S.P., Z. Alemseged, C.W. Marean, J.G. Wynn, D. Reed, D. Geraads, R. Bobe & H.A. Béarat (2010), Evidence for stone-tool assisted consumption of animal tissues before 3.39 million years ago at Dikika, Ethiopia. *Nature* 466: 857-860.

Tocheri, M.W., C.M. Orr, M.C. Jacofsky & M.W. Marzke (2008), The evolutionary history of the hominin hand since the last common ancestor of *Pan* and *Homo*. *Journal of Anatomy* 212: 544-562.

Skinner, M.M., N.B. Stephens, Z.J. Tsegai, A.C. Foote, N.H. Nguyen, T. Gross, D.H. Pahr, J.-J. Hublin & T.L. Kivell (2015), Human-like hand use in *Australopithecus africanus*. *Science* 347: 395-399.

The Neolithic transition

Akey, J.M., A.L. Ruhe, D.T. Akey, A.K. Wong, C.F. Connelly, J. Madeoy, T.J. Micholas & M.W. Neff (2010), Tracking footprints of artificial selection in the dog genome. *Proceedings of the National Academy of Sciences of the United States of America* 107: 1160-1165.

Axelsson, E., A. Ratnakumar, M.-L. Arendt, K. Maqbool, M.T. Webster, M. Perloski, O. Liberg, J.M. Arnemo, Å. Hedhammar & K. Lindblad-Toh (2013), The genomic signature of dog domestication reveals adaptation to a starch-rich diet. *Nature* 495: 360-364.

Bosse, M., H.J. Megens, L.A.F. Frantz, O. Madsen, G. Larson, Y. Paudel, N. Duijvesteijn, B. Harlizius, Y. Hagemeijer, R.P.M.A. Crooijmans & M.A.M. Groenen (2014), Genomic analysis reveals selection for Asian genes in European pigs following human-mediated introgression. *Nature Communications* 5: 4392.

Broushaki, F., M.G. Thomas, V. Link, S. López, L. Van Dorp, K. Kirsanov, Z. Hofmanová, Y. Diekman, L.M. Cassidy, L. Díez-del-Molino *et al.* (2016), Early Neolithic genomes from the eastern Fertile Crescent. *Science* 353: 499-503.

Curry, A. (2013), The milk revolution. *Nature* 500: 20-22.

Diamond, J. (2002), Evolution, consequences and future of plant and animal domestication. *Nature* 418: 700-707.

Evershed, R.P., S. Payne, A.G. Sherratt, M.S. Copley, J. Coolidge, D. Urem-Kotsu, K. Kotsakis, A.E. Özdogan, O. Nieuwenhuyse et al. (2008), Earliest date for milk use in the Near East and southeastern Europe linked to cattle herding. *Nature* 455: 528-531.

Frantz, L.A.F., V.E. Mullin, M. Pionnier-Capitan, O. Lebrasseur, M. Ollivier, A. Perri, A. Linderholm, V. Mattiangeli, M.D. Teasdale, E.A. Dimopoulos et al. (2016), Genomic and archeological evidence suggests a dual origin of domestic dogs. *Science* 352: 1228-1231.

Gerland, P., A.E. Raftery, H. Sevcikova, N. Li, D. Gu, T. Spoorenberg, L. Alkema, B.K. Fosdick, J. Chunn, N. Lalic, G. Bay, T. Buettner, G.K. Heilig & J. Wilmoth (2014), World population stabilization unlikely this century. *Science* 346: 234-237.

Grimm, D. (2015), Dawn of the dog. *Science* 348: 274-279.

Lazaridis, I., D. Nadel, G. Rollefson, D.C. Merrett, N. Rohland, S. Mallick, D. Fernandes, M. Novak, B. Gamarra, K. Sirak et al. (2016), Genomic insights into the origin of farming in the ancient Near East. *Nature* 536: 419-424.

Matsuoka, Y., Y. Vigouroux, M.J. Goodman, J. Sanchez G., E. Buckler & J. Doebley (2002), A single domestication for maize shown by multilocus microsatellite genotyping. *Proceedings of the National Academy of Sciences of the United States of America* 99: 6080-6084.

Piperne, D.R., A.J. Ranere, I Holst, J. Iriarte & R. Dickau (2009), Starch grain and phytolith evidence for early ninth millennium B.P. maize from the Central Balsas River Vally, Mexico. *Proceedings of the National Academy of Sciences of the United States of America* 106: 5019-5024.

Ranere, A.J., D.R. Piperno, I. Holst, R. Dickau & J. Iriarte (2009), The cultural and chronological context of early Holocene maize and squash domestication in the Central Balsas River Valley, Mexico. *Proceedings of the National Academy of Sciences of the United States of America* 106: 5014-5018.

Skoglund, P., H. Malmström, M. Raghavan, J. Storå, P. Hall, E. Willerslev, M.T.P. Gilbert, A. Götherström & M. Jakobsson (2012), Origins and genetic legacy of neolithic farmers and hunter-gatherers in Europe. *Science* 336: 466-469.

Language: early or late?

Bolhuis, J.J., I. Tattersall, N. Chomsky & R.C. Berwick (2014), How could language have evolved? *PLoS Biology* 12: e1001934.

Bradley, B.J. (2007), Reconstructing phylogenies and phenotypes: a molecular view of human evolution. *Journal of Anatomy* 212: 337-353.

Dediu, D. & S.C. Levinson (2013), On the antiquity of language: the reinterpretation of Neandertal linguistic capacities and its consequences. *Frontiers in Psychology* 4: 397.

Fitch, W.T. (2012), Evolutionary developmental biology and human language evolution: constraints on adaptation. *Evolutionary Biology* 39: 613-637.

Hoffmann, D.L., D.E. Angelucci, V. Villaverde, J. Zapata & J. Zilhão (2018), Symbolic use of mineral shells and mineral pigments by Iberian Neandertals 115,000 years ago. *Science Advances* 4: eaar5255.

Hoffmann, D.L., C.D. Standish, M. García-Diez, P.B. Pettitt, J.A. Milton, J. Zilhão, J.J. Alcolea-González, P. Cantalejo-Duarte, H. Collado, R. De Balbín et al. (2018), U-Th dating of carbonate crusts reveals Neandertal origin of Iberian cave art. *Science* 359: 912-915.

Konopka, G., J.M. Bomar, K. Winden, G. Coppola, Z.O. Jonsson, F. Gao, S. Peng, T.M. Preuss, J.A. Wohlschlegel & D.H. Geschwind (2009), Human-specific transcriptional regulation of CNS development genes by FOXP2. *Nature* 462: 213-218.

Krause, J., C. Lalueza-Fox, L. Orlando, W. Enard, R.E. Green, H.A. Burbano, J.-J. Hublin, C. Hänni, J. Fortea, M. de la Rasilla *et al.* (2007), The derived *FOXP2* variant of modern humans was shared with Neanderthals. *Current Biology* 17: 1908-1912.

Lombard, M. & P. Gärdenfors (2017), Tracking the evolution of causal cognition in humans. *Journal of Anthropological Sciences* 95: 219-234.

Mellars, P. (2010), Neanderthal symbolism and ornament manufacture: the bursting of a bubble? *Proceedings of the National Academy of Sciences of the United States of America* 107: 20147-20148.

Renfrew, C. (1999), Reflections on the archaeology of linguistic diversity. In: *The Human Inheritance: Genes, Language, and Evolution*, ed. B. Sykes. Oxford University Press, Oxford, pp. 1-32.

Living in groups: altruistic behaviour

Derex, M., M.-P. Beugin, B. Godelle & M. Raymond (2013), Experimental evidence for the influence of group size on cultural complexity. *Nature* 503: 389-391.

Gilbert, P. (1997), The evolution of social attractiveness and its role in shame, humiliation, guilt and therapy. *British Journal of Medical Psychology* 70: 113-147.

Gintis, H. (2000), Strong reciprocity and human sociality. *Journal of Theoretical Biology* 206: 169-179.

Gintis, H. (2011), Gene-culture coevolution and the nature of human sociality. *Philosophical Transactions of the Royal Society of London. B. Biological Sciences* 366: 878-888.

Gintis, H., S. Bowles, R. Boyd & E. Fehr (2003), Explaining altruistic behavior in humans. *Evolution and Human Behavior* 24: 153-172.

Gluth, S. & L. Fontanesi (2016), Wiring the altruistic brain. *Science* 351: 1028-1029.

Happé, F. (2003), Theory of mind and the self. *Annals of the New York Academy of Sciences* 1001: 134-144.

Heyes, C.M. & C.D. Frith (2014), The cultural evolution of mind reading. *Science* 344: 1357.

Lehmann, J., A.H. Kortsjens & R.I.M. Dunbar (2007), Group size, grooming and social cohesion in primates. *Animal Behaviour* 74: 1617-1629.

Pruitt, J.N. & C.J. Goodnight (2014), Site-specific group selection drives locally adapted group compositions. *Nature* 514: 359-362.

Richerson, P. (2013), Group size determines cultural complexity. *Nature* 503: 351-352.

Szathmáry, E. (2011), To group or not to group. *Science* 334: 1648-1649.

Van der Steen, W.J. (1999), Evolution and altruism. *The Journal of Value Inquiry* 33: 11-29.

West, S.A., A.S. Griffin & A. Gardner (2007), Social semantics: altruism, cooperation, mutualism, strong reciprocity and group selection. *Journal of Evolutionary Biology* 20: 415-432.

Wilson, D.S. & E.O. Wilson (2007), Rethinking the theoretical foundation of sociobiology. *The Quarterly Review of Biology* 82: 327-348.

Cultural evolution

Franklin, A., L. Bevis, Y. Ling & A. Hurlbert (2009), Biological components of colour preference in infancy. *Developmental Science* 13: 346-354.

Hurlbert, A. & Y. Ling (2007), Biological components of sex differences in color preference. *Current Biology* 17: R623-R625.

Van Straalen, N.M. & J. Stein (2003), Evolutionary views on the biological basis of religion. In: *Is Nature Ever Evil? Religion, Science and Value*, ed. W.B. Drees. Routledge, London, pp. 321-329.

Varki, A., D.H. Geschwind & E.E. Eichler (2008), Explaining human uniqueness: genome interactions with environment, behaviour and culture. *Nature Reviews Genetics* 9: 749-763.

Do humans still evolve?

Quantitative characters and heritability
Bohren, B.B., H.E. McKean & Y. Yamada (1961), Relative efficiencies of heritability estimates based on regression of offspring on parent. *Biometrics* 17: 481-491.
Boomsma, D.I., E.J.C. de Geus, G.C.M. van Baal & J.R. Koopmans (1999), A religious upbringing reduces the influence of genetic factors on disinhibition: evidence for interaction between genotype and environment on personality. *Twin Research* 2: 115-125.
Breen, M.S., C. Kenema, P.K. Vlasov, C. Notredame & F.A. Kondrashov (2012), Epistasis as the primary factor in molecular evolution. *Nature* 490: 535-538.
Burt, C. (1963), Is intelligence distributed normally? *The British Journal of Statistical Psychology* 16: 175-190.
Eaves, L., A. Heath, N. Martin, H. Maes, M. Neale, K. Kendler, K. Kirk & L. Corey (1999), Comparing the biological and cultural inheritance of personality and social attitudes in the Virginia 30 000 study of twins and their relatives. *Twin Research* 2: 62-80.
Galis, F. (1999), Why do almost all mammals have seven cervical vertebrae? Developmental constraints, *Hox* genes and cancer. *Journal of Experimental Zoology (Mol. Dev. Evol.)* 285: 19-26.
Hill, W.G. (1971), Design and efficiency of selection experiments for estimating genetic parameters. *Biometrics* 27: 293-311.
Kelsoe, J.R. (2010), A gene for impulsivity. *Nature* 468: 1049-1050.
Phillips, P.C. (2008), Epistasis – the essential role of gene interactions in the structure and evolution of genetic systems. *Nature Reviews Genetics* 9: 855-867.
Visscher, P.M., W.G. Hill & N.R. Wray (2008), Heritability in the genomics era – concepts and misconceptions. *Nature Reviews Genetics* 9: 255-266.
Wei, W.-H., G. Hemani & C.S. Haley (2014), Detecting epistasis in human complex traits. *Nature Reviews Genetics* 15: 722-733.

Ecogeographic variation in human body form
Betti, L., S.J. Lycett, N. Von Cramon-Taubadel & O.M. Perason (2015), Are human hands and feet affected by climate? A test of Allen's rule. *American Journal of Physical Anthropology* 185: 132-140.
Cowgill, L.W., C.D. Eleazer, B.M. Auerbach, D.H. Temple & K. Okazaki (2012), Developmental variation in ecogeographic body proportions. *American Journal of Physical Anthropology* 148: 557-570.
Hagen, J.B. (2017), Bergmann's rule, adaptation, and thermoregulation in arctic animals: conflicting perspectives from physiology, evolutionary biology, and physical anthropology after World War II. *Journal of the History of Biology* 50: 235-265.
Kasabova, B.E. & T.W. Holliday (2015), New model for estimating the relationship between surface area and volume in the human body using skeletal remains. *American Journal of Physical Anthropology* 156: 614-624.
Roseman, C.C. & B.M. Auerbach (2015), Ecogeography, genetics and the evolution of human body form. *Journal of Human Evolution* 78: 80-90.
Savell, K.R.R., B.M. Auerbach & C.C. Roseman (2016), Constraint, natural selection, and the evolution of human body form. *Proceedings of the National Academy of Sciences of the United States of America* 113: 9492-9497.

Tilkens, M.J., C. Wall-Scheffer, T.D. Weaver & K. Steudel-Numbers (2007), The effects of body proportions on thermoregulation: an experimental assessment of Allen's rule. *Journal of Human Evolution* 53: 286-291.

Evolution of biomedical traits

Bersaglieri, T., P.C. Sabeti, N. Patterson, T. Vanderploeg, S.F. Schaffner, J.A. Drake, M. Rhodes, D.E. Reich & J.N. Hirschhorn (2004), Genetic signatures of strong recent positive selection at the lactase gene. *American Journal of Human Genetics* 74: 1111-1120.

Balter, M. (2005), Are humans still evolving? *Science* 309: 234-237.

Byars, S.G., D. Ewbank, D.R. Govindaraju & S.C. Stearns (2010), Natural selection in a contemporary human population. *Proceedings of the National Academy of Sciences of the United States of America* 107 (Suppl. 1): 1787-1792.

Deng, L. & S. Xu (2018), Adaptation of human skin color in various populations. *Hereditas* 155: 1.

Gerbault, P., A. Liebert, Y. Itan, A.J. Powell, M. Currat, J. Burger, D.M. Swallow & A.D. Thomas (2011), Evolution of lactase persistence: an example of human niche construction. *Philosophical Transactions of the Royal Society of London. B. Biological Sciences* 366: 863-877.

Kirk, K., S.P. Blomberg, D.L. Duffy, A.C. Heath, I.P.F. Owens & N.G. Martin (2001), Natural selection and quantitative genetics of life-history traits in Western women: a twin study. *Evolution* 55: 423-435.

Lachance, J. & S.A. Tishkoff (2013), Population genomics of human adaptation. *Annual Review of Ecology, Evolution and Systematics* 44: 123-143.

Llardo, M.A., I. Moltke, T.S. Korneliussen, H. Cheng, A.J. Stern, F. Racimo, P. De Barros Damgaard, M. Sikora, A. Seguin-Orlando, S. Rasmussen *et al.* (2018), Physiological and genetic adaptations to diving in sea nomads. *Cell* 173: 569-580.

Perry, G.H., N.J. Dominy, K.G. Claw, A.S. Lee, H. Fiegler, R. Redon, J. Werner, F.A. Villanea, J.L. Mountain, R. Misra *et al.* (2007), Diet and the evolution of human amylase gene copy number variation. *Nature Genetics* 39: 1256-1260.

Ségurel, L. & C. Bon (2017), On the evolution of lactase persistence in humans. *Annual Review of Genomics and Human Genetics* 18: 297-319.

Stearns, S.C., S.G. Byars, D.R. Govindaraju & D. Ewbank (2010), Measuring selection in contemporary human populations. *Nature Reviews Genetics* 11: 611-622.

Swallow, D.M. (2003), Genetics of lactase persistence and lactose intolerance. *Annual Review of Genetics* 37: 197-219.

Wooding, S., U.-k. Kim, M.J. Bamshad, J. Larsen, L.B. Jorde & D. Drayna (2004), Natural selection and molecular evolution in PTC, a bitter-taste receptor gene. *American Journal of Human Genetics* 74: 637-646.

Evolution of the life cycle

Bogin, B. & B.H. Smith (1996), Evolution of the human life cycle. *American Journal of Human Biology* 8: 703-716.

Dean, C., M.G. Leakey, D. Reid, F. Schrenk, G.T. Schwartz, C. Stringer & A. Walker (2001), Growth processes in teeth distinguish modern humans from *Homo erectus* and earlier hominins. *Nature* 414: 628-631.

DeSilva, J.M. (2018), Comment on 'The growth pattern of Neandertals, reconstructed from a juvenile skeleton from El Sidrón (Spain)'. *Science* 359: eaar3611.

Kachel, A.F. & L.S. Premo (2012), Disentangling the evolution of early and late life history traits in humans. *Evolutionary Biology* 39: 638-649.

Lahdenperä, M., V. Lummaa, S. Helle, M. Tremblay & A.F. Russell (2004), Fitness benefits of prolonged post-reproductive lifespan in women. *Nature* 428: 178-181.

Rosas, A., L. Ríos, A. Estalrrich, H. Liversidge, A. García-Tabernero, R. Huguet, H. Cardoso, M. Bastir, C. Laluela-Fox, M. De la Rasilla & C. Dean (2017), The growth pattern of Neandertals, reconstructed from a juvenile skeleton from El Sidrón (Spain). *Science* 357: 1282-1287.

Partner choice and sexual selection

Chaix, R., C. Cao & P. Donnelly (2008), Is mate choice in humans MHC-dependent? *PLoS Genetics* 4: e1000184.

Dixson, B.J., G.M. Grimshaw, W.L. Linklater & A.F. Dixson (2011), Eye-tracking of men's preferences for waist-to-hip ratio and breast size of women. *Archives of Sexual Behavior* 40: 43-50.

Dixson, B.J.W. & M.J. Rantala (2016), The role of facial and body hair distribution in women's judgments of men's sexual attractiveness. *Archives of Sexual Behavior* 45: 877-889.

Gaulin, S.J.C. & J.S. Boster (1992), Human marriage systems and sexual dimorphism in size. *American Journal of Physical Anthropology* 89: 467-475.

Keller, A., H. Zhuang, Q. Chi, L.B. Vosshall & H. Matsunami (2007), Genetic variation in a human odorant receptor alters odour perception. *Nature* 449: 468-472.

Lee, A.J., S.L. Dubbs, A.J. Kelly, W. von Himpel, R.C. Brooks & B.P. Zietsch (2013), Human facial attributes, but not perceived intelligence, are used as cues of health and resource provision potential. *Behavioural Ecology* 24: 779-787.

Mautz, B.S., B.B.M. Wong, R.A. Peters & M.D. Jennions (2013), Penis size interacts with body shape and height to influence male attractiviness. *Proceedings of the National Academy of Sciences of the United States of America* 110: 6925-6930.

Platek, S.M. & D. Singh (2010), Optimal waist-to-hip ratios in women activate neural reward centers in men. *PLoS ONE* 5: e9042.

Prause, N., J. Park, S. Leung & G. Miller (2015), Women's preferences for penis size: a new research method using selection among 3D models. *PLoS One* 10: e133079.

Singh, D. (1993), Body shape and women's attractiveness: the critical role of the waist-to-hip ratio. *Human Nature* 4: 297-321.

Wedekind, C., T. Seebeck, F. Bettens & A.J. Paepke (1995), MHC-dependent mate preferences in humans. *Proceedings of the Royal Society of London, Series B* 260: 245-249.

Wedekind, C. & S. Füri (1997), Body odour preferences in men and women: do they aim for specific MHC combinations or simply heterozygosity? *Proceedings of the Royal Society of London, Series B* 264: 1471-1479.

Wilson, M.L., C.M. Miller & K.N. Crouse (2017), Humans as model species for sexual selection research. *Proceedings of the Royal Society B* 284: 20171320.

Evolutionary medicine

Low, F.M., P.D. Gluckman & M.A. Hanson (2012), Developmental plasticity, epigenetics and human health. *Evolutionary Biology* 39: 650-665.

Williams, G.C. & R.M. Nesse (1991), The dawn of Darwinian medicine. *The Quarterly Review of Biology* 66: 1-22.

Wang, G. & J.R. Speakman (2016), Analysis of positive selection at single nucleotide polymorphisms associated with body mass index does not support the 'thrifty gene' hypothesis. *Cell Metabolism* 24: 531-541.

Epilogue

Bhullar, B.-A.S., J. Marugán-Lobón, F. Racimo, G.S. Bever, T.B. Rowe & M.A. Norell (2012), Birds have paedomorphic dinosaur skulls. *Nature* 487: 223-226.

Bhullar, B.-A.S., Z.S. Morris, E.M. Sefton, A. Tok, M. Tokita, B. Namkoong, J. Camacho, D.A. Burnham & A. Abzhanov (2015), A molecular mechanism for the origin of a key evolutionary innovation, the bird beak and palate, revealed by an integrative approach to major transitions in vertebrate history. *Evolution* 69: 1665-1677.

Callaway, E. (2016), CRISPR's hopeful monsters: gene editing storms evo-devo labs. *Nature* 536: 249.

Eisen, M. (2012), Darwin's tangled bank in verse. http://www.michaeleisen.org/blog/?p=1245.

Laland, K.N., T. Uller, M.W. Feldman, K. Sterelny, G.B. Müller, A. Moczek, E. Jablonka & J. Odling-Smee (2015), The extended evolutionary synthesis: its structure, assumptions and predictions. *Proceedings of the Royal Society B* 282: 20151019.

Nakamura, T., A.R. Gehrke, J. Lemberg, J. Szymaszek & N. Shubin (2016), Digits and fin rays share common developmental histories. *Nature* 537: 225-228.

Credits

We have done our best to trace the original sources of all figures and obtain permissions. Those who hold copyright of material reproduced in this book that is not acknowledged below should contact the publisher.

Fig. Preface, original, N.M. van Straalen.
Fig. 1.1, from Boyd, R. & J.B. Silk (1997), *How Humans Evolved*, 6th ed. W.W. Norton & Company, New York, © 1997 W.W. Norton & Company Inc., reproduced with permission from W.W. Norton & Company Inc.
Fig. 1.2, from Boyd, R. & J.B. Silk (1997), *How Humans Evolved*, 6th ed. W.W. Norton & Company, New York, © 1997 W.W. Norton & Company Inc., reproduced with permission from W.W. Norton & Company Inc.
Fig. 1.3, from Lewin, R. & R.A. Foley (1998), *Principles of Human Evolution*. Blackwell Science, Malden, reproduced with permission from Wiley Global Permissions, ©1998 Blackwell Science Inc.
Fig. 1.4, original, N.M. van Straalen.
Fig. 1.5, © Bone Clones 2012.
Fig. 1.6, from Lovejoy, C.O., B. Latimer, G. Suwa, B. Asfaw & T.D. White (2009), Combining prehension and propulsion: the foot of *Ardipithecus ramidus*. *Science* 326: 72, reproduced with permission from AAAS.
Fig. 1.7, Private Collection/Bridgeman Images.
Fig. 1.8, from Lewin, R. & R.A. Foley (1998), *Principles of Human Evolution*. Blackwell Science, Malden, reproduced with permission from Wiley Global Permissions, ©1998 Blackwell Science Inc.
Fig. 1.9, © Naturalis Biodiversity Center, Leiden, The Netherlands.
Fig. 1.10, photograph by Kees Verhoef, August 2006.
Fig. 1.11, from Lordkipanidze, D., M.S. Ponce de Leon, A. Margvelashvili, Y. Rak, G.P. Rightmire, A. Vekua & C.P.E. Zollikofer (2013), A complete skull from Dmanisi, Georgia, and the evolutionary biology of early Homo. *Science* 342: 326-331, reproduced with permission from AAAS.
Fig. 1.12, from Lewin, R. & R.A. Foley (1993), *Principles of Human Evolution*. Blackwell Science, Malden, reproduced with permission from Wiley Global Permissions, ©1998 Blackwell Science Inc.
Fig. 1.13, from *Pour la Science* 64(2) (1983) © Grégoire Soberski.
Fig. 1.14, original, N.M. van Straalen.
Fig. 2.1, from Duboule, D. (1994), Temporal colinearity and the phylotypic progression: a basis for the stability of a vertebrate Bauplan and the evolution of morphologies through heterochrony. *Development Supplement*: 135-142, reproduced with permission from the Company of Biologists.
Fig. 2.2, from Lewin, R. & R.A. Foley (1993), *Principles of Human Evolution*. Blackwell Science, Malden, reproduced with permission from Wiley Global Permissions, ©1998 Blackwell Science Inc.
Fig. 2.3, from Slack, J.M.W. (2006), *Essential Developmental Biology*. Blackwell Publishing, Malden, reproduced with permission from Wiley Global Permissions, © 2006 Blackwell Science Inc.
Fig. 2.4, from Wikipedia, by Zephyris (Richard Wheeler), licensed under the Creative Commons Attribution-Share Alike 3.0 Unported license.
Fig. 2.5, from Wolpert, L., R. Beddington, T. Jessel, P. Lawrence, E. Meyerowitz & J.L. Smith (2002), *Principles of Development*, 2nd ed. Oxford University Press, Oxford.

Fig. 2.6, original, N.M. van Straalen.
Fig. 2.7, from Wolpert, L., R. Beddington, T. Jessel, P. Lawrence, E. Meyerowitz & J.L. Smith (2002), *Principles of Development*, 2nd ed. Oxford University Press, Oxford.
Fig. 2.8, from Wolpert, L., R. Beddington, T. Jessel, P. Lawrence, E. Meyerowitz & J.L. Smith (2002), *Principles of Development*, 2nd ed. Oxford University Press, Oxford.
Fig. 2.9, original, N.M. van Straalen.
Fig. 2.10, from Slack, J.M.W. (2006), *Essential Developmental Biology*. Blackwell Publishing, Malden, reproduced with permission from Wiley Global Permissions, © 2006 Blackwell Science Inc.
Fig. 2.11, from Wolpert, L., R. Beddington, T. Jessel, P. Lawrence, E. Meyerowitz & J.L. Smith (2002), *Principles of Development*, 2nd ed. Oxford University Press, Oxford.
Fig. 2.12, from Slack, J.M.W. (2006), *Essential Developmental Biology*. Blackwell Publishing, Malden, reproduced with permission from Wiley Global Permissions, © 2006 Blackwell Science Inc.
Fig. 2.13, original, N.M. van Straalen
Fig. 2.14, from Strickberger, M.W. (2000), *Evolution*, 3rd ed. Jones and Bartlett Publishers, Sudbury, reproduced with permission from Jones & Bartlett Learning.
Fig. 2.15, original, N.M. van Straalen.
Fig. 2.16, from Wolpert, L., R. Beddington, T. Jessel, P. Lawrence, E. Meyerowitz & J.L. Smith (2002), *Principles of Development*, 2nd ed. Oxford University Press, Oxford.
Fig. 2.17, modified after Alberts, B., A. Johnson, J. Lewis, M. Raff, K. Roberts & P. Walter (2002), *Molecular Biology of the Cell*, 5th ed. Garland Science, New York, reproduced with permission from Paul Martin, Bristol (photo top left).
Fig. 2.18, from Wikipedia, by James Heilman, MD, licensed under the Creative Commons Attribution-Share Alike 3.0 Unported license.
Fig. 3.1, kindly provided by Atlas van Stolk, Rotterdam.
Fig. 3.2, © Sebastiaan Donders, Fallen Serenety Productions.
Fig. 3.3, from Reed, D.L., J.E. Light, J.M. Allen & J.J. Kirchmann (2004), Pair of lice lost or parasites regained: the evolutionary history of anthropoid primate lice. *BMC Biology* 5: 7, reproduced with permission from David L. Reed.
Fig. 3.4, from Wu, D.D., D.M. Irwin & Y.-P. Zhang (2008), Molecular evolution of the keratin associated protein gene family in mammals, role in the evolution of mammalian hair. *BMC Evolutionary Biology* 8: 241, reproduced with permission from Ya-ping Zhang.
Fig. 3.5, from LeGros Clark, W.E. (1964), *The Fossil Evidence for Human Evolution*. Revised ed. University of Chicago Press, Chicago, reproduced with permission from University of Chicago Press.
Fig. 3.6, from Aiello, L. & C. Dean (2006), *An Introduction to Human Evolutionary Anatomy*. Elsevier Scientific Press, Amsterdam, reproduced with permission from Elsevier.
Fig. 3.7, from DeSilva, J.M., K.G. Holt, S.E. Churchill, K.J. Carlson, C.S. Walker, B. Zipfel & L.R. Berger (2013), The lower limb and mechanics of walking in *Australopithecus sediba*. *Science* 340: 163-165, reproduced with permission from AAAS.
Fig. 3.8, from Bennet, M.R., J.W.K. Harris, B.G. Richmond, D.R. Braun, E. Mbua, P. Kiura, D. Olago, M. Kibunjia, C. Omuombo, A.K. Behrensmeyer et al. (2009), Early hominin foot morphology based on 1.5-million-year-old footprints from Ileret, Kenya. *Science* 323: 1197-1201, reproduced with permission from AAAS.
Fig. 3.9, from Wolpert, L., R. Beddington, T. Jessel, P. Lawrence, E. Meyerowitz & J.L. Smith (2002), *Principles of Development*, 2nd ed. Oxford University Press, Oxford.
Fig. 3.10, C.E. Stevens & I.D. Hume (1998) Contributions of microbes in vertebrate gastrointestinal tract to production and conservation of nutrients. *Physiological Reviews* 78: 393-427, reproduced with permission of The American Physiological Society.
Fig. 3.11, from https://www.geol.umd.edu/~jmerck/honr219d/notes/12a.html.

Fig. 3.12, from Portmann, A. (1969), *Einführung in die vergleichende Morphologie der Wirbeltiere*. 4te, überarbeitete und ergänzte Auflage. Schwabe & Co. Verlag, Basel/Stuttgart, reproduced with permission from Schwabe AG.

Fig. 3.13, from Strickberger, M.W. (2000), *Evolution*, 3rd ed. Jones and Bartlett Publishers, Sudbury, reproduced with permission from Jones and Bartlett Learning.

Fig. 3.14, from Strickberger, M.W. (2000), *Evolution*, 3rd ed. Jones and Bartlett Publishers, Sudbury, reproduced with permission from Jones and Bartlett Learning.

Fig. 3.15, from BC Open Textbooks, *Anatomy and Physiology*, Chapter 28.3, Fetal Development.

Fig. 3.16, from Gerard M. Doherty: Current Diagnosis & Treatment: Surgery, 14th edition, © 2015 McGraw-Hill Education, New York.

Fig. 3.17, from Portmann, A. (1969), *Einführung in die vergleichende Morphologie der Wirbeltiere*. 4te, überarbeitete und ergänzte Auflage. Schwabe & Co. Verlag, Basel/Stuttgart, reproduced with permission from Schwabe AG; image on the right from Pioneer 10 plaquette.

Fig. 3.18, Reprinted by permission from Springer Nature Customer Service Centre GmbH: Gilad, Y., A. Oshlack, G.K. Smyth, T.P. Speed & K.P. White (2006), Expression profiling in primates reveals a rapid evolution of human transcription factors. *Nature* 440: 242-245, © 2006.

Fig. 3.19, from O'Bleness, M.S., C.M. Dickens, L.J. Dumas, H. Kehrer-Sawatzki, G.J. Wyckoff & J.M. Sikela (2012), Evolutionary history and genome organization of DUF1220 protein domains. *G3 Genes|Genomes|Genetics* 2: 977-986, reproduced under Creative Commons with permission from the author.

Fig, 4.1, reproduced from page 332 in De Vries, H. (1905), A visit to Luther Burbank. *The Popular Science Monthly* 67: 329-347, made available by Wikimedia Commons.

Fig. 4.2, original, N.M. van Straalen.

Fig. 4.3, from Iltis, H.H. (1983), From teosinte to maize: the catastrophic sexual transmutation. *Science* 222: 886-894, reproduced with permission from AAAS.

Fig. 4.4, original, N.M. van Straalen.

Fig. 4.5, original, N.M. van Straalen.

Fig. 4.6, DNA analysis by Tjalf de Boer, Vrije Universiteit Amsterdam, 2006.

Fig. 4.7, from Dennis, M.Y., X. Nuttle, P.H. Sudmant, F. Anotonacci, T.A. Graves, M. Nefedov, J.A. Rosenfield, S. Sajjadian, M. Malig, H. Kotkiewicz *et al.* (2012), Evolution of human-specific neural SRGAP2 genes by incomplete segmental duplication. *Cell* 149: 912-922, reproduced with permission from Elsevier.

Fig. 4.8, from Yunis, J.J. & P. Prakash (1982), The origin of man: a chromosomal pictorial legacy. *Science* 215: 1525-1529, reproduced with permission from AAS.

Fig. 4.9, from Sipek Jr., A., A. Panczak, R. Mihalova, L. Hrckova, E. Suttrova, V. Sobotka, P. Lonsky, N. Kasprikova & V. Gregor (2015), Pericentric inversion of human chromosome 9 epidemiology study in Czech males and females. *Folia biologica* 61: 140-146, reproduced with permission from Folia Biologica.

Fig. 4.10, original, N.M. van Straalen.

Fig. 4.11, original, N.M. van Straalen.

Fig. 4.12, from Shifman, S., J. Kuypers, M. Kokoris, B. Yakir & A. Darvasi (2003), Linkage disequilibrium patterns of the human genome across populations. *Human Molecular Genetics* 12: 771-776, reproduced with permission from Oxford University Press.

Fig. 4.13, original, N.M. van Straalen.

Fig. 4.14, original, N.M. van Straalen.

Fig. 4.15, from Ramachandran, S., O. Deshpande, C.C. Roseman, N.A. Rosenberg, M.W. Feldman & L.L. Cavalli-Sforza (2005), Support from the relationship of genetic and geographic distance in human populations for a serial founder effect originating in Africa. *Proceedings of the National Academy of Sciences of the United States of America* 102: 15942-15947.

Fig. 4.16, from Waddington, C.H. (1957), *The Strategy of the Genes: A Discussion of Some Aspects of Theoretical Biology*. George Allen & Unwin Ltd., London, reproduced with permission from Allen & Unwin Publishers.

Fig. 4.17, from *Protocols and Applications Guide*. Chapter 16, Epigenetics, Promega Corporation.

Fig. 4.18, from Smallwood, S.A. & G. Kelsey (2012), De novo DNA methylation: a germ cell perspective. *Trends in Genetics* 28: 33-42, reproduced with permission from Elsevier.

Fig. 5.1, original, N.M. van Straalen.

Fig. 5.2, original, N.M. van Straalen.

Fig. 5.3, original, N.M. van Straalen.

Fig. 5.4, original, N.M. van Straalen.

Fig. 5.5, original, N.M. van Straalen.

Fig. 5.6, original, N.M. van Straalen.

Fig. 5.7, modified from Fitch, W.M. & C.H. Langley (1976), Protein evolution and the molecular clock. *Federation Proceedings* 35: 2092-2097.

Fig. 5.8, Reprinted by permission from Springer Nature: Springer: Where Do We Come from? The Molecular Evidence for Human Descent, by J. Klein and N. Takahata ©2002.

Fig. 5.9, from Lewin, R. & R.A. Foley (1998), *Principles of Human Evolution*. Blackwell Science, Malden, reproduced with permission from Wiley Global Permissions, ©1998 Blackwell Science Inc.

Fig. 5.10, original, N.M. van Straalen.

Fig. 5.11, modified from Hammer, M.F., T. Karafet, A. Rasanayagam, E.T. Wood, T.K. Altheide, T. Jenkins, R.C. Griffiths, A.R. Templeton & S.L. Zegura (1998), Out of Africa and back again: nested cladistic analysis of human Y chromosome variation. *Molecular Biology and Evolution* 15: 427-441.

Fig. 5.12, background map © d-maps.com.

Fig. 5.13, original, N.M. van Straalen.

Fig. 5.14, Reprinted by permission from Springer Nature Customer Service Centre GmbH: Moreno-Mayar, I., B.A. Potter, L. Vinner, M. Steinrücken, S. Rasmussen, J. Terhorst, J.A. Kamm, A. Albrechtsen, A.-S. Malaspinas, M. Sikora *et al.* (2018), Terminal Pleistocene Alaskan genome reveals first founding population of Native Americans. *Nature* 553: 203-207, © 2018.

Fig. 5.15, original, N.M. van Straalen.

Fig. 5.16, from Lalueza-Fox, C. & M.T.P. Gilbert (2011), Paleogenomics of archaic hominins. *Current Biology* 21: R1002-R1009, reproduced with permission from Elsevier.

Fig. 6.1, from Boyd, R. & J.B. Silk (1997), *How Humans Evolved*, 6th ed. W.W. Norton & Company, New York, © 1997 W.W. Norton & Company Inc., reproduced with permission from W.W. Norton & Company Inc.

Fig. 6.2, original source unknown.

Fig. 6.3, from Zilhão, J. & F. d'Errico (2003), *The Chronology of the Aurignacian and of the Transitional Technocomplexes: Dating, Stratigraphies, Cultural Implications: Proceedings of Symposium 6.1 of the XIVth Congress of the UISPP, University of Liège, Belgium, 2001*, reproduced with permission from Nicholas Conrad, University of Tübingen.

Fig. 6.4, from Boyd, R. & J.B. Silk (1997), *How Humans Evolved*, 6th ed. W.W. Norton & Company, New York, © 1997 W.W. Norton & Company Inc., reproduced with permission from W.W. Norton & Company Inc.

Fig. 6.5, photograph by N.M. van Straalen.

Fig. 6.6, Reprinted by permission from Springer Nature Customer Service Centre GmbH: Diamond, J. (2002), Evolution, consequences and future of plant and animal domestication. *Nature* 418: 700-707, © 2002.

Fig. 6.7, data from Wikipedia.org.

Fig. 6.8, photograph by Sander Collet.

CREDITS 279

Fig. 6.9, from Gluth, S. & L. Fontanesi (2016), Wiring the altruistic brain. *Science* 351: 1028-1029, reproduced with permission from AAAS.

Fig. 6.10, Drawing by Sudha Premnathe, reproduced from R. Gadagkar (2000) The origin and resolution of conflicts in animal societies. *Resonance* 5(4): 62-73.

Fig. 6.11, from West, S.A., A.S. Griffin & A. Gardner (2007), Social semantics: altruism, cooperation, mutualism, strong reciprocity and group selection. *Journal of Evolutionary Biology* 20: 415-432, reproduced with permission from John Wiley and Sons.

Fig. 6.12, from Lehmann, J., A.H. Kortsjens & R.I.M. Dunbar (2007), Group size, grooming and social cohesion in primates. *Animal Behaviour* 74: 1617-1629, reproduced with permission from Elsevier.

Fig. 6.13, © Cyril Ruoso/Biosphoto/SteveBloom.com.

Fig. 6.14, from Szathmáry, E. (2011), To group or not to group. *Science* 334: 1648-1649, © Mauricio Anton, Photo Researchers.

Fig. 7.1, from Burt, C. (1963), Is intelligence distributed normally? *The British Journal of Statistical Psychology* 16: 175-190, reproduced with permission from John Wiley and Sons.

Fig. 7.2, from Stearns, S.C. (1992), *The Evolution of Life Histories*, Oxford University Press, Oxford, p. 48, after Falconer, D.S. (1981), *Introduction to Quantitative Genetics*, Longman, New York, reproduced with permission from Oxford University Press.

Fig. 7.3, Reprinted by permission from Springer Nature Customer Service Centre GmbH: Kelsoe, J.R. (2010), A gene for impulsivity. *Nature* 468: 1049-1050, © 2010.

Fig. 7.4. from Kasabova, B.E. & T.W. Holliday (2015), New model for estimating the relationship between surface area and volume in the human body using skeletal remains. *American Journal of Physical Anthropology* 156: 614-624, reproduced by permission from John Wiley & Sons.

Fig. 7.5, from Byars, S.G., D. Ewbank, D.R. Govindaraju & S.C. Stearns (2010), Natural selection in a contemporary human population. *Proceedings of the National Academy of Sciences of the United States of America* 107 (Suppl. 1): 1787-1792.

Fig. 7.6, from Gerbault, P., A. Liebert, Y. Itan, A.J. Powell, M. Currat, J. Burger, D.M. Swallow & A.D. Thomas (2011), Evolution of lactase persistence: an example of human niche construction. *Philosophical Transactions of the Royal Society of London. B. Biological Sciences* 366: 863-877, reproduced with permission from The Royal Society.

Fig. 7.7, Reprinted by permission from Springer Nature Customer Service Centre GmbH: Lahdenperä, M., V. Lummaa, S. Helle, M. Tremblay & A.F. Russell (2004), Fitness benefits of prolonged post-reproductive lifespan in women. *Nature* 428: 178-181, © 2004.

Fig. 7.8, from Wedekind, C., T. Seebeck, F. Bettens & A.J. Paepke (1995), MHC-dependent mate preferences in humans. *Proceedings of the Royal Society of London*, Series B 260: 245-249, reproduced with permission from The Royal Society.

Fig. 7.9, from Dixson, B.J., G.M. Grimshaw, W.L. Linklater & A.F. Dixson (2011), Eye-tracking of men's preferences for waist-to-hip ratio and breast size of women. *Archives of Sexual Behavior* 40: 43-50, and Dixson, B.J.W. & M.J. Rantala (2016), The role of facial and body hair distribution in women's judgments of men's sexual attractiveness. *Archives of Sexual Behavior* 45: 877-889, reproduced with permission from Springer.

Fig. 7.10, original, N.M. van Straalen.

Fig. 7.11, original, D. Roelofs.

Fig. 7.12, original, N.M. van Straalen, images from http://walkingwithdinos.wikia.com.

Fig. Epilogue, from Laland, K.N., T. Uller, M.W. Feldman, K. Sterelny, G.B. Müller, A. Moczek, E. Jablonka & J. Odling-Smee (2015), The extended evolutionary synthesis: its structure, assumptions and predictions. *Proceedings of the Royal Society* B 282: 20151019, reproduced with permission from The Royal Society.

Index

(page numbers printed in bold refer to occurrences in figures or tables)

abdA 62
abdominal A 62
abdominal cavity 43, 94
abductor 80, 84
Aboriginals 173
acetabulum 79, **80**
acetylation 147
Acheulean culture **185**, **188**
Achilles tendon 81
achondroplasia 128
adaptation 12, 26, **34**, 49, 71, 77, 82, **84**, 87, 109, 114, 121, 148, 172, 183, 185, 212, 221, 230, 238, **239**, 242
adaptationism 113
additive genetic variance 217
ADHD 118
adjusted Out of Africa model 171, **172**
AER 65
Afar 13
Africa
  agriculture 192, 196
  apes 11, **195**
  fossil finds **13**, 19, 23, 29, 35, 88
  hybridisation 177
  malaria 134
  migrations 26, 36, 138, 141, 164, 166, **168**, **169**, 173
  savanna 12, 75, 212
  stone tools 185, **189**
Africans 152, 166, **169**, 172, **173**, 177, 178, 196, 222
Afrotheria 98
Agouti 148
agriculture 195, 226
AIDS **195**, 228
Aiello, L. 88
Alexander, R.D. 211
alignment 158, **159**
alisphenoid 85
allantois 44, **58**, 96
allele
  adverse 128, 134, 137
  codominant 128, **130**
  cystic fibrosis 236
  dominant 128, 129
  frequency 113, 124, **127**, **130**, 135, **137**
  Hardy-Weinberg 119, 125, **126**, 140
  inbreeding 140, 203
  lactase persistence 226, **227**, 228
  LDL receptor 117
  linkage 132
  mutation 115, 116
  neanderthal 177

obesitas 240
recessive 128, 134
VNTR 118
Allen's rule 221, **223**
allometry 101
allopatric speciation 19
almond 192, **193**
alpha-lactalbumin 71
Altaï 35, 179
Altamira 189, **191**
altruism 201, **203**, **204**, 208
Altschul, S. 159
Alu-elements 114
Amazonia 192
Amish 138
amnion 44, 45, **58**
Amniota 45
Amphioxus 84
amygdala 100
anagenesis 17, **18**, 112, 164
Anatolia 196
anatomical language equipment 198, **199**
ancient DNA 7, 35, 161, 170, 174, 176, 227, 242
androstenone 231
angioblast 91, **92**
angular 86
animal experiments 55, 58
animal pole 48, **49**, 56
animal-vegetal axis 48, **49**, 50
Annelida 51
Antennapedia 60, **61**
anterior cingulate cortex 202, **203**
anterior insula 202
anterior-posterior axis 45, 50, 59, 62, 64
anteroventral nucleus 102
antibiotic resistance 237, **238**
antibiotics 237
antioxidant 236
anus 43, **51**
aorta 91, **93**, 95
ape 11, **12**, 15, 37, 41, 73, 75, 80, **102**, 106, 122, **123**, 159, 162, 200, 229, 234
ape-man 12, 17, 22, 37, 184
aphrodisiac 231
apical ectodermal ridge 65, **66**
apomorphic 23, 154, 181, 199
apoptosis 59, 65
appendix 69, 89
aquatic ape 75
Aramis 20
arboreal 12, 21
archenteron 42, **43**, 51

archinephric duct 94, **96**, 97
Arcy-sur-Cure 188
Ardipithecus ramidus **20**, 22, 81
argon dating 15, **16**, 28
Armenia 173
art 189, **191**
artefact 14, 160, 170, 173, 183, **187**, 188
Arthropoda 51
articular 85, **86**
artificial selection 193
Ashkenazi Jews 133
assimilation, genetic 144, 210
Atapuerca 29
atavism 154
Antp 60, **61**, **62**
atrium 92, 93, **94**
Aurignacien 186, 187
Australians 166, 171, 172, **173**, 178
Australopithecus
    afarensis 13, 17, **23**, **30**, 79, **80**, 83, 102, 185
    africanus 13, 17, 23, 24, **30**, **199**
    anamensis 13, 17, 24, 30
    bahrelghazali 13, 17, 24
    deyiremeda 13, 17, 24, 30
    garhi **13**, **17**, **24**, **30**
    sediba 13, 17, 24, 26, 30, 79, 82
autapomorphic **154**, 156
autosome 123
axis 45, 48, **49**, **50**, 52, 56, **58**, **62**, 64, **65**, 91
axochord 46, 52
azoospermia factors 236
Azores 120, 138

Bächler, E. 33
background, genetic 215, 218
balancing selection 131
Baldwin effect 144, 145
Baldwin, J.M. 143
basicranium 84, 199
Basks 138
bat 59, 89, 122, **195**
beads **187**, 191
Behe, M.J. 72
Beijing 28
Berger, L.R. 24, 28
Bergmann's rule 221
Bering strait 141, 173
Beringia 172, 174
Bhullar, B.-A. 243
bilaminar disk **44**, 45
bilateral symmetry 48, 51, 62, 144
bilirubin **235**, 236
biliverdin **235**
biogenetic law 38
biometric traits 215, **216**
bipedalism 11, **12**, 36, 76, 79, 81, 104
Bithorax complex 62
Black Eve 167
BLAST 159

blastocoel **42**, **43**, 44
blastocyst **44**, 45, 54, 58
blastopore **42**, **43**, 50, **51**
blastula 41, **43**, 45, 50, 56
blind watchmaker 71
Blombos 191
blood circulation 91, 93, **95**
blood pressure 224
blood vessel 46, 76, 84, 91
BMI 239
BMP2 **65**
BMP4 52, 66
BNC 181
boar smell 231
body axis 56
body cavity 43, 91
body decoration 199, 201
body hair 73, 75, 77, **234**
body louse 74
body mass index 239
body plan 8, 22, 37, 40, 46, 51, 59, 66, 111, 115, 145
body surface 223
body weight 80, 100, 224, 239
Boeing 747 72
bonobo 11, **30**
bony fish 90
bootstrap 159, **160**
Bosporus 196
bottleneck 135, 138, **139**
Boule, M. 33
Bowlby, J. 212
Boyer, P. **211**
brachial index 222
brain
    behaviour 99, 202, **203**, **207**, 210, 239
    development 58, 64, 99, 104, **105**, **181**
    evolution 88, 99, 100, **106**, 115, 122
    genes **103**, **105**, **106**, 121, 221
    heat regulation 76
    neoteny 40, 121
    sex 233
    sexual dimorphism 102, 149
    volume 18, 21, **23**, 25, **26**, 29, **30**, 99, 101, **102**, 228
Brazilians 170
breast cancer 66, 218
breast size 233, **234**
BRCA 218
Britton, R. 120
Brno 109
Broca's area 199
bronchi 90
brow ridge 27, 32
Brown, M. 118
Brunet, M. 21
Buffalo **193**
Burns, J. 86
Buttermilk Creek 174
by-product evolution 113, 134, 149

INDEX

caecum 87, 89
Caenorhabditis elegans 51, 53, 55
calcaneus 81, 84
calf muscle 81
canalisation 144, **145**
cancer 55, 66, 218
Canis lupus 194
Cann, R. 166, 170
caput femoris 80, 84, **222**
carbon dating 16
cardiovascular disease 118, 224, 236
Carnivora **160**
carnivory 88, **89**, 195
Cavalli-Sforza, L. 141
cave bear 33
cave drawing 33, 183, 189, **191**, 199, 201
cavitation **44**
centimorgan 132
central dogma 143, 146
central limit theorem 215
centromere **115**, 118, **124**
cerebral hemisphere 99
cerebral isthmus 100
Cerebral Rubicon 25
cervical vertebrae 219
cesium chloride 118
CFTR 236
Chad 14, 21
character state 153, **154**, **155**, 157
Châtelperonian 186
Chauvet 189
chiasma opticum 102
chicken 39, 50, **51**, 57, **63**, 66, 192, **195**, 243
childbirth 79
chimpanzee
    behaviour 183, **184**, 229
    brain 21, **30**, 100
    chromosomes 123
    classification 12, **153**
    DNA 77, 104, **105**, 106, 120, **159**, **177**
    knuckle walk 12
    limbs **80**, 81
    pelvis **78**, 79
    phylogeny 11, 75, 107, 120, 152, **154**, **155**
    skull **40**
chin 27
China 14, 27, 36, 165, 171, 173, 192, 239
Chinese 166, 172, **173**, 213
cholesterol 117, 148, 224, **225**,
Chordata 12, 38, 45, **51**, 61, 84, 91
Chordin 52
chorion **44**, 58
chromatin 146, **147**
chromatin condensation 116
chromosome
    chart **123**, 124
    fusion 123
    Hox genes 61
    unstable region 1q21.1 **105**, 107, 121

linkage 131
mutation 115, 122, 124, 236
pairing 109, **115**, **123**
X and Y 133, 134, 136, 167, **169**, 181, 236
church 32, 212
circumcision 143
cis-vaccenylacetate 231
clade 152, **154**, 159, **160**
cladistics 154
cladogenesis 17, **18**, 112, 164
Clark, G. 188
claviform signs 189, **191**
clawed frog 43, **49**, **51**, 55, 57
cleavages 38, 41, 44, 49, **51**, 53, 56
climate 12, **20**, **34**, 75, 172, 174, 192, 195, 221
clock
    fossil dating 15
    molecular 161
clothing 75, 186
Clovis 173
coalescence 162, **163**, 166, 168
co-dominance 128
coelom 43, **51**, 91, **92**, 95
co-evolving gene complex 133
cognition 33, 102
cognitive capacity 101, 103, 228
co-linear 62
collum femoris 80
colon 87
colonisation 14, 135, 138, 141, 172, 175, 282
Columbus 141, 188
communication
    cellular 181, 202
    chemical 232
    language 149, 198, 209
complexity
    body 61
    molecular 72, 146
    social 195
condyles 80
conservative gene 104, 116
convergence 156, 172
co-optation 71, 84, 91, 94, 97, 115
copulation 231
corkscrew manoeuvre 79
corpus luteum 57
Correns, C. 109
cortical rotation 49, **50**
cow **193**, 194
creationism 73
cremaster 97
Crick, F. 143
cricoides 85
CRISPR-Cas9 237, 243
crocodile **154**, 155
Cro-magnon **34**, 99, 173, 189, 197
crossing-over 114, **115**, 120
crural index 222
cultural drift 208, **209**

cultural diffusion 196
cultural explosion 119, 189, 191
culture 119, 139, 173, 183, **185**, **187**, **188**, **191**, 196, 198, 208, **209**, **210**, 233
cVA 231
cycling, antibiotics 238
cystic fibrosis 236
cytochrome c 161
cytosine 146, **147**, 176

Dali Man **165**
Danio rerio 53, **54**, **56**
Daoxian 171
Dart, R. 24
Darwin, C. 17, 32, 38, 49, 71, 86, 109, 127, 142, 145, 149, 164, 193, 205, 209, 211, 233, **243**
Darwinian evolution 49, 71, 72, 109, 145, 149, 205, 209
Darwinism 164
dating fossils 15, **16**, 201
Dawkins, R. 71, 73, 114, 204, 208
De Knijff, P. 167
De Vos, J. 31
De Vries, H. 109, **110**
De Waal, F. 200
deamination, cytosine 176
decoration 170, **187**, 199, 201
deletion **103**, **115**, 120
delivery 58, 79
demography 136, 138, 169, **196**, 230
dendritic spines 121
Denisova Man 17, **35**, 36, 121, 179
Dennett, C. 114, 211
dental 86
dental arch 23
derived character 23, **155**, 156
descensus testiculorum 97
determinate cleavages 51, 53
Deuterostomia 47, **51**, **52**, 91
development
 body axes 48, 51, 60, **65**
 body plan 37, **38**, 39, 51, 59
 cognitive 200
 ear 86,
 embryo 42, **43**, **44**, **50**, 56, 219
 epigenetics **145**, 146
 evolution 8, 37, 39, 111, 144
 genetics 59, 63, **103**, **106**, 149
 gill arches 84, **85**
 heart 91, **92**, 95
 heterochrony **40**, 77, 107, 121, 229
 kidney 94, **96**
 limbs **65**, 66
 lung 90
 model animals 53, **54**, 56, **58**, **63**
 mutation 69, 111, 145
 nervous system 46, **47**, 100, **103**, 121, 181
 plasticity 144, **145**, 222
 testis 97, **98**

developmental genes 37, 52, 59, 64, **145**
developmental genetics 37, 54, 66
developmental homeostasis 144
diabetes 239
Diamond, J. 197
diastema 23
diencephalon 58, 99
density centrifugation 118
diffusion
 cultural 196
 demic 197
Dikika 184
dilatation 199
dimorphism 22, 102, 233
Diptera 60
directional selection 111, **131**, 219, 227
disease 55, 64, 89, 93, 105, 117, 120, 127, 134, 138, 140, 194, **195**, 208, 215, 229, 232, 236
dispersal 14, 141
disruptive selection **131**
distal 13, 48, 64, **65**, 66, 80, 223
distribution range 14, 19, 165, 179, **186**
divergence 11, 21, 37, 151, **162**, 163, 169, 179
divergence time **54**, **74**, 75, 151, 161
diverticulum 90
dizygotic twin 218
Dmanisi 14, 29, 180
DNA
 ancient 7, 30, 35, 141, 161, 170, 174, 176, 181, 197, 227, 242
 accelerated evolution 104
 binding 59
 homology 158, **159**, 161, 164
 methylation 146, **147**, 148
 mutations 114, **115**, 116, **117**, 119, 158, 162
 non-coding 77
 Neanderthal 35, 181, 176, **177**, 179
 mitochondrial 136, 152, 166, 178
 phylogenetic reconstruction 161, 166, 170
 polymorphic **115**, **117**, **119**, 166
 recombination 133
 relatedness 230
 repair 161
dog 87, 194, **195**
Dollo, L. 157
Dollo's rule 157
dolphin 200
domain amplification 103, **106**
domestic animal 192, **193**, 194
domestication 192, 194, 211
dominant allele 128, 226
dopamine 118, **220**
dopamine receptor 118
dorsal blood vessel 52
dorsalising centre 50
dorso-ventral axis 49, **50**, 65
double circulation 92, **94**
Drachenloch cave 33
DRD4 119, 220, **228**

Dreyer, T.F. 35
drift
 continental 13
 cultural 208, **209**
 genetic 8, 124, 135, **137**, 140, 142, 162, 242
Drosophila 51, 53, **54**, 59, **61**, **62**, **63**, 112, 231
D-statistic 177
Dubois, E. 27, 191
Duboule, D. 37, **38**
ductus arteriosus 93, **95**
ductus Botalli 93
DUF1220 103, 105, **106**, 220
Dunbar, R. 206
duplication 59, **60**, 62, 73, 103, **106**, 107, 115, 120, 121, 193
dwarfism 31, 128
dystrophin 181

Echinodermata 51
ectoderm **42**, 43, 45, 47, 65
ectodermal ridge **65**, **66**
ecogeographic rules 221
EEA **212**, 213, 239
effective population size 135, 136, **163**
Eichler, E. 121
einkorn 193
Eldredge, N. 111
elephant 31, 89
elk 193
embryonic development
 body axis 48, 51
 C. elegans 53
 clawed frog 49, 51, **55**, **57**
 cleavages 46
 ear 86
 epigenetics 146
 evolution 8, 39
 human **44**
 genetics 66, 103, 219
 gill arches 84, 86
 heart 51, 91, **92**, **93**
 kidney 43, 94, **96**
 lung 90
 mouse 58
 nervous system 47
 zebra fish 56
empathy 200, **203**
encephalisation quotient 101, **102**
endoderm **42**, 43, 45
endometrium 44, 46, 58
Engrailed 63, 65
Enterobius vermicularis 53
entotaxony 82
environment of evolutionary adaptedness **212**, 239
EPAS1 181, 228
epiblast **44**, 45
epiboly 56
epicard 92

epididymis 95
epigenetic inheritance 149
epigenetic landscape 144, **145**
epigenetic marker 146, **147**, 148
epigenetics 144, 146, 148
epistasis 218, 220
Equus 192, **193**
Euarchontoglires 98
Europe
 agriculture 194, 196
 cultural explosion 119, 185, 191
 migration 14, 19, 35, 141, 172, 197
 Neanderthal 33, **34**, 36, 177, 179, **180**, 222
 stone tools 185, 188, 189
Europeans 141, 152, 166, 171, **173**, 174, **177**, 197, **222**, 227, 240
Eustachian tube 93
evolutionary rate 112, 160
exaptation 69, 71, 185
expensive tissue hypothesis 88
expression
 art 100, 183, 189, 201, 209
 gene **50**, 52, 60, **62**, **63**, 66, 104, **105**, 115, 134, 144, **147**, 218, 226, **228**
 self 98, 191
expression domain **62**, **63**
Extended Synthesis 241
extensor 80, **84**

feathers 154, **155**, 233
Feldhofer 27, 31
Felsenstein, J. 159
femur 21, 27, 79, **80**, **84**, 222
fermentation 88, **89**, 227
fertile crescent 192
fertility 97, 149, 178, 190, 228, 233
FGF8 **65**
fibrinopeptide 161
fibula **222**
filter feeding 84
fire 34, 88, 189, 211
Fisher, R. 111, 140
Fitch, W.M. 161
fitness 74, 122, 128, 130, 139, 144, 202, 208, 218, 229, 236
fitness, inclusive 202, 204, 230
fixation 136, 163
fixation time 136, 163
flagellum **73**
flea 75
Fleming, A. 237
Flores Man 31
fluctuating asymmetry 145
foetus 45, 58, 77, 85, 90, 96, 116
forebrain 99
foregut fermenter 88
foot **20**, 25, 81, **82**, 84
foot prints 82, **83**
foramen magnum 21, 77, **84**

foramen ovale 93, **95**
Fore people 208
fossilisation 14, 75, 176
FOXP2 103, 199
Framingham Heart Study 224, **225**, 230
frequency-dependent selection 130, **131**
Freud, S. 100
fruit fly 53, **54**, 60, 64
F_{ST} 140, 142, 175
Fuhlrott, J. 32
Funnel Beaker culture 197
Fuyan cave 171

G6PD 134, 226, **228**
galactose 226
Galis, F. 219
Gallus gallus 53, 57
Gärdenfors, P. 199
gastrula 42, **43**
gastrulation 42, **43**, 44, **49**, **50**, 53, 91
Geissenklösterle 186, **187**
gene conversion 103, 107, 114, **115**
gene duplication 59, **60**, 120
gene expression
  brain 105
  developmental genes 52, 60, 62, 64, 66, 219
  domain **63**, 64
  epigenetics 146
  genome-wide 46
  mutation 115
gene-environment interaction 217
gene family 76, 106, 121
gene genealogy **163**, 175
gene loss 106, 115, 120, 122
gene therapy 236
generation time 53, 57, 218
genetic assimilation 144, 210
genetic background 197, 215, 218
genetic distance 132, 140, **142**, 223
genetic drift 8, 124, 135, **137**, 140, 142, 162, 242
genetic linkage 132, **133**, 226, 232
genetic mapping 117
genome
  ancient DNA 174, 178, 197
  association studies 31, 117, 215, 240
  domestication 193, 194
  duplication 59, 105, 107
  editing 237, 243
  epigenetics 146, **147**
  epistasis 218, 220
  linkage 132, **133**
  migration 197
  mitochondrial 158, **159**, 176
  model animal 53, 55, 57, **62**
  mutation 114, 115, 120, 122, 161, 169, 210, 241
  Neanderthal 122, 176, **177**, 178, 179, **181**
  phylogeny 7, 157, 162
  polymorphism 116, 166, 177
  selection 103, 105, 107, 113, 134

structure 114, 121, 133
genotype 127, 129, 132, 139, 144, 217, 220, 232
genotype frequency 124, **126**, **128**, **131**
genotyping 119, 133, 231
Georgia Man 14, 27, 29, 199
germ disc 45
germ layer 41, **42**, **43**, 45
germline 134
gibbon 22, 158, **159**, 198
Gilbert's syndrome 235
gill arch 69, 84, **85**, **93**, 238
gill slit 84, **85**
gizzard **154**, **155**
globin 161
glucuronic acid 235
gluteus maximus 80, 84
gluteus medius 80
Goldschmidt, R. 111
Goldstein, J. 118
Gona 183
gonocoel 43
Good Samaritan 202
gorilla 11, **12**, 21, 74, 77, 97, **102**, 107, **123**, 153, 158, **159**
Gould, S.J. 69, 71, 86, 111, 113, 164
Greenberg, J. 174
gracile hominins 18, 19
gradualism 111, 164
grandmother hypothesis 229
Greenlanders 33
group selection 202, 205, **206**, 212
group size 206, **207**
guinea pig 122
GULO 122

HaA 77
Hadar 13, 183
Haeckel, E. 38, **39**, 41
haem **235**
haemoglobin 235, 236
hair 41, 73, 75, 77, 234
Haldane, J.B.S. 204
half sib 218
hallux abduction 23, 82, **83**
halter 59
Hamilton's rule 202, **204**
Hammer, M. 167, **169**
hammer stone 183, **188**
Han Chinese 172, **173**
hand 11, 25, 66, 173, 184, 185, 222
hand axe 26, 173, 185, **188**
Haplorhini 12
haplotype 133, 167, 178, **181**, 226, 228
haplotype tree **169**
HAR **104**
Hardy, A. 75
Hardy, G.H. 125
Hardy-Weinberg equilibrium **126**, **127**, 135, 139, 140

INDEX

harem 22, 135, 136
Harpending, H. 138
hCG 57
HDL 117
head louse 74
heart 43, 46, 48, **51**, 66, 91, **92**, **93**, **95**, 235
heat regulation **12**, 75
Hedgehog **63**, 64
Heidelberg Man **19**, 34, **35**, 36, 41, 166, 177, 179, 185, **188**
hemichordial placentation 46
Hennig, W. 154
herbivory 18, 87, **89**
heritability **217**, 219, 225
Herto 37
heterochrony 31, 39, **40**
heterogametic 134
heterosis 129, **130**, 135
heterozygosity 125, 232
heterozygote advantage 129, **130**, **131**
heterozygous **119**, 129, 232
heel bone 20, 81
Heyes, C. 200
hindbrain 100
hindgut fermenter 88, **89**
hip 66, 79, 82, 222, 233, 234
hip dysplasia 66
hippocampus 100, 102
histone 146, **147**
histone modification **147**
HIV 195
HLA Class 1 **181**
Hofman, M. 102
Hogben, L. 56
Hohle Fels Venus 190
Holt Oram syndrome 66
holoblastic cleavages 44
Holocene 192, 195
homeobox 59, 112
homeosis 59
Hominidae 11, **12**, 102, **153**
Homininae 11, **12**, **20**
hominins
 ancient DNA 175
 behaviour 198, 239
 bipedalism **12**, 15, 75, 77, 81, **83**
 brain 27, 30, 76, 78, 88, 99, 101, **102**
 culture 183, 185, 189, 198, **199**
 fossil **13**, 17, 18, 20, 21, 24, 28, 31, 34, 36, 165
 hybridisation 172, 180
 neoteny 41, 229
 sexual dimorphism 22, 233
 speciation 18, 30, 112, 123, 164
 tree **17**, 19, 21, 74, 179, 180
Hominoidea **12**, **153**
Homo antecessor **17**, 29
Homo erectus
 behaviour 88, 191, 198
 brain volume 26, 99, **102**

life cycle 41, 229
migrations 14, 28, 31, **165**, 167, 185
morphology 27, 28, **30**, 74, 79, **83**, 165, 199
origin **17**, **19**, 24, 26, 164
skull 29, 181
stone tools 185, **188**
Homo
 ergaster 19, 26, 28, **30**, 34, 185, **188**, **212**
 floresiensis 31
 georgicus 30
 habilis **17**, 24, 26, 29, **30**, 82, 99, **102**, 164, 184, 185, **188**, 198, **199**
 heidelbergensis **17**, **19**, 35, 41, 166, 177, 179, 185, **188**, **199**
 naledi **15**, **17**, 29, **30**, 177
Homo neanderthalensis
 behaviour 34, 189, 198, **199**, 201, 209
 brain 99
 DNA 35, **103**, 121, 128, 176, 179, **181**
 fossil 16, 27, 31, 33
 hybridisation 172, **177**, 178, **180**
 morphology **33**, **34**, 74, 79, **222**
 phylogenetic position **17**, **19**, 32, **35**, **139**, 164, 166
 skull 32, 35
 stone tools 185, **188**
Homo rudolfensis **17**, 24, 26, 30
Homo sapiens
 behaviour 34, 189, **190**, **191**, 201, 213, 234
 bipedalism 11, 81, **83**
 brain volume 31, 99, **102**
 classification 12
 DNA 61, 77, 167, **173**, 177, 179, **181**
 hybridisation 172, 178, **180**
 language 198, **199**
 life-cycle 234
 migrations 75, 141, **142**, 170, **172**, 174
 neoteny 31, **40**, 41
 origin 14, **17**, 29, 32, **35**, 164, **165**, **168**, **169**
 pelvis 66, 78, 79
 population 138, **142**, 173, 179
 skull 15, 31, **33**
 stone tools 173, 186, **188**
homogametic 134
homology 52, 54, 56, 58, **60**, 64, 121, 154, 157, 159, 164
homoplasy 153, 156
homosexuality 102, 149
homozygous **119**, **126**, **127**, **128**, **130**, 134, **139**, 232
homozygosity 125, 139
hoofed animal 59
hopeful monster 111
horse 88, **193**
Hortus Botanicus 109
hospital 166, 237, **238**
hourglass model **38**, 44
Hox genes 54, 59, **60**, 62, 64, 66, 112, 133, 219
Hoyle, F. 72
Hublin, J.-J. 36

humerus 222
Hunger Winter 148
hunter-gatherers 194, 196
hunting 26, 34, 88, 99, 189, 192
hypertension 228
Hutchinson's disease 107
Huxley, J. 111
Huxley, T. 32
hybridisation 177, 179, **180**, 193
hyoid 85, 199
hypermethylation 147
hypervariable locus 116
hypoblast **44**, 45
hypoglossal canal **199**
hypomethylation 147, 148
hypothalamus 100, 102

IBD 141, 171, 223
ICSI 236
IgA 89
ileum 87
ilium 78, 79, 84
imaginal discs 63
immune system 89, 129, **181**, 194, 224, **232**
implantation **44**, 54, 58
in vitro fertilisation 236
inbreeding 139
inbreeding coefficient 140, 203
inclination angle 23, 81, **80**, 84
incus **85**, **86**
indel **115**, 120
indeterminate cleavages 51, 53
Indochina 173
Indonesia 27, 30
inner cell mass **44**, 45, 54
ingroup 151
inguinal canal 97
insertion **115**, 120
insula anterior 202, **203**
insulin 240
intelligence 215, 220
intelligent design 71, 73
interference RNA 53, 147
intermediary inheritance 129
intersexual selection 211
intestine **42**, **44**, 49, **51**, 57, **86**, **93**, 218
intrasexual selection 233
introgression 174, 178, 181, 194
invagination **42**, 43
inversion 106, 115, 120, 122, **124**
inner world **198**, **201**, 209
IQ 107, **216**, 219
irreducible complexity 72
irreversible character 157
ischium 78, 79
isolate breaking 140
isolation by distance 141, **142**
isometry 101
isotope ratio 16

Jablonka, E. 148
Jacob, F. 69, 107
Jacobson's organ 231
Jacob, T. 31
Japan 173
Japanese macaque 209, **210**
jaundice 235
Java Man 27, 28, 164, 191
jaw
  development 85, 103, 234
  fossil 18, 24, 34
  joint **85**, **86**,
  phylogenetic character **154**, **155**
Jebel Irhoud 36, 171
Jenner, E. 194
Johanson, D. 22
juvenile traits 31, 39, **40**, 77, 229

K/Ar dating 16, 28
kangaroo 87
karyotype 124
Kennewick Man 174
Kenyanthropus platyops 17, 24
keratin 76, 122, **155**, 156
kidney 43, 94, **96**
Kimura, M. 135
Kimura's two-parameter model 158
kin selection 202, 204, 206
kinship 202
knee **23**, **80**, **82**, 84
knuckle walk 12
koala 87
Koro Toro 13
Koryak 33, **175**
Koshima 209, 210
Krause, J. 176
KRTAP **76**, 122
kuru 208
kyphosis 78

La Chapelle-aux-Saints 33, 209
lactase persistence 226, **227**, **228**
lactose 225, 227, 240
lactose intolerance 226
lactose synthase 71
Laetoli 82, **83**
Lamarck, J.-B. de 142
Lamarckism 143, 145
lamprey **154**, **155**
language 24, 33, 149, 167, 174, 198, 200, **201**, 209
lanugo 77
larva 39, 45, 55, 62, 64
laryngeal nerve 92, **93**
larynx **85**, 92, **93**, 199
Lascaux 189
Laurasiatheria 98
LCTp 226
LDL receptor **117**
Leakey, L. 25, 36, 183

INDEX

Leakey, M. 24
leg 11, 21, **23**, 48, 59, **61**, **63**, 77, 79, 81, **82**, 112, **222**, 243
leptokurtosis 216
leukaemia 55, 66
Levallois technique 186
Lewis, E. 59
Lewontin, R. 113
Liang Bua 30
Liang Timpuseng 191
life-cycle 25, 148, 201, 219, 229, 240
life-time reproductive success 224
limb 48, 64, **65**, **66**, 153, 156, **222**
limbic system 100
LINE 114
linkage 131, **133**, 226, 232
liver **43**, **235**
lizard **154**, **155**
llama 192
Lmx1 **63**, **65**
Lokalalei 167
Lomekwi 184
long branch attraction 160
Lordkipanidze, D. 29
lordosis 78
Lorenz, K. **211**
louse 74
LRS 224
Lucy 22
lumbar vertebrae **79**, **84**
lumping **30**
lung 43, 87, **90**, 93, **154**, **155**
lung artery **93**, **95**
lungfish 90
lung vesicle 90
Lynch, M. 113
lysozyme 71

Macaca fuscata 209, **210**
macaque 77, **105**, 209
Machado-Joseph disease 120, 138
macro-mutation 110, **111**, 112, **115**, 193
Magdalenian culture **191**
magpie 200
maize 112, 192, 194
Mal'ta cave 174
Malapa cave 24
malaria 134, 215, 228
Malay archipelago 31, **172**, 173
malleus 85, **85**, **86**
mammal
    behaviour 22
    brain 88, 100, 103, **106**
    characteristics 75, 91, 93, **98**, 114, 122, 157, 219
    classification 12
    development **44**, 50, 84, 91, **96**,
    evolution 31, 39, **86**, 98, **154**, **160**, **162**
    intestine **87**, 88, 89
    reproduction 46, 71, 97, 134, 226, 234
Mammalia 12
mammoth 189
Marmara sea 196
Marsupialia **162**
Mata Menge 31
maternal effect 49
maternal inheritance 166
mathematical capacity 107
Mauer jaw 34
Mayas 141
Maynard Smith, J. 73, 203, 205
Mayr, E. 164
Mbuti 141
MCM6 226
Meckel's cartilage 84, **85**
meerkat 198
Melanesians 171
meme 208, **209**
memory 100
menarche 225
Mendel, G. 109
Mendelian inheritance 192
menopause 215, 224, 229
menstruation 225, 232
mental capacity 101, 105, 107, **228**
mental retardation 106
meroblastic cleavages 45, **56**
MERS 195
mesoderm 42, **43**, **45**, 51, 91, **92**
mesonephros 95, **96**, **97**
metanephros 95, **96**, **97**
mesencephalon **58**, 99
metencephalon **58**, 99
methylation 146, **147**, 148
metric characters 215, **216**
MHC-II 129, 133, 231, **232**
microcephalin 181
microcephaly 31, 105
microliths 186
micro-mutation **115**
microsatellite **115**, 118, 120, 126, 141, **142**
midbrain 99
Middle Awash 13
Middle East
    agriculture 192, 194, 196, 226
    migration *H. sapiens* 14, 34, 141, 165, 171, **172**, 173, 178
    stone tools 185
middle-ear ossicles **85**, 86
migration
    America **172**, **174**, **175**
    Australia 171, **172**
    cell **43**, **44**, 57, 103, 121
    genetic variation 126, 133, 135, 141, **142**, 197, 229
    Homo erectus 26, 167, 185, 199
    Out of Africa 138, 142, 166, 171, **172**, 178
milk 71, 194, 196, 225, 227, **228**

milk gland 69, 141, **142**, 143
millet 192
mind reading **201**
mirror test 200
mismatch 239
mitochondrial DNA 152, 161, 166, 168, 178
mitochondrial Eve 166
MJD 120
Modern Synthesis 111, 124, 128, 140, 143, 164, 203, 221, 241
Moldavia 189, **190**
molecular clock 161, **162**
Mollusca 51
monogamy 22
monomorphic 135
monophyletic group 152, **153**, 176
monozygotic twin 54, 218
moral behaviour 200, **211**
moratorium on genome editing 237
Morgan, T.H. 54, 110, 132
Morgan, E. 75
morphogen 56
Morris, D. 73
morula 41, **43**, 51
mosaic evolution 22, 69
mother's milk 226
Mount Improbable 71
mouse 44, 50, **51**, 53, 57, **58**, 103, 143, 148, **154**, 155
Mousterian culture 185, 187, **188**
mouth 51, **52**
Movius, M.L. 185
MRI 185
mRNA 48, **49**, 53, 63
mtDNA 166, 168, 176, 178
Müllerian duct 95, **96**, 97
multilevel selection 205
multiregional evolution 36, **165**, 167, **168**, 170
Mus musculus 53, 57
musculus
    cremaster 97
    gluteus 80
    obturatorius **80**, 81
    sartorius 80
    temporalis 18, 103
mutant 55, 58, 60, **61**, 63, 111, 134, 193, 226, 228, 236
mutation
    behaviour 118, 210, **220**
    chromosome 122, **123**, **124**, 135
    culture 208, **209**
    development 37, 59, 62, 69, 112, 144
    disease 64, 66, 106, **117**, 120, 122, 218, 238
    divergence 162, 169
    evolution 8, 71, 103, 109, **110**, **111**, 113, 116, 242
    fixation 137, 140
    indels 115, 120
    pseudogene 76, 122
    selection 124, 128, **131**, 143, 226
    SNPs 115
    types 103, 115, 130
    virus 114
    VNTRs 115, 117, **118**, 120
mutation frequency 115, 117, 161, 163, 169
mutation-selection equilibrium 131, 135
mutation theory 113, 140
mutational equilibrium 116, 135
mutationism 110, 113
Mutationstheorie 110
myelencephalon 58, 99
MYH16 103
myosin 103

nails **154**, **155**
nakedness 41, 74, 76, 234
Naked Ape 74
Napier, J. 25
Nariokotome boy 28
Naturalis 27, 191
natural selection 8, 71, 83, 109, **111**, 113, 140, 142, 144, 193, 223, 225, 229, 236, 241
Native American 172, 174, **175**
NBPF 103, **106**, 220
Neander 31
Neanderthal Man
    behaviour 34, 189, 198, **199**, 201, 209
    brain 99
    DNA 35, **103**, 121, 128, 176, 179, **181**
    fossil 16, 27, 31, 33
    hybridisation 172, **177**, 178, **180**
    morphology **33**, **34**, 74, 79, 222
    phylogenetic position 17, **19**, 32, **35**, 139, 164, 166
    skull 32, 35
    stone tools 185, **188**
near-neutral evolution 137
Nei, M. 113, 140
Nematoda 53
Nematostella vectensis 53
neocortex 100, 121, 206
neocortex ratio **207**
Neolithic 171, **189**, 191, 195, 197, 227
neolithic transition 176, 181
neoteny 31, **40**, 77, 122, 229
nephrocoel 43
nervous system 43, 46, 47, 51, 52, 99, 231
nervus
    laryngeus **93**
    vagus 92, **93**
neural crest 47
neural crest cells 47, 57, 91
neural fold 47
neural plate 47, 99
neural tube 47, 48, 99
neuroblastoma 103, 107
neurulation 47, 99
neutral evolution 8, 113, 135, 138, 140, 164
New Guinea 166, 179, 192, 208
Nieuwkoop, P. 56

INDEX

Nieuwkoop centre 56
node, phylogeny 151
nomadic existence 192
non recombining region of Y 134
non-synonymous mutation 103, 115, 162, 199
normal distribution 215, **216**
NOTCH2NL **103**, 107
notochord 46, 52
NRY 134, 167, **169**
nucleosome 146, **147**
nucleus suprachiasmaticus 102
Nüsslein-Volhard, C. 59
Nutcracker Man 19

oak 192, **193**
obesity 148, 239
obturator groove 80
obturatorius 81
Ockam, William of 156
Ockam's razor 156
Oenothera erythrosepala 109
oesophagus 90
Ohta, T. 137
Oldowan culture 183, **185**, 188
Olduvai 13, 25, 183
olfactory receptor 122, 231
Omo 13, 36, 183
omnivory 19, 87, 89
ontogeny 38, **39**
open reading frame 76, 104, 107, 115, 122
operational taxonomic unit 152
opposable 20, 184
orang-utan **105**, **123**, **153**, **159**
ordered character 157
organogenesis 42, 57, 67, 95
Orrorin tugenensis 13, 17, 21, **81**
orthognathic 29, 41
orthologous genes 59, **60**, 77, 121, 161
OTU 152, **154**, 156
Ötzi 194
Out of Africa model 166, **168**, **169**, 171, 172, 176, 178
Out of Beringia model 174
outgroup 151, **152**, **154**
ovarium 57, 97
overdominance **128**, 129
overpronation 81, **82**
oviduct 95, 97
oxytocin 102

Pääbo, S. 35, **177**, 179
palaeogenetics 176
Paleolithic **185**, **187**, **189**, **190**, 192, 196
palaeontology 14
Paley, W. 71, **72**
Pan paniscus 11, **30**
Pan troglodytes
    behaviour 183, **184**, 229
    brain 21, **30**, 100

chromosomes 123
classification 12, **153**
DNA 77, 104, **105**, 106, 120, **159**, 177
knuckle walk 12
limbs 80, 81
pelvis 78, 79
phylogeny 11, 75, 107, 120, 152, **154**, **155**
skull 40
Pangloss, dr. 113
panmixia 126
paracentric inversion 115
paralogous genes 59, **60**, 61, 103, 121, 122
paramesonephric duct 95, 96
Paranthropus
    aethiopicus 13, 17, 18
    boisei 13, **17**, 102
    robustus 13, 17
parapatric speciation **19**
paraphyletic group 152, **153**
parasites 53, 75, **195**, 215, 224
parent-offspring regression **217**, 218, 220
Parlevar 211
parsimony 156, 160
partner choice 215, 230, **232**, **234**, 240
PCR 119
PDE4DIP **106**
Pearson Type IV distribution 216
Pediculus humanus 74
paedomorphosis 40
Peking Man 27
pelvic floor muscle 81
pelvis 14, 23, 66, 78, 80, 82, 84, 96
penicillin 237
penis 234
peramorphosis 40, 41
perch **154**, 155
pericard 91, **92**
pericentric inversion 106, **115**, **124**
peritoneum 43
periventricular nucleus 102
pharyngeal basket 84
pharyngeal arch 84
pharynx 84, 90, 91, 87, 92, **93**
phenotype
    behaviour **118**, 119, 221
    development 77
    epigenetics 146, 148
    mutation 64, 144
    Neanderthal 181
    plasticity 144, 146
    relation with genotype 53, 66, 113, 128, 143, 170
    selection 124, **131**, 226, 228
phenotypic plasticity 144, 210, 222
pheromone 231
philanthropy 202
phosphorylation 147
phyletic evolution 18, 111
phyletic gradualism 111

phylogenetic reconstruction 151, 153, **154**, **155**, 156, 159, 161, 164
phylogenetic tree 17, **154**, 157, 159, 161, 166, **180**,
phylogeny 17, **39**, 76, 152, 155, **160**, 162, 180
phylotypic stage 37, **38**, 84
phylum 12, 37, **38**, 46, 53
Pickford, M. 21
pig 89, 160, 191, 194
pigeon **154**, **155**
Pigliucci, M. 241
pigmentation 172
pinworm 53
Pithecanthropus erectus 27
Pitx1 112
placenta **44**, 46, 96, 115, 157
Placentalia 12
plasticity 144, 146, 210
pleiotropy 218
Pleistocene 33, **34**, 192, 194
plesiomorphic character 23, 154, 226
pneumatic duct **90**, 91
point mutation 115
polar body 42, 50
polygyny 168
polymorphism 113, **115**, 117, **119**, 131, 135
polyphyletic group 152
Pongidae 153
population genetics 109, 125, 128, 130, 133, 135, 142, 151, 164
population growth 169, 195, **196**
population size 113, 136, 139, **163**, 167
population pressure 195
Portmann, A. 98
postcranium 29
postorbital constriction **33**
potassium-argon dating 15, **16**
potato 194
POU2F3 181
PPAR 240
pre-adaptation 71
predation 206, **207**
prefrontal cortex 100, 221
pregnancy 45, 95, 134, 230, 233
pregnancy test 56, 57
primate slowdown 161, **162**
primates 11, **12**, 15, 46, 74, 76, 98, 100, **102**, 104, 105, 107, 122, **160**, 162, 195, 206, 207
primitive character 23, 36
primitive streak **44**, 45
primrose 109
prisoner's dilemma 208
progesterone 57
prognathic 23, 27, 41
pronation 81, 82
pronephros 95, **96**
protein domain **103**, 106
Protostomia 47, **51**, **52**, 91
proximal 48
pseudo-sampling 159

pseudogene 76, 77, 103, 107, **121**, 122
psychology
    attractive ideas 208, **209**, 211
    behaviour 200, **212**, 213
    evolutionary 204
    language 199
Pthirus pubis 74
pubic hair 75
pubic louse 74
pubis 78, 84
public goods game 208
punctuated equilibria 111, 164
purifying selection 131, 236
purine 158
pyrimidine 158

Q fever 195
quadrate 85, 86
quantitative characters 215
quantitative genetics 203, 215, 217

rabies 195
rachitis 32
radial cleavage **51**
radial glia cells **103**, 105, 107
radioactive decay 15, **16**
radius 13, 222
Rampasasa 31
rate
    birth 195
    death 195
    development 40, 229
    evolution 112, 160
    metabolic 88
    mutation 115, 130, 169
    recombination 132
    selection 130
    substitution 104
recessive 127, **128**, **130**, 134, 139
reciprocal altruism 202, 208
reciprocity 202, **203**, 208
recombination 8, 58, 109, 114, **115**, 120, 131, 133, 135, 232, 242
recombination frequency 132, 134, 232
recurrent laryngeal nerve 93
regional continuity 165
regulatory mutation **104**, 105, **115**
Reich, D. 170, 176
Reichert, K.B. 86
Reichert's cartilage 85
relatedness 55, 223
religion 149, **211**, 212
repetitive DNA 118
reproductive isolation 19, 124, 145
reptile 39, 85, 93
resistance
    antibiotics 237, **238**
    malaria 134, **228**
response to selection 217, 218, 222, 225

rete mirabile 76
retromolar space **33**
retrotransposon 114
reverse transcription 115
reversion 155
rhesus macaque 77, 105
rice 192, 194
Riddle, R. 64
Rift valley **13**, 14, 171
RNA 53, **103**, 114, 143, **147**, 237
RNA interference 53
Robertsonian translocation 123
robust hominins 15, **18**, **19**, 24, 32, 35
Roebroeks, W. 34
Romania 178
root, phylogeny 151, **154**
Rubicon 25
rudimentory organ 89

sacral vertebrae 79
sacrum 78, 79
sagittal crest 18
sagittal keel 27, 35, 165
Sahel 192
Sahelanthropus tchadensis **13**, **17**, 21
salamander 39, **154**, **155**
saltation 110
Sami 33
San Marco cathedral 113
Sangiran 28
SARS **195**
sartorius muscle 80
scales **154**, **155**
scavenger **89**
Schaaffhausen, H. 32
schizophrenia 104
sciatic notch **78**, 79
scrotum 97, **98**, 234
SDN-POA 102
sea squirt 48, 61
sedentary 192
Selbstdarstellung 98
selection
    adaptation 31, 71, 83, 88, 223, **228**, 240
    artificial 113, 192, **193**
    biomedical traits 224, **228**, 230
    cultural **209**
    coefficient 127, **128**
    drift 135, 137, 140
    evolution 8, 109, 124, 145, 209, 241
    frequency dependent **130**, 134
    group 202, 205, **206**, 212
    Hardy-Weinberg equilibrium 126, **128**
    heritability 217, 219, 222
    kin 202, **204**, 206, 208
    modes **131**
    Modern Synthesis 110, 124
    molecular 71, 103, 107, 119, 178, 226, 229, 236
    mutation 71, 111, 113, 131, 142

    natural 8, 31, 71, 83, 88, 109, 111, 114, 142, 144, 149
    sexual 22, 74, 98, **232**, 233, **234**
selection differential 218, 225
selection gradient **225**
selective sweep 134, 226
selectionism 110
self-consciousness 200
self-transcendence 119, 212
seminal vesicles 95
Senut, B. 21
Sephardic Jews 120
sex chromosome 134
sex ratio 22, 136
sexual behaviour 100, 102, 149, 230
sexual dimorphism 22, 102, 233
sexual selection 22, 74, 98, 233, **234**
sexual transmutation 112
sheep 88, 124, **160**, **193**, **195**
SHH 64
shinbone 33
shoulder 23, **78**, 84
Shubin, N. 243
siamang **153**
Siberia 35, **172**, 174, 177, 179, 186, 222
signal transduction 50
Sima de los Huesos 178
SINE 114
sinus venosus **94**
siRNA 147
sister clade 152
sister group 160, 173
sister species 17, 152, 179
Skoglund, P. 176
skull 14, **18**, 21, **23**, 25, 27, 29, **30**, 32, **33**, 35, 40, 48, 77, **84**, **86**, **103**, **165**, 181, **199**
smell 102, 122, 231
SNP 115, **117**, 132, **133**, **177**, 181
social behaviour 205, 208
social learning 149, 209,
social interaction 98, 206, **207**
social environment 99, 209, 211
sociobiology 205
Sog 52
Solo river 28
soma 143
somatic development 39, **40**
somite 56
somatopleura 92
Sonic Hedgehog 64
South-East Asia 179, **180**
spandrels 113,
spatial orientation 102
speciation **19**, 59, **60**, 145, 151
Spemann centre **50**
sperm 49, **50**, 94, 97, 166, 236
sphenoid 84, **85**
spina bifida 48

spine 79
spiral cleavage 51
spiracle 93
spirituality 119
Spencer, H. 128
splanchnopleura 92
splitting 18, 30
squamosum 86
SRGAP2 103, 121
SRY 134
SSR 118
stabilizing selection 131
stalk-eyed flies 233
stapes 85
starch digestion 194, 228
STAT2 181
Stearns, S. 225
stickleback 112
stone age 188, 189, 192, 239
stone tools 25, 33, 88, 99, 173, 183, **185**, **187**, 199
STR 118
stratigraphy 15, 188
Strickberger, M. **211**
Stringer, C. 165, 167
strong reciprocity 208
Strongylocentrotus purpuratus 51
structural mutation 115
stylohyoid ligament 85
sublimation 100
substitution
    nucleotide 104, 115, 117, 124, **158**
    amino acid 117
supra-angular 86
Swaab, D. 102
sweaty T-shirt test 231, **232**
swim bladder **90**
Swiss army knife 186
symbolic thinking 149, 188, 191, 201
sympatric speciation **19**
symphysis pubica 78
symplesiomorphic **154**
synapomorphic 153, **154**, 156, 158
synaps 121
synonymous mutation 115
synpolydactyly 66
synteny 62

Tabin, C. 64
tangled bank 241, **243**
Tasmania 211
Tattersall, I. 29, 32, 164
Taung 24
Tbx5 66
TDF 134
Teleostei **90**, 91
Templeton, A. 165
temporal bone of skull 85
temporal muscle 18, **103**
temporal lobe of brain 100, 102

teosinte 112, 193
testicles 97, **98**, 234
terrestrial 12, 75
testicular descent 97
testis 94, **98**, 134
testosterone 95, 149, 231, 234
tinkering 69, 85, 107, 241
thalamus 100
theory of mind 200
Therapsida 86
thighbone 21, 33, **80**
Thompson, d'A.W. 101
thoracic segment 60, 63
thoracic vertebrae 199
thorax 61, 84
thrifty gene 148, 239
thyroid 85, 228
Tibetans 181
tibia 84, **222**
Tirolean iceman 194
Toba volcano 138
Tobias, P. 25
tongue 122, **199**
toe 20, 66, 81, **84**
toolbox 59, 64, 69
tools 25, 33, 36, 88, 99, 170, 173, 183, **185**, 187,
    188, **199**
torus supraorbitalis 21, 27, **33**, 165
toxoplasmosis 195
trachea 90
transcription 66, 115
transcription factor 56, 59, 63, 66, **104**, **105**, 239
transcriptional regulation 103, 104
transgenerational transfer 148
transition
    bipedalism 12
    cultural 185
    neolithic 171, 192, 197
    nucleotide 157, **158**
    species 21, 25, 121, **165**
translocation 115, 120, 122
transmission
    cultural 197, 209, 211, 186
    genetic 59, 163
transmutation 112
transplantation 55, 57
transposon 114
transversion **158**
tree length 156
Tribolium castaneum 51, 53, 55
Trinil 27, **28**
trisomy 21 122
Triticum 193
trochanter
    insect **61**
    femur 80
trophoblast 44, 46
Tugen Hills 13
Turkana 13, 28, 83, 184

INDEX

Turkey 192, 196
twin 54, 211, 218, 220
type III secretion system 72, 73

ubiquitin **147**
Ubx 60, **62**, 63
UDPGT 235
Ultrabithorax 60, 62
umbilical cord 44, 46, **58**, 96
ungulate 76, **160**
unordered character **157**
unrooted tree 151
upright walking 11, **12**, 21, **23**, 26, **33**, 41, 75, 78, 81, **84**
urachus 96
Uraha **13**
Urbilateria 51
ureter 95, **96**, 97
urges 100, 208, 211
urine 56, 94, 96
urogenital system 96, **97**,
uterus 45, 54, 58, 95, **96**
UV radiation 114

Vacca 194
vaccine 194
vagina 95
valgus angle 23, **80**, 81, **84**
Van den Bergh, G. 31
vas deferens 95, **96**
vegetal pole 48, **49**, 50, 57
VegT 48
ventral 48, 51, **52**, 65, **90**, **92**
ventral aorta 52
ventral nervous system 46, **52**
ventral striatum 202
ventricle, heart 92, **94**, **95**
Venus figurines 189, **190**
Verhaegen, M. 75
vermiform appendix 89
vertebrae
  cervical 219
  lumbar 79
  sacral 79
vertebrates **12**
Vestigial 63
vestigial organ 89
Vg1 48, 49, 63
Virchov, R. 32
virus 46, 114, **115**, 158, **195**, 237
viscosity, blood 181
vitamin C 122
VNTR 115, 118, 119
volcano **16**, 138
Voltaire 113
vomeronasal organ 231
Von Baer, K.E. 37
Von Humboldt, A. 241

Von Koenigswald, R. 28
Von Tschermak, E. 109

Waddington, C.H. 144, 146, 210, **211**
Wahlund effect 140
waist-to-hip ratio 233, **234**
watchmaker analogy 71, **72**
Watson, J. 143
Weak Garden of Eden 167
Wedekind, K. 231
Weinberg, W. 125
Weismann, A. 143
Weismann barrier 143, 146
West, S. 205
West-Eberhard, M.-J. 145
whale 153, **160**
wheat 192, 194
Wheeler, P. 88
White, T. 18, 20, 36, 165
Wieschaus, E. 59
Willerslev, E. 176
Williams, G.C. 205
Wilson, D.S. 205, **211**
Wilson, E.O. 205, 211
Willendorf Venus 189
wing 59, 63, 65
Wnt **50**, 64, 104
wolf 194
Wolffian duct **96**
Wolpoff, M. 165, 167
Wood, B. 26
world population 195, **196**
Wrangham, R. 34
Wright, R. 191
Wright, S. 111, 128
Wynne Edwards, V.C. 205

X chromosome 134, 181
Xenopus 49, **50**, 53, 55
Xi'an 165
Xwnt 48

Y chromosome 133, 135, 167, 169, 236
yolk 38, 45, 48, **50**, 56
yolk sac **44**, 56
Yoruba 172, **173**, 175

Zagros 197
zebra 192, **193**
zebra fish 43, 48, **51**, 53, 55, **56**, 243
Zhoukoudian 28
zona pellucida **44**
zone of polarising activity 64, **65**
zoonosis 194, **195**
Zoroaster 197
ZPA 64, **65**
zygomatic bone 18
zygote 41, **42**, **43**, 55, 149